Gravitational Waves in Physics and Astrophysics

An artisan's guide

AMERICAN ASTRONOMICAL SOCIETY IOP | ebooks™

AAS Editor in Chief

About the program:

AAS-IOP Astronomy ebooks is the official book program of the American Astronomical Society (AAS) and aims to share in depth the most fascinating areas of astronomy, astrophysics, solar physics, and planetary science. The program includes publications in the following topics:

Books in the program range in level from short introductory texts on fast-moving areas, graduate and upper-level undergraduate textbooks, research monographs, and practical handbooks.

For a complete list of published and forthcoming titles, please visit iopscience.org/books/aas.

About the American Astronomical Society

The American Astronomical Society (aas.org), established 1899, is the major organization of professional astronomers in North America. The membership (~7,000) also includes physicists, mathematicians, geologists, engineers, and others whose research interests lie within the broad spectrum of subjects now comprising the contemporary astronomical sciences. The mission of the Society is to enhance and share humanity's scientific understanding of the universe.

Editorial Advisory Board

Gravitational Waves in Physics and Astrophysics

An artisan's guide

M Coleman Miller
University of Maryland, College Park, MD 20742-2421, USA

Nicolás Yunes
University of Illinois at Urbana-Champaign, Urbana, IL 61801-3008, USA

IOP Publishing, Bristol, UK

ISBN 978-0-7503-3051-0 (ebook)
ISBN 978-0-7503-3049-7 (print)
ISBN 978-0-7503-3052-7 (myPrint)
ISBN 978-0-7503-3050-3 (mobi)

DOI 10.1088/2514-3433/ac2140

Version: 20211201

AAS–IOP Astronomy
ISSN 2514-3433 (online)
ISSN 2515-141X (print)

British Library Cataloguing-in-Publication Data: A catalogue record for this book is available from the British Library.

Published by IOP Publishing, wholly owned by The Institute of Physics, London

IOP Publishing, Temple Circus, Temple Way, Bristol, BS1 6HG, UK

US Office: IOP Publishing, Inc., 190 North Independence Mall West, Suite 601, Philadelphia, PA 19106, USA

For Cecilia, and for my parents.

For the Graviteers, past and present. For Jessica, for Penny, and for my parents. You have all taught me much.

Contents

Preface

Nico: Before I let you go, we should probably talk about the preface of our book. I mean, we should chat about what we want to include in there. I'm guessing we wanna tell the reader what the book is about, who it is intended for, and also thank all the people that helped us along the way.

Cole: Yeah, that's a good idea. But we should also explain how the "characters" come into play in the book, why we've decided to use them, and what their purpose is. But first things first. Who is this book for?

If you recall, when we first started talking about writing this book, we had senior undergraduate students, and first- or second-year graduate students in mind. The idea was that these students would have already seen undergraduate quantum mechanics and classical electrodynamics, say at least at the level of Griffiths' textbooks, and also classical mechanics, say at the level of Goldstein's book. Students would also benefit from having seen general relativity at the undergraduate level, say from parts of Hartle's or Schutz's textbooks. But the whole point of this book is that these students wouldn't need to have taken a formal, graduate course (or four) in general relativity in order to make sense of our book. Of course, if the student has taken general relativity at the graduate level, say from a book like Carroll's or Wald's, then that's great! The more you know, the better because the more certain statements in the book will make sense to you.

Nico: That's right. And we also wanted to make the book accessible to a variety of communities, from the gravity student to the gravitational-wave or data analysis student, from the astronomy student to the nuclear physics or high-energy physics student. That's why we tried to present the material from "first principles," even if the beginning bits may seem a bit simple to the more advanced student.

But we also had a different driver for this book: we wanted to train the student in the art of the Fermi estimate. This is the idea that many times we can estimate the answer to a question, without doing a bunch of complicated math. Of course, in the end, those complicated calculations are often needed to make a precise prediction. But much insight can be gained by thinking about the scales of the problem and the main physics effects at play.

Cole: Yeah, it's funny how sometimes complicated calculations can lead us astray, to an answer that, on second thought makes no sense! A Fermi estimate reveals this. And also, when you do the Fermi estimate first, many times it may help you devise a calculational strategy, give you hints on how to attack the problem.

There is this story about how Fermi, when he was at Los Alamos, helped estimate the yield of an atomic bomb with nothing more than a few strips of paper and wind. The story is that in 1945, he was observing the Trinity test at a safe location, and he needed to estimate the nuclear yield. So when the blast wave was coming toward him, he threw a bunch of strips of paper up in the air, to let the blast wave blow them away. He then measured how far away the paper flew from where he was (by walking to the paper that now laid on the floor!), and with this, he calculated the yield. His estimate was off only by a factor of 2!

Nico: I've always found that learning by solving problems is the most effective way of learning. When you can see example after example of problems related to the complex concepts you are trying to learn, things just become easier. So in addition to the Fermi estimates that we have popped in throughout the book, every chapter also has a bunch of homework problems for the students to do. There are some that are very easy and require only Fermi estimates. There are some that are more complicated and require a bit more calculation. There are even some in which the student has to respond to an accusation of Dr I. M. Wrong!

Cole: Ah, yes, the characters. That was a fun little addition to this book. The idea was to introduce three characters that would accompany the reader throughout the book. We have Captain Obvious, who is always there for the reader to provide an alternative (and often simpler) explanation than our own! And we also have Major Payne, who brings more mathematical and theoretical rigor to the table, attempting to correct what Captain Obvious says. And then there is Dr. Wrong, who really seems to dislike us, the authors, and is always getting confused. Fortunately, Captain Obvious usually steps in to smooth things out. None of the characters represent particular people; instead, they have a blend of characteristics of people that you and I have met throughout our lives.

Nico: Yeah, these characters have helped us in a variety of ways. It was nice to have these interludes to break the flow of content a bit. Sometimes books can get a bit too dense, and it's nice to stop for a while to "smell the roses," so to speak, by considering alternative explanations or additional rigor. Also, the characters allowed us to bring in that extra level of math, without making the entire book too math-y, which would not have served our intended audience well. And then, there is also the fun that Dr. Wrong brings in!

Cole: We should probably also mention that the alternative explanations typically introduced by the Captain, but also some of our own explanations, are purposely nonrigorous. Their raisons d'être are to build intuition and a physical picture of typically complicated mathematical results, sometimes at the cost of letting go of rigor a bit. We have to make it clear to the reader that these analogies, like any analogies in general relativity, are to be taken with a grain of salt.

Nico: Yeah, the characters were great fun and allowed us to go through a bunch of material without making the book too monotonous. I mean, we cover a ton of stuff here! We go from the basics of gravitational radiation and its sources in the first two chapters to a more detailed description of how to model these sources and how to detect them in Chapters 3 and 4. These four chapters, the first half of the book, cover what we could call the "basics" of gravitational-wave science. The remaining four chapters then cover applications, from astrophysics in Chapter 5 to cosmology in Chapter 6, nuclear physics in Chapter 7, and tests of general relativity in Chapter 8.

By no means is this a complete treatment though. If we had attempted that, we would have needed to write way more than just 200 or 300 pages worth of content! The idea here was just to present the key ingredients that students who wish to specialize in this topic would need to then do a deep dive in whatever subarea they choose. For this reason, we moved some important, but more specialized, material to the appendices. We have one on Bayesian statistics, one on dynamics, and one on calculation methods for neutron stars. Each of these appendices could, by itself, be a book (or a series of books), but the content presented here is enough (we hope!) to help the chapters make sense.

Cole: Exactly. All in all, the material we present in this book would be ideal for a one-semester course in gravitational-wave science for senior undergrads and beginning graduate students. It's not clear that all chapters can be covered in detail in just one semester. But because of the structure of the book, the instructor has the freedom to choose which of the topics in the second half of the book will be covered. And of course, there is ample room to add material in any one of the chapters presented here.

Nico: Hopefully this book will be useful to the students and to the instructors. But we should also mention in the preface that we can't take all the credit for this work. All of the material presented here has appeared in the literature in some way or another because obviously none of it is original research! Some of this material is in scientific papers, broken down here to its elementary bits for easier digestion. Some of the material is in other textbooks, like the introduction to post-Newtonian theory by Poisson and Will, or the presentation of gravitational waves and cosmology by Maggiore, or the treatise of dynamics in the book by Binney and Tremaine.

At the end of each chapter, we provide a list of references for the interested student to begin her or his deep dive, but for sure this list is not going to be exhaustive. I mean, there are just too many good references out there to list, and if we attempted something exhaustive, we would certainly leave some references out.

Cole: True! We're exclusively citing books and we have tried to include a wide variety, from introductory to advanced and from classics to more recent texts—some in multiple chapters if they apply over several subjects—but even so, we have certainly missed other excellent books. And even when we mention certain people in the book, the intention is not here to give the sole and unique credit for the entirety of the field to that person. The people we mention in the book did play a key role in the development of the field, but so did many other people whom unfortunately we do not have room to discuss. After all, this is not a historical study of the field of gravitation!

Nico: Indeed. But we do owe a tremendous debt to many of these people, either because the material presented in our book builds on their work, or sometimes because they helped in the process of writing the book itself. I guess we should mention the people who helped us with the contents first: Cecilia Chirenti and Emanuele Berti gave us good insights into quasi-normal modes and Eric Poisson was instrumental in helping us understand tidal deformabilities better.

Next, but not least, there is a large group of people who spent time looking over our book and making suggestions or pointing out typos (you're gonna break some eggs when you make an omelet!). Special thanks must go to Peter Adshead, Chris Belczynski, Floor Broekgaarden, Rohit Chandramouli, Katerina Chatziioannou, Cecilia Chirenti, Neil Cornish, Gil Holder, Jorge Jose, Tushar Nagar, Jacquelyn Noronha-Hostler, Caroline Owen, Christopher Plumberg, Alexander Saffer, Kristen Schumacher, Alejandro Vigna Gomez, and Kent Yagi.

Cole: And I guess that's about it, right? Maybe we should think about how to end the preface. I mean, it's a preface, so it doesn't have to be too long. But maybe we should end with some inspiring words. How about the following, from Walt Whitman, which I have always considered to be representative of the spirit of scientific exploration as a wonderful journey rather than a destination?

"This day before dawn I ascended a hill and look'd at the crowded heaven,

And I said to my spirit *When we become the enfolders of those orbs, and the pleasure and knowledge of every thing in them, shall we be fill'd and satisfied then?*

And my spirit said *No, we but level that lift to pass and continue beyond."*

About the authors

Cole Miller

Cole Miller is a Professor of Astronomy at the University of Maryland and a member and former director of the Joint Space-Science Institute.

He received his PhD in 1990 from the California Institute of Technology and was a postdoctoral research associate at the University of Illinois at Urbana-Champaign and then the University of Chicago before starting as an Assistant Professor at Maryland. He has authored numerous review articles on gravitational-wave astrophysics. He has also given several dozen popular presentations to audiences ranging from kindergartners to assisted home residents, he was the lead scriptwriter for the 2000 Adler Planetarium show "Black Holes: Into the Dark Abyss," and he was a consultant for the 2012 University of Maryland dance performance "Gravity."

He has published papers on many aspects of compact objects and gravitational waves, especially dynamical interactions and fast variability in accreting neutron stars. He was recently awarded a Radboud Excellence Professor position.

He is currently on the Council of the American Physical Society and is on the Executive Committee of the Division of Astrophysics. He has also been on the Executive Committee of the High Energy Astrophysics Division of the American Astronomical Society and was the chair of the LIGO Program Advisory Committee from 2010–2014. Dr. Miller has published over 160 articles in peer-reviewed journals and has given more than 270 professional presentations on his work.

Dr. Miller lives in College Park, Maryland, with his wife, Dr. Cecilia Chirenti.

Nicolás Yunes

Nicolás Yunes is a Professor of Physics at the University of Illinois Urbana-Champaign and founding director of the Illinois Center for Advanced Studies of the Universe.

He received his PhD in 2008 from The Pennsylvania State University and has been a Research Associate at Princeton University and a NASA Einstein Fellow at MIT and Harvard University. Before moving to the University of Illinois, he was an Assistant and then Associate Professor of Physics at Montana State University, where he cofounded the eXtreme Gravity Institute in 2015.

Dr. Yunes is the author of a popular physics book, *Is Einstein Still Right?: Black Holes, Gravitational Waves, and the Quest to Verify Einstein's Greatest Creation*, which he coauthored with Dr. Clifford M. Will. Dr. Yunes is also the producer of the "Celebrating Einstein" science festival, the "Rhythms of the Universe" spoken word event, the "Einstein Gravity Playlist" planetarium show, and he has appeared regularly in a weekly segment of local Montana radio.

He is an expert in the use of gravitational-wave observations to systematically test Einstein's theory of general relativity and in other aspects of gravitational physics, including black hole and neutron star theory. Dr. Yunes has published over 200 original articles in peer-reviewed scientific journals, including *Science* and *Nature* magazine. He has also received numerous grants from NSF and NASA, including the prestigious NSF CAREER award, and awards such as the Young Scientist Prize from the International Society for General Relativity and Gravitation.

Dr. Yunes has served on several national and international committees run by NASA and the NSF and has been part of various collaborations, including the LIGO Scientific Collaboration and the International LISA Consortium. He has served as Chair of the Division of Gravitational Physics of the American Physical Society, and he is a member of the Editorial Board of Classical and Quantum Gravity.

Dr. Yunes lives in Champaign, Illinois, with his wife and daughter.

About the Characters

S. O. Obvious was recruited by the US Army at age 18 and rose through the ranks rapidly to become an Army Captain at age 29, after taking some time off to complete a Bachelor's degree and then a PhD degree in Applied Physics. After the Army, S. O. Obvious became a professor at a top-tier research university, where S. O. Obvious is still affectionately referred to as "captain." Captain Obvious is very intelligent and for the most part friendly, but can also come off as a bit standoffish.

I. M. Payne graduated from high school at age 16, completed an Ivy league Bachelor's degree in Physics in 2 years, obtained a PhD in Mathematics and another in Theoretical Physics by age 24, and then joined West Point and rapidly ascended to the rank of Major. Major Payne is extremely intelligent, but also quite arrogant and condescending. The Major's relationship with Captain Obvious is not good, because of what Major Payne perceives as Captain Obvious' incessant need to explain the basics over and over again.

I. M. Wrong obtained a PhD at age 30 and retired at age 60 after a career in industry. Dr. Wrong has always been interested in physics and was prompted by the recent discovery of gravitational waves to revisit this subject after first seeing it over 30 years ago. Dr. Wrong currently spends a great deal of time at different physics blogs and websites, trying to originate ideas that will overturn the world of establishment physics.

Symbols

List of symbols used throughout this book, together with their physical meaning and their units in SI and in geometric ($c = 1 = G$) units. Units are listed via the notation SI = geometric. Greek indices represent spacetime quantities, while Latin indices in the middle of the alphabet represent spatial quantities. Bold-faced mathematical symbols represent three-dimensional spatial vectors.

$\approx$	An approximate value including a numerical factor, e.g., $c \approx 3 \times 10^8$ m s^{-1}
$\sim$	A rough value or a dependence, sometimes without numerical factors or constants, e.g., there are $\sim$300 days in a year, and a circular orbital speed is $v \sim r^{-1/2}$
G	Newton's gravitational constant (m^3 s^{-2} kg^{-1} = 1)
c	Speed of light in vacuum (m s^{-1} = 1)
ρ	Energy density (kg m^{-3}= m^{-2})
$\mathbf{r}$ and $r^i = x^i$	3D (spatial) distance vector and components (m = m)
$\mathfrak{M}$	Characteristic mass scale (kg = m)
$\mathfrak{L}$	Characteristic length scale (m = m)
$\mathfrak{R}$	Characteristic radius of a compact object (m = m)
$\mathfrak{T}$	Characteristic timescale of a compact object (s = m)
m_1 and m_2	Individual mass of the components of a binary system (kg = m)
$m = m_1 + m_2$	Total mass of a binary system (kg = m)
$\mu = m_1 m_2/m$	Reduced mass of a binary (kg = m)
$\eta = \mu/m = m_1 m_2/m^2$	Symmetric mass ratio of a binary (1)
$\mathcal{M} = \eta^{3/5} m$	"Chirp mass" (kg = m)
R_1 and R_2	Individual radii of the components of a binary system (m – m)
$C_{1,2} = m_{1,2}/R_{1,2}$	Individual compactnesses of the components of a binary system (kg m^{-1} = 1)
a	Semimajor axis of a binary system (m = m)
ω	Angular orbital frequency of a binary (rad s^{-1}= m^{-1})
f	Orbital frequency of a binary (s^{-1} = m^{-1})
$f_{\rm GW}$	Gravitational-wave frequency (s^{-1} = m^{-1})
P	Orbital period of a binary (s = m)
r_{12}	Orbital separation of a quasi-circular binary system (m = m)
v	Orbital speed of a quasi-circular binary system (m s^{-1} = 1)
a	Semimajor axis of an elliptical binary system (m = m)
e	(Newtonian) Orbital eccentricity of an elliptical binary system (1)
$r_p = a(1 - e)$	Pericenter distance of an elliptical binary system (m = m)
$\omega_{\rm QNM}$	Quasi-normal frequencies of a perturbed black hole (rad s^{-1}= m^{-1})

Chapter 1

Overview of Gravitational Radiation

Gravitational radiation was first detected directly in 2015. Since that time, the rate of discoveries has increased to the point that we can soon expect ~100 compact object (for us, black hole and neutron star) coalescences per year. With such rates, gravitational-wave detections have expanded beyond physics and into the realms of astronomy, cosmology, nuclear physics, and particle physics, and have opened new observational windows onto some of the most dynamic phenomena in the universe. These include merging neutron stars and black holes, supernova explosions, and possibly echoes from the very early history of the universe as a whole.

Gravitational-wave astrophysics, however, has important differences from standard astronomy and nuclear and particle physics. In electromagnetic observations, every waveband contains sources that are so strong that they can be detected without knowing anything about them. You don't need to understand nuclear fusion in order to see the Sun! Similarly, particle colliders detect hundreds of trillions of decay products when energetic beams crash onto a particle detector, but plenty of particles can be detected even if we don't know quantum chromodynamics. In contrast, as we will see, many gravitational radiation events are so weak that sophisticated statistical techniques are required to extract them properly from the detector noise, and then to characterize them. These techniques involve matching templates of expected waveforms against the observed data stream. Maximum sensitivity therefore requires a certain understanding of what the waves look like and thus of the characteristics of the sources that produce them.

In this book, you will encounter a survey of many aspects of gravitational radiation, starting with the anticipated sources. Before discussing the different types of sources, though, we first need to have some general perspective of how gravitational radiation is generated and how strong it is. We will thus begin by discussing radiation in such a general context.

doi:10.1088/2514-3433/ac2140ch1

1.1 Radiation in General

By definition, a radiation field must be able to carry energy to infinity. If the amplitude of the field is a distance r from the source in the direction (θ, ϕ) is $A(r, \theta, \phi)$, then the flux at r is $F(r, \theta, \phi) \propto A^2(r, \theta, \phi)$. If, for simplicity, we assume that the radiation is spherically symmetric, $A(r, \theta, \phi) = A(r)$, then the luminosity at a distance r must be $\mathcal{L}(r) = \text{Flux} \times \text{Area} \propto A^2(r) \times (4\pi r^2)$.

Captain Obvious: The fact that $F(r, \theta, \phi) \propto A^2(r, \theta, \phi)$ is "clearly" true because flux is energy per time per area, and the energy of a field is proportional to the square of the field itself. In electrodynamics, for example, the energy contained in an electromagnetic wave is proportional to the square of the electric field plus the square of the magnetic field.

Major Payne: Sigh. Already I have to step in to clear some things up. "Flux" is used in different ways by people in different fields. For example, in relativity, it is common to say that the flux is proportional to the integral of the stress–energy tensor of a field contracted onto certain normal 4-vectors and integrated over a two-dimensional sphere at spatial infinity. That would mean that to relativists, "flux" is what the authors are calling "luminosity"! The authors assure me that in this book, "flux" will mean energy per area per time, and luminosity will mean energy per time, but when you read papers you need to understand what convention is being used.

To set things up, let's talk about multipole moments. There are various definitions, but for our purposes, we will suppose that we have a distribution of mass and energy over some finite and bounded volume V, so that the mass–energy density is $\rho(\mathbf{r})$ at a location $\mathbf{r}$, where it is convenient to locate the origin of the system at the center of mass–energy. You could do something similar with the net density of electric charge $\rho_e(\mathbf{r})$. It is convenient for many applications to represent $\rho(\mathbf{r})$ using a sort of series. The lowest-order term, which is called the monopole moment, is $\int_V \rho(\mathbf{r}) d^3r$ and is therefore the total mass–energy (or the total electric charge if we are integrating $\rho_e(\mathbf{r})$). The next-order term, which is called the dipole moment, is $\int_V \rho(\mathbf{r})\mathbf{r} d^3r$, which we can also write as $\int_V \rho(\mathbf{r})\, r^i\, d^3r$ for component i of the dipole moment. Thus, the dipole moment would have three terms. The next-order term, which is called the quadrupole moment, likewise has as its (i, j) component $\int_V \rho(\mathbf{r}) r^i r^j\, d^3r$, and so on up to higher moments. Because these reference only the location of the mass–energy (or electric charges) rather than their motion, these sometimes are prefixed with "electric" in reference to the generation of electric field by charges; thus, the "electric dipole" and "electric quadrupole," for example, even if we're considering mass–energy rather than electric charge. One can also introduce multipoles related to motion, which are therefore sometimes called "magnetic" (again even if we are thinking about mass–energy distributions). For example, the "magnetic dipole moment" is $\int_V \rho(\mathbf{r})\, [\mathbf{r} \times \mathbf{v}(\mathbf{r})] d^3r$ if the velocity at location $\mathbf{r}$ is $\mathbf{v}(\mathbf{r})$. Please keep these definitions in mind as we proceed with our discussion.

Major Payne: No, no, no! That's not the real reason these multipole moments are called "electric" or "magnetic." Sure, maybe it makes sense to explain it this way, but the real definitions have to do with parity. The latter is a (discrete) transformation in which one flips the direction of the coordinate axis at a fixed time (i.e., on a so-called time-like hypersurface) in the "passive transformation" view. In the "active" view, one can think of a parity transformation as acting on the components of a geometric object by flipping its signs. Yet another way to think about a parity transformation, and probably the view favored by Captain Obvious, is as a rotation $\theta \to \pi - \theta$ and $\phi \to \phi + \pi$ using standard spherical polar coordinates because then you can think of it as a "reflection on a mirror."

No matter how you think about it, when you act on a geometric object, like a vector or a tensor, with a parity transformation, then there are different possible outcomes. One possibility is that the entire vector or tensor picks up a minus sign. Objects that transform in this way are called odd or axial. The electric field of Maxwell's theory transforms in this way, and so, in analogy, tensors that transform similarly are called of "electric type." Another possibility is that the vector or tensor is left unchanged (invariant) under a parity transformation. Objects that transform in this way are called even or polar. The magnetic field is an example of a polar vector, and so, again in analogy, tensors that transform in this way are called of "magnetic type."

When dealing with multipole moments, one has to be more careful and clear. Consider a geometric object that you expand far away from its source. In many cases, such an object decomposes into the product of a set of functions of the spherical polar angles (θ, ϕ) and another set of functions of time and radius, which are related to the multipole moments, e.g., for a scalar $S = F^{\ell m}(t, r)G^{\ell m}(\theta, \phi)$. For a scalar, the functions of polar angles $G^{\ell m}(\theta, \phi)$ are typically spherical harmonics $Y^{\ell m}(\theta, \phi)$, but for other more complicated tensor objects, they can be other functions. Regardless of their nature, how these angular functions transform under parity depends on the ℓ number, and therefore, the transformation properties of the nonangular functions must also depend on ℓ. Nonangular functions, and therefore multipole moments, that transform as $\hat{\mathcal{P}}[F^{\ell m}] = (-1)^{\ell+1}F^{\ell m}$ under a parity transformation $\hat{\mathcal{P}}$ are said to be odd or axial, while those that transform as $\hat{\mathcal{P}}[F^{\ell m}] = (-1)^{\ell}F^{\ell m}$ are said to be even or polar.

Let us temporarily focus on electrodynamics in our discussion of radiation. When one expands the static electromagnetic field of a source in multipole moments, the slowest-decreasing moment is the monopole, and the electric field from a static charge decreases with distance like $E(r) \propto 1/r^2$. This is generally true of amplitudes from static monopoles: $A(r) \propto 1/r^2$. This implies that $L(r) \propto A(r)^2 r^2 \propto 1/r^2$, so a static electromagnetic field does not carry energy away from the source out to infinity, because $L(\infty) \propto \lim_{r\to\infty} 1/r^2 = 0$.

If, on the other hand, we consider a time-varying source that produces an electromagnetic wave, then the amplitude of the latter would decay as $A(r) \propto 1/r$ (see any textbook on electromagnetism for derivations of this fact), implying that for time-varying sources, energy is carried out to infinity because $L(r) \propto A(r)^2 r^2 \propto (1/r^2) r^2 \propto$ constant. This tells us two things that are true in general, regardless of the nature of the radiation (e.g., electromagnetic or gravitational). First, radiation (i.e., a field that is able to carry energy to infinity) requires time variation of the source. Second, the amplitude of the radiation field must decay as $1/r$ far from the source.

We can now explore what types of source variations will produce radiation. We'll start with electromagnetic radiation. The electric monopole moment $\int \rho_e(\mathbf{r}) d^3r$ is simply the total charge Q. For an isolated source, this charge cannot vary, so there is no electromagnetic monopolar radiation. For the electric dipole moment, $\int \rho_e(\mathbf{r})\, \mathbf{r}\, d^3r$, there is no applicable conservation law, so electric dipole radiation is possible. Figure 1.1 shows how the center of charge moves around in a classical electron or positron trajectory. Usually, if there is no symmetry to prevent a process from happening in nature, it will happen! One can also look at the variation of internal currents. The magnetic dipole is $\int \rho_e(\mathbf{r})\, [\mathbf{r} \times \mathbf{v}(\mathbf{r})]\, d^3r$, where $\mathbf{v}$ is the three-dimensional velocity vector field associated with the body, and here the cross product is the usual one in three-dimensional flat space. Once again, this can vary because there is no conservation law that prevents it, so magnetic dipole radiation is possible. The lower-order moments will typically dominate the field unless their variation is reduced or eliminated by some special symmetry.

Now consider the gravitational radiation emitted by an isolated source with mass–energy density $\rho(\mathbf{r})$. The monopole moment $\int \rho(\mathbf{r})\, d^3r$ is simply the total mass–energy of the source. This quantity is constant, because mass–energy is conserved in the absence of external perturbations, so there cannot be monopolar gravitational radiation. The static dipole moment $\int \rho(\mathbf{r})\, \mathbf{r}\, d^3r$ is just the center of mass–energy of the system. In the center of mass frame, therefore, this moment does not change, so there cannot be gravitational radiation of the electric dipole type; and because the existence of radiation is frame independent, this statement is true in

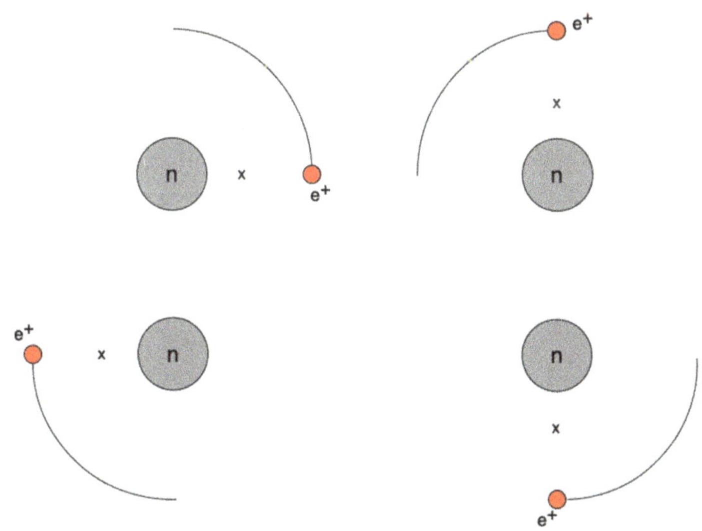

Figure 1.1. A sketch showing a positron in red in "orbit" around a neutron in gray (not drawn to scale!). In reality, quantum mechanics tells us that positrons or electrons are not classical objects in classical trajectories, but if the positron is far away enough from the neutron (such that, e.g., quantization can be ignored), we can treat the situation classically. In this picture, the center of mass stays fixed (as most of the mass is concentrated in the neutron), but the center of charge moves around (shown by an "x" in the diagram).

general.[1] The gravitational equivalent of the magnetic dipole moment is $\int \rho(\mathbf{r})\, [\mathbf{r} \times \mathbf{v}\, (\mathbf{r})] d^3r$. This quantity, however, is simply the total angular momentum of the system, so its conservation means that there is no magnetic dipole gravitational radiation either. The next static moment is quadrupolar: $I_{ij} = \int \rho(\mathbf{r})\, r_i r_j\, d^3r$, where $r_j = \delta_{ij} r^i$ and r^i are the components of the $\mathbf{r}$ vector (and here we just raise and lower indices with the flat spatial metric). This quantity need not be conserved and therefore there can be quadrupolar gravitational radiation.

Major Payne: If the authors are not going to define things, I guess it's up to me to do so. A vector V can be raised or lowered using the metric tensor $g_{\alpha\beta}$, which is to say that $V_\alpha = g_{\alpha\beta} V^\beta$ and $V^\beta = g^{\alpha\beta} V_\alpha$. A vector is a type of tensor, where the latter is defined as a geometric object that obeys certain transformation rules. The indices of a tensor are raised and lowered similarly to how they are for vectors. For example, if you have a tensor T whose (α, β) component is $T^{\alpha\beta}$, then $T^\alpha{}_\gamma = T^{\alpha\beta} g_{\beta\gamma}$. The metric is a special tensor that is defined such that the invariant interval between events in spacetime separated by dx^α is $ds^2 = g_{\alpha\beta} dx^\alpha dx^\beta$ (and $g^{\alpha\beta}$ is the matrix inverse of $g_{\alpha\beta}$). Here, and I'm assuming in much of the rest of the book, the authors will use the Einstein summation convention, in which there is an implied summation over all of the components when there is a repeated index. For example, in four-dimensional spacetime $A^\alpha B_\alpha = A^t B_t + A^x B_x + A^y B_y + A^z B_z$ if we are using Cartesian coordinates. Authors, you are welcome!

Of course, there is also an octupolar contribution, a hexadecapolar contribution, and so on, but these are typically suppressed by powers of the ratio of the light-crossing time to the characteristic timescale of temporal variation of the system. Why is that? Well, because when we talked about source variations above, what we really meant was time derivatives of the source multipole moments. For example, gravitational radiation emitted by an isolated source is produced by the time variation of the quadrupole moment. Now, a simple argument reveals how many derivatives must act on any given multipole moment to lead to gravitational radiation. Because gravitational waves are perturbations of spacetime, and these must be dimensionless (as we will see later in this chapter), and because they must decay as one over distance, we must have the same number of time derivatives as the multipolar order of the moment these derivatives are hitting! That is, the amplitude of gravitational waves must be proportional to

$$h \sim \frac{1}{r}[\partial_{tt} I + \partial_{ttt} O + \partial_{tttt} H + \ldots], \tag{1.1}$$

where I is the quadrupole moment, O is the octupole, H is the hexadecapole, and so on. Every time a time derivative hits these multipole moments, it introduces a factor of one over the characteristic timescale of temporal variation of the system.

[1] We can justify this statement heuristically by thinking about radiation as consisting of individual particles, such as photons or gravitons. The directions and energies of the particles are frame dependent, but the total number of such particles emitted over the history of the source is a scalar and thus all observers have to agree that radiation has been produced.

Therefore, the quadrupole term has two reciprocal powers of this timescale, the octupole three and the hexadecapole four and so on. Because we are assuming these moments are varying slowly (for example, on the orbital period of a binary system), the higher moments are suppressed more and more the higher the multipole order one considers.

Captain Obvious: There is an intuitive way to understand that it is actually something like the trace-free quadrupole that needs to vary. Recall that in Newtonian gravity, the gravitational field outside a spherical distribution of mass is the same as it would be if the mass were all concentrated at a point at the center of the sphere. A similar statement, called Birkhoff's theorem, applies in general relativity in a vacuum outside a spherical mass distribution with no net mass currents. Thus, if you had a "gravity meter" outside a spherical object able to sense any variation in the gravitational field (wave like, or not!), and if the object were to collapse or explode spherically, then your meter would not register any changes. Thus, spherical expansions and collapses cannot emit gravitational waves. The original formula the authors gave above for the quadrupole moment I_{ij} would change for an expansion or collapse, and thus it needs to be modified. Note, by the way, that if there is a net rotation of the object, then expansion or collapse can produce gravitational waves even if the mass distribution is spherically symmetric, due to the changing mass currents.

Gravitational waves and electromagnetic waves are both predominantly of a particular multipolar form (quadrupolar in the gravitational case and dipolar in the electromagnetic case), but the similarities don't end there. Both types of waves are also transverse to the direction of propagation. That is, the gravitational-wave perturbations of the gravitational field oscillate perpendicular to the direction of propagation, just like electric and magnetic fields. Heuristically, one can imagine that this must be the case because of the fundamental limitation of both Maxwell's and Einstein's theory: the fields of the theory cannot travel faster than the speed of light. Imagine an electromagnetic wave traveling in a vacuum at the speed of light and argue by contradiction. If electromagnetic radiation had a component that oscillated in the direction of propagation of the wave, then after boosting to a frame comoving with the wave, one would find that parts of the wave are at times traveling faster and at times slower than the speed of light. This is clearly forbidden, and thus no such component of electromagnetic (or gravitational) radiation is possible.

Major Payne: Heuristics, bah humbug! The reason for the transverse nature of electromagnetic and gravitational radiation is the existence of certain constraints in the field equations of Maxwell's and Einstein's theory. Maxwell's theory says that the electric and the magnetic field have to be divergence free in the absence of sources. These equations act as constraints that force the nontransverse component of electromagnetic waves to vanish. As in the electromagnetic case, the Einstein equations also contain constraints that force the part of the metric tensor that represents propagating degrees of freedom, a. k. a. gravitational waves, to be transverse.

But like a midnight commercial on television, there is more! The Einstein equations force the waves to not only be transverse but also traceless. That means that the relevant part of the quadrupole moment that sources gravitational waves must be proportional to

$$I_{ij}^{\text{traceless}} = \int \rho(\mathbf{r})\left(r_i r_j - \frac{1}{3}r^2\delta_{ij}\right)d^3r, \tag{1.2}$$

where δ_{ij} is the "Kronecker delta," also known as the identity matrix in linear algebra. And of course, as in the above case, only the traceless part of higher multipole moments (mass and current) contribute to gravitational waves. Incidentally, this is why most gravitational-wave calculations are typically done in "transverse-traceless gauge," i.e., in a coordinate system that isolates the transverse and traceless part of the metric perturbation, which represents gravitational waves.

Where are you going? I'm not done! The Einstein equations also reduce the possible number of gravitational-wave polarizations from six (the maximum allowed in a metric theory) to just two, as in electromagnetic theory. These constraints are important not only because they force the waves to oscillate transversely to their propagation, but also because the traceless condition implies that the perfectly spherical expansion of a distribution of mass cannot produce gravitational waves. Note that in other theories of gravity, other polarizations are possible; for example, if gravity were to be represented by a massive particle, a "massive graviton," and therefore travel at less than the speed of light, then longitudinal gravitational-wave polarizations would be allowed. I'm sure the authors will get back to this later, probably in Chapter 8.

1.2 What Can Generate Gravitational Radiation?

Now that we know that gravitational radiation must be transverse and traceless, we can draw some general conclusions about the type of motion that can generate such radiation. A perfectly spherically symmetric variation is only monopolar and does not change the trace-free quadrupole, so it does not produce radiation. No matter how violent an explosion or a collapse (even into a black hole!), no gravitational radiation is emitted if spherical symmetry is maintained. In addition, a rotation that preserves axisymmetry (without contraction or expansion) does not generate gravitational radiation because the quadrupolar and higher moments are unaltered. Therefore, for example, a neutron star can rotate rigidly arbitrarily rapidly without emitting gravitational radiation as long as it maintains axisymmetry about its rotation axis.

This immediately allows us to focus on the most promising types of sources for gravitational-wave emission. The general categories are binary systems, continuous-wave sources (e.g., rotating stars with nonaxisymmetric lumps), bursts of radiation (e.g., asymmetric collapses), and stochastic sources (i.e., individually unresolved sources with random phases; arguably the most interesting of these would be a background of gravitational waves from the early universe). We will discuss each of these later in the book, but for now, you can take a sneak peek at what these waves should look like in Figure 1.2.

Can you come up with an approximate expression for the dimensionless amplitude h of a gravitational wave (which can be considered to be a perturbation of the metric) some large distance r from a source?

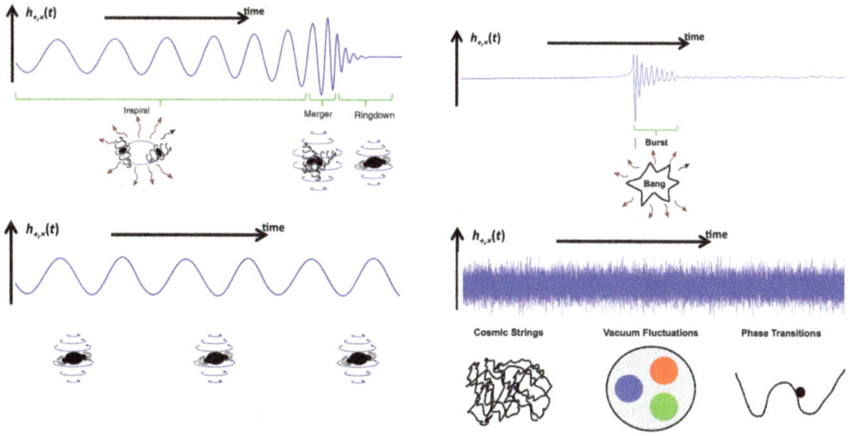

Figure 1.2. Sketches of the gravitational waves emitted by the main four classes of sources: binary coalescence (top left), bursts (top right), continuous (bottom left), and stochastic (bottom right). Adapted from Schutz (2005). Copyright © 1969, Springer Nature. With permission of Springer.

Major Payne: Now wait a minute. I've noticed that the authors are being sloppy (no surprise there) in an important way. If we want to be precise (and we should!), we need to draw a clear distinction between a spacetime (or geometry) and a metric. A spacetime can be considered a geometric entity; it might have flat portions (where, e.g., the sum of the interior angles of a triangle is 180°) or positive-curvature portions or negative-curvature portions. The geometric reality of a spacetime does not require a particular coordinate system. But to get to a metric, we need to specify our coordinates. For example, the Minkowski spacetime is flat. You can represent this using Cartesian coordinates, in which case the square of the invariant interval between two nearby events, represented by a line element, is $ds^2 = -c^2dt^2 + dx^2 + dy^2 + dz^2$ if for a particular observer the time separation between the events is dt and the x, y, and z separations are respectively dx, dy, and dz (let me note in passing that the authors are using a "(−, +, +, +)" metric signature in which the time component has a negative coefficient). Or if you represent this using spherical polar coordinates, the invariant interval between the same events would be $ds^2 = -c^2dt^2 + dr^2 + r^2(d\theta^2 + \sin^2\theta d\phi^2)$. When we look at the line elements, we see that they are different; that is, the metrics are different. But the spacetime hasn't changed! So careful scientists, unlike the authors, draw a distinction between spacetime and metric. But, like the authors, many people will say "metric" when they actually mean "spacetime," so you should be alert to the difference.

Anyways, let's go back to figuring out how to calculate wave-like perturbations of spacetime far away from a source. One of the main raisons d'être of this book is to teach you how to do order-of-magnitude estimates, i.e., estimates that provide the scaling and size of a given quantity. You may think you cannot possibly do this, that you cannot possibly answer the question above, because you are perhaps not an expert in gravitational-wave astrophysics yet (this is just Chapter 1 of this book!). Well, you would be wrong. There is immense power in knowing how to do a back-of-the-envelope, order-of-magnitude calculation, and much of this book will be devoted to teaching you how to do just that. And the great thing about Fermi estimates, as they are sometimes called after the great physicist Enrico Fermi, is that what matters are the physical principles rather than the details. When possible and when done right, Fermi estimates are almost magical.

Let us then argue our way to a Fermi estimate of the amplitude of a gravitational wave. We said before that gravitational waves had to be predominantly quadrupolar, and hence they must depend on the quadrupole moment I. In classical mechanics, you may have learned that this moment is just one of the (principal) components of the moment of inertia matrix (actually a tensor) $I_{ij} = \int \rho \, r_i r_j \, d^3x$. Clearly, because ρ is nonzero only inside the body, this quantity has dimensions of $\mathfrak{M}\,\mathfrak{L}^2$, where $\mathfrak{M}$ is some characteristic mass and $\mathfrak{L}$ is some characteristic size of the object. We also argued that the amplitude of any radiative field that carries energy out to infinity must decay as $1/r$, so we must have

$$h \sim \frac{\mathfrak{M}\,\mathfrak{L}^2}{r}. \tag{1.3}$$

We will later see that h has to be dimensionless because it represents fractional changes in length and time, so how do we determine what else goes in here? In general relativity, we usually set $G = 1 = c$, and these are called "geometric units." In these units, mass, distance, and time all have the same effective "units," but we can't, for example, turn a distance squared into a distance. Our current expression has effective units of distance squared (or mass squared, or time squared, all of which are the same when $G = 1 = c$). We note that time derivatives have to be involved as a static system can't emit any waves. As we pointed out back in Equation (1.1), two time derivatives will cancel out the current units, making h properly dimensionless, so that seems like it would work. Let's try it out! We then have

$$h \sim \frac{1}{r}\frac{\partial^2(\mathfrak{M}\mathfrak{L}^2)}{\partial t^2}. \tag{1.4}$$

But now what? To get back to physical units we have to restore factors of G and c. It is useful to remember certain conversions: for example, if M is a mass, GM/c^2 has units of distance, and GM/c^3 has units of time; and while we are at it, it is useful to remember the conversion factors $1 M_\odot \approx 1.5$ km $\approx 5\ \mu$s in geometric units (and also while we're at it, everybody knows that 1 yr $\approx \pi \times 10^7$ s). Playing with this a little bit gives finally

$$h \sim \frac{G}{c^4}\frac{1}{r}\frac{\partial^2(\mathfrak{M}\mathfrak{L}^2)}{\partial t^2}. \tag{1.5}$$

Because G is small and c is large, the prefactor is tiny! That tells us that unless $\mathfrak{M}$ and $\mathfrak{L}$ are large, the system is changing fast, and r is not too large, the gravitational-wave metric perturbation is dramatically minuscule.

Captain Obvious: General relativists and particle physicists often fight about which units to use. The general relativists like "geometric" units in which $G = 1 = c$, but the particle physicists like "natural" units in which $\hbar = 1 = c$. In both cases, they don't write the constants explicitly in their expressions, and their unit system is not the same! For example, general relativists will tend to convert all quantities to lengths or times, while particle physicists typically prefer units of energy (like electron Volts)—much fun arises when these two groups of scientists try to work together! Something to remember to convert between the two is that 1 eV $= 2 \times 10^{-7}$ m $= 6 \times 10^{-16}$ s. This takes some getting used to, but you'll be happy to learn that there is no dimensionless combination of G and c (or $\hbar$ and c), and so there is always a unique way to get the units to work out. By the way, you can get even more insight from this approach: c only appears if high speeds (approaching or equaling the speed of light) are important, G emerges only when gravity is important, and you only get $\hbar$ when quantum mechanics is important. So you can intuit which fundamental constants might play a role even before you start the problem!

Using this approach can lead to amazing leaps in physical understanding. Consider, for example, Hawking radiation. If we suggest (correctly) that the characteristic wavelength of Hawking radiation must be of order the size of the event horizon of a black hole, which is proportional to its mass M, then because the energy is proportional to the reciprocal of the wavelength in natural units, and because the energy of thermally emitted waves is also proportional to the temperature, this means that $E_H = kT_H \sim M^{-1}$, where k is Boltzmann's constant. With constants put in, $T_H \sim \hbar c^3/(GMk)$ (the actual factor in front is $1/8\pi$), so by physical intuition and dimensional analysis we have "derived" a profound result about black holes!

Let's make a very rough estimate for a circular binary. Suppose the total mass is $m = m_1 + m_2$, the reduced mass is $\mu = m_1 m_2/m$, the semimajor axis is a, and the orbital frequency ω is therefore given by $\omega^2 a^3 \sim Gm$ by Kepler's third law (after all, Newtonian gravitation is a good approximation to general relativity in many cases). Without worrying about precise factors, we can imagine that the quadrupole moment will vary with the orbital frequency, so that $\partial^2/\partial t^2 \sim \omega^2$. Then, because $\mathfrak{M}\mathfrak{L}^2 \sim \mu a^2$, we have[2]

$$h \sim \frac{G}{c^4}\frac{\mu a^2 \omega^2}{r} = \frac{G^2}{c^4}\frac{\mu m}{r\, a}. \tag{1.6}$$

This can also be written in terms of the orbital period $P = (1/2\pi)(Gm/a^3)^{1/2}$, and with the correct factors put in we get, for example, for an equal-mass system

$$h \approx 10^{-22}\left(\frac{m}{2.8\, M_\odot}\right)^{5/3}\left(\frac{0.01\ \text{s}}{P}\right)^{2/3}\left(\frac{100\ \text{Mpc}}{r}\right), \tag{1.7}$$

[2] You may wonder: why μ and not the total mass m? After all, m also has the right units and is symmetric between m_1 and m_2. We can break the tie by appealing to the limit in which one mass is much less than the other (think of a dust particle orbiting around a black hole). Then because the overwhelming majority of the total mass isn't moving much (it's in the big mass, which is nearly centered on the center of mass), there can't be much gravitational radiation. If we used m, however, we would conclude that there would be lots of gravitational radiation. That's wrong, and that's why we use μ instead.

which we have scaled to a double neutron star system. This is a really, really, really small number: it corresponds to less than the ratio of the radius of an atomic nucleus to the size of Earth. That's why it is so challenging to detect these waves! More on this in a little bit.

Remarkably, though, the flux of energy is not tiny. To see this, let's calculate the flux given some dimensionless gravitational-wave amplitude h. We can take a Fermi approach to this by noting that the flux has normal units of energy per area per time, which in geometrized units is $1/\text{time}^2$. We know that the flux has to go like the amplitude squared, or h^2, but because h^2 is dimensionless we need to put in the square of the frequency f of the wave, so that $F \sim h^2 f^2$.

Captain Obvious: Remember that flux has a direction: for example, if you think about the mass flux in a wind (i.e., the mass per area per time that flows by you), you also need to specify the direction of the wind!

Major Payne: Yes, well, if you want a deeper physical understanding of gravitational-wave flux, it is useful to compare with electromagnetic energy flux. For an electromagnetic wave, the energy flux is $e/(4\pi)\mathbf{E} \times \mathbf{B}$, where $\mathbf{E}$ and $\mathbf{B}$ are, respectively, the electric and magnetic fields. For a vacuum electromagnetic wave, $\mathbf{E} \perp \mathbf{B}$ and $|\mathbf{E}| \propto |\mathbf{B}|$ (indeed, in cgs units the magnitudes are equal). Thus, in a vacuum, the energy flux is proportional to the square of the electric field (or of the magnetic field). In contrast, a gravitational wave is more like a potential than a force field. We can motivate this by noting that in electromagnetic theory the Lorentz force is directly proportional to the electric field, while in gravitation the Newtonian force is directly proportional to the gradient of the gravitational potential, or equivalently, the gradient of certain components of the metric tensor. The gravitational-wave flux then has to be proportional to the square of the derivative of the gravitational wave, which is thus proportional to the square of the amplitude times the square of the orbital frequency f, namely $F \sim h^2 f^2$.

If we want to be even more quantitative (and we should!), we can approach this from another direction. The energy density in gravitational waves is

$$\rho_{\rm GW}\, c^2 = \frac{c^2}{32\pi G}\langle \dot{h}_{ij}\dot{h}^{ij}\rangle, \tag{1.8}$$

where h_{ij} is the metric perturbation (in the transverse-traceless gauge), the overdot indicates a time derivative, and the angle brackets indicate a time average (or equivalently, an average over multiple wavelengths).

Given that there are two gravitational-wave polarizations, which we label h_+ and $h_\times$ (see Section 1.3, and note that h_+ and $h_\times$ are intrinsically similar but just rotated 45° with respect to each other), we can rewrite this as

$$\rho_{\rm GW}\, c^2 = \frac{c^2}{16\pi G}\langle \dot{h}_+^2 + \dot{h}_\times^2\rangle. \tag{1.9}$$

If the + component is $h_{+,0}\sin(2\pi f t)$ and the × component is $h_{\times,0}\cos(2\pi f t)$, then because the time averages of $\sin^2$ and $\cos^2$ are both 1/2, we get

$$\rho_{\rm GW}\, c^2 = \frac{\pi c^2}{8G} f^2 (h_{+,0}^2 + h_{\times,0}^2). \tag{1.10}$$

If we define $h_{c,0}^2$ as the average of h_+^2 and $h_\times^2$, i.e., $h_{c,0}^2 = (h_{+,0}^2 + h_{\times,0}^2)/2$, then our final expression for the energy density is

$$\rho_{\rm GW}\, c^2 = \frac{\pi}{4}\frac{c^2}{G} f^2\, h_{c,0}^2. \tag{1.11}$$

Our last step is to note that for gravitational waves moving radially outward from a point source[3], the energy flux is related to the energy density by $F = \rho_{\rm GW} c$, and therefore

$$F = \frac{\pi}{4}\frac{c^3}{G} f^2\, h_{c,0}^2. \tag{1.12}$$

Because $F \sim (c^3/G) f^2 h^2$ the prefactor is enormous! For the double neutron star system above, with $h \sim 10^{-22}$ and $f \sim 100$ Hz, this gives a flux of a few times 10^{-2} erg cm^{-2} s^{-1}. For comparison, the electromagnetic flux from Sirius, the brightest star in the night sky, is about 10^{-4} erg cm^{-2} s^{-1}! That means that if you could somehow absorb gravitational radiation perfectly with your eyes, you would find a huge number of events with peak fluxes greater than every star except the Sun. What this really implies, of course, is that gravitational radiation interacts very weakly with matter (and in particular with our eyes), which again means that it is mighty challenging to detect.

1.3 How Can We Detect Gravitational Radiation?

To answer this question, the first thing we need to know is how a test particle reacts to an impinging gravitational wave. As we mentioned before, the physical effect of such a wave is to stretch and compress spacetime. The usual picture (see, e.g., Figure 1.3) is to imagine a circle of test particles (i.e., particles with masses so small that we can ignore their curvature of spacetime), and a gravitational wave that impinges on it. If you study how these test particles behave (through somcthing called the geodesic deviation equation), you'll find that they move up and down and left and right, producing either a plus "+" pattern or a "×" pattern, which represent the two polarization modes of gravitational waves, as shown in Figure 1.3. This behavior is indeed very similar to what happens in electromagnetism, where an electromagnetic wave will push charged particles in directions consistent with the wave's polarization. The answer to the question above is simple then: just use a sufficiently accurate "ruler" to measure how the distance between two objects changes as gravitational waves go through!

But this sounds fishy, or "a little bit strange," like some colleagues would say when they are trying to be polite. How can we detect gravitational radiation at all when your "ruler" will also stretch and contract as the wave goes through? This

[3] Note that caveat. Consider a point near to, but not exactly at, the center of the Sun; the photon energy density is huge but the energy flux is comparatively small because photons are traveling in all directions.

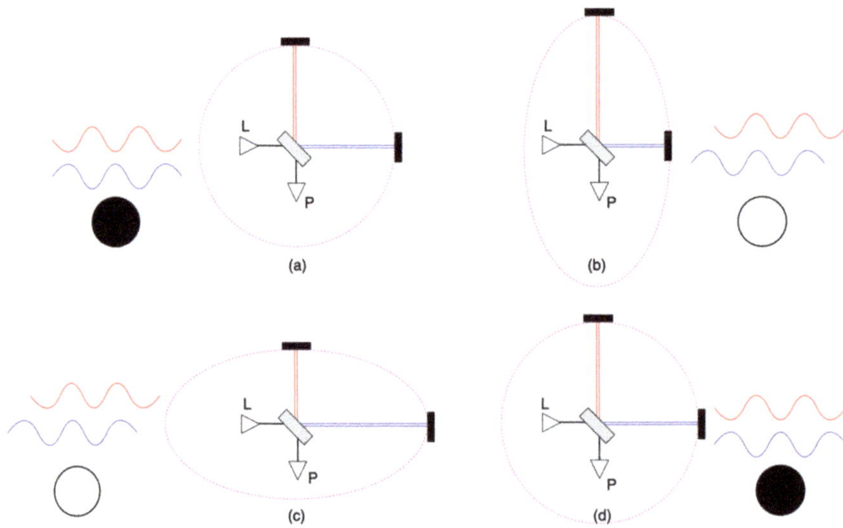

Figure 1.3. Schematic depiction of the response of a laser interferometer to an impinging gravitational wave. In this picture, a laser is emitted (L) to a beam splitter that separates the beam into two orthogonal directions; these beams then bounce off a mirror and recombine at the beam splitter to be detected by a photodetector (P). A gravitational wave is assumed to be traveling into the page with + polarization so that the response of the detector is to stretch in the up–down direction while shrinking in the left–right direction (top-right panel) and then to stretch in the left–right direction while shrinking in the up–down direction (bottom-left panel). We also show the up–down laser link in red and the left–right laser link in blue, together with a bright or dark port. The dotted circle represents how a circle of test particles would react to such an impinging gravitational wave.

question is rooted in the belief that all lengths are changed by the same factor when a gravitational wave comes by, including any rulers you may use to measure lengths. To be perfectly honest, this isn't quite true. If you try to stretch an atom, it will resist the stretching, and because not all atoms are created equal, all bodies do not react in exactly the same way. Having said that, for an idealized detector without an internal structure (that is, a detector where all the parts can be thought of as moving freely and not curving spacetime), it's a pretty good approximation to argue that all lengths are changed by the same factor when a wave goes through. If, then, your ruler changes in the same way as the distances you're trying to measure, how can you measure any change in the distance?

To get a good heuristic understanding of how measurements are possible, let's think about a laser interferometer. A laser interferometer works (in the most basic sense!) by having light bounce from a corner station along each of two arms to end stations and then having the light bounce back to interfere with itself (see, e.g., Figure 1.3). When the crests of the two beams coincide, there is a lot of light, and when the nulls add there is very little light. Typically, we build the interferometer such that when unperturbed it sits on a "dark fringe," meaning with complete destructive interference. This is because if the arm lengths are somehow perturbed (at an extreme, imagine "delicately" hitting one of the mirrors with a hammer), then the effect of the perturbation will be to generate some constructive interference, and therefore, allow some light to go through. And it turns out that it is much easier to detect light with a photodetector than it is to detect its absence!

Let's now consider a laser interferometer as a gravitational-wave detector. When a gravitational wave comes along, the lengths of the two arms change in different ways, depending on the phase of the wave and the orientation of the wave relative to the detector (see again Figure 1.3). But, working on the special relativistic axiom that light in a vacuum has a fixed speed, the crests of each of the light waves travel at the speed of light, regardless of any gravitational wave going through the detector. The wavelengths of the photons are stretched ever so slightly, but the crests of the waves travel at c. And before anybody complains, yes, it is true that the cavities inside which the lasers travel are not in a perfect vacuum, but they are evacuated to a very high degree, and this is close enough to vacuum for our purposes. Because the arm lengths are changed as the gravitational wave goes through, the crests return to the corner station to interfere at slightly different times than they would without being stretched or shrunk. As a result, the interference pattern changes as the gravitational wave goes through and the arm lengths shrink and expand. In this way of explaining it (and there are other, equivalent but more mathematical, explanations), the fixed speed of light in a sense gives you an "invariant ruler."

Okay, but you may think that things are still a little bit strange ⋯ even if we were able to build an "invariant" ruler, how can we detect gravitational waves when the shifts in lengths are so ridiculously tiny? Indeed, as we mentioned in passing before, the gravitational-wave amplitude is suppressed by a factor of G/c^4, so, for example, for two objects separated by a 4 km ruler, the change in distance is only one part in a thousand the size of the nucleus! A smart experimentalist, in fact, may argue that the wavelength of light used in the detectors is roughly 1 μm, which means that the resolution of our ruler is only $\lambda = 10^{-4}$ cm. However, the total length shift of the arms is about 10^{-17} cm, at a signal-to-noise ratio of 1 and a frequency of 200 Hz for 4 km arms. How can we then measure this tiny distance change with such a poor ruler?

The answer is that lasers have a lot of photons in them, and this comes to our rescue! If there are N photons over the period of a gravitational wave, then you can roughly get a measurement precision of $\Delta d \sim N^{-1/2}\lambda/(2\pi)$, where $1/(2\pi)$ can be pictured as being because you can measure distances to roughly one radian of the wavelength. One way to motivate the $N^{-1/2}$ factor is to imagine that you have a Gaussian with a width of w, and you have N samples from the Gaussian. You can estimate the location of the peak of the Gaussian to a precision of $\approx N^{-1/2}w$ (this is a good exercise to perform on a computer!). Thus, in the same way that we can determine the direction to a bright star with a precision many times better than the diffraction limit would suggest (otherwise high-precision astrometry missions such as Gaia would fail), we can measure the length shifts due to gravitational waves to much better than the wavelength of the light.

Let's make an illustrative estimate for the advanced Laser Interferometer Gravitational-Wave Observatory (advanced LIGO) as follows. Say that we are looking for gravitational waves at a frequency $f_{\rm GW}$, which means that the duration of one wavelength is $T_{\rm GW} = 1/f_{\rm GW}$. Advanced LIGO has an instantaneous laser power in its cavities that is roughly half a megawatt, or 5×10^{12} erg s^{-1}; the initial injected power is a lot less than this, but LIGO is not just a Michelson interferometer. It actually also contains a Fabry–Perot cavity, which allows the laser light to bounce around many times, building up the power in the cavity. The energy of one 10^{-4} cm photon is roughly $hc/\lambda \sim 2 \times 10^{-12}$ erg, so half a megawatt of power is about $\dot{N} \sim 2 \times 10^{24}$ photons per second. Suppose we're trying to detect a wave with a frequency of 200 Hz, so that $T_{GW} = 5 \times 10^{-3}$ s. There are then $N = \dot{N}T \approx 10^{22}$ photons interacting with the mirrors during one gravitational-wave wavelength, and thus we might expect a measurement uncertainty of $\Delta d \approx 10^{-11} \times 10^{-4}$ cm$/2\pi \approx 2 \times 10^{-16}$ cm.

That's pretty good! But we're still a factor of 10 away from what LIGO and other ground-based detectors can do. The missing element is that the light actually travels significantly farther than 4 km, despite that being the length of the arms. As we mentioned before, the light actually bounces many times inside the Fabry–Perot cavities, traveling a few hundred kilometers in total. Therefore, our calculation earlier about the length change induced by a gravitational wave in a 4 km detector is incorrect; we really need to multiply that number (10^{-17} cm) by the ratio of a few hundred kilometers to 4 km. This turns out to give us about the right answer at higher frequencies (several hundred Hertz and above), where the finite number of photons dominates the noise (this is called shot noise).

There are, of course, other sources of noise, and generations of clever instrument builders have worked and continue to work extremely hard to reduce that noise. However, this is the topic for a different book, so we won't discuss it further here.

1.4 Exercises

1. Let's play a bit with dimensional analysis.
   (a) Demonstrate that there is no combination of G and c, and of $\hbar$ and c, that is dimensionless.
   (b) What combination of G and c must multiply with a mass M to get a length and to get a time?
   (c) What combination of G and c yields a luminosity? Hint: this means that luminosity is "dimensionless" in $G = c = 1$ units.
   (d) Calculate the value of that luminosity in cgs or SI units, and compare this value with the luminosities of stars, quasars, and the peak luminosity of gravitational-wave sources (search the Web for characteristic values).

2. Dr. I. M. Wrong is 10^4 km from you, waving both arms rapidly in the course of a vigorous argument. The waving frequency is an amazing 100 Hz and is remarkably sinusoidal. Given plausible assumptions about the mass and radius of the doctor's arms, could advanced LIGO detect the waving if the detector's 1σ dimensionless strain sensitivity at 100 Hz is 5×10^{-23}?

3. Compare the total energy ever emitted in gravitational waves with the total emitted in starlight. To do this, assume the following: (i) that gravitational-wave energy is dominated by mergers of supermassive black holes, (ii) that a

typical current supermassive black hole has 10^{-4} of the total mass of the stars in a galaxy, and (iii) that in their lifetimes most supermassive black holes have one merger with a comparable-mass supermassive black hole, and in doing so emit 5% of their mass–energy in gravitational waves. For the starlight, assume that within a Hubble time (the current age of the universe; you should round this to 10^{10} years), 5% of the mass in stars has converted from hydrogen to helium, with an efficiency of 0.7%; that is, the energy released has been $E = 0.007\Delta M$, where ΔM is 5% of the total stellar mass. So what is greater? The total energy ever emitted in gravitational waves or that in starlight?

4. Demonstrate that for a spherical distribution of mass, the trace-free quadrupole (Equation (1.2)) is identically zero in a frame at rest with respect to the mass such that the origin of the system is at the center of the distribution. Therefore, spherical expansion or spherical contraction does not change the value of the quadrupole. Do the calculation for a general spherical distribution without, for example, requiring that the mass have constant density.
5. Dr. I. M. Wrong is deeply worried about orbital inspiral as expected from gravitational waves because it means that Earth will spiral into the Sun! The good doctor has run for political office based on the platform of doing something about this vital problem. Use Equations (1.6) through (1.12) to determine whether this is a problem. Hint: you haven't been given exact coefficients, so you should do a quick calculation and then determine whether greater precision is needed to answer the overall question of whether we need to take care of Earth's orbit!
6. Say that there is a neutron star with mass 1.4 $M_{\odot}$ and radius 12 km. Imagine that a black hole magically appears at its center, with mass $M < 1.4$ $M_{\odot}$, so that it is completely inside the neutron star.
   (a) As a warm-up for this problem, estimate how long it would take for all the neutron star mass that is outside of the black hole to be swallowed by the black hole, assuming that the star freefalls radially into the black hole. Use Newtonian gravity. Hint: yes, you can integrate the equations to get the time, but there is a cleverer method that uses Kepler's laws, where we note that a freefall from a stationary start is like half of an orbit with an eccentricity of about 1.
   (b) For the head-scratching part of the problem, think about what effects might prevent the collapse from being a freefall. For example, would it matter if the initial black hole mass were much less than the neutron star mass?

Useful Books

Ashtekar, A., Berger, B. K., Isenberg, J., & MacCallum, M. 2015, General Relativity and Gravitation: A Centennial Perspective (Cambridge: Cambridge Univ. Press)

Blair, D. G. 2012, Advanced Gravitational Wave Detectors (Cambridge: Cambridge Univ. Press)

Carroll, S. M. 2019, Spacetime and Geometry: An Introduction to General Relativity (Cambridge: Cambridge Univ. Press)
Dewitt-Morette, C. M., & Rickles, D. 2018, The Role of Gravitation in Physics—Report from the 1957 Chapel Hill Conf. (Berlin: Pro Business)
Maggiore, M. 2007, Gravitational Waves: Volume 1: Theory and Experiments (Oxford: Oxford Univ. Press)
Maggiore, M. 2018, Gravitational Waves: Volume 2: Astrophysics and Cosmology (Oxford: Oxford Univ. Press)
Misner, C. W., Thorne, K. S., & Wheeler, J. A. 1973, Gravitation (Princeton, NJ: Princeton Univ. Press)
Reitze, D., Saulson, P., & Grote, H. 2019, Advanced Interferometric Gravitational-wave Detectors (Singapore: World Scientific)
Rybicki, G., & Lightman, A. 1985, Radiative Processes in Astrophysics (New York: Wiley)
Schutz, B. F. 2003, Gravity from the Ground Up: An Introductory Guide to Gravity and General Relativity (Cambridge: Cambridge Univ. Press)
Schutz, B. F. 2005, Growing Black Holes: Accretion in a Cosmological Context (Berlin: Springer), 321
Schutz, B. F. 2009, A First Course in General Relativity (Cambridge: Cambridge Univ. Press)
Saulson, P. 2017, Fundamentals of Interferometric Gravitational Wave Detectors (Singapore: World Scientific)
Taylor, E. F., & Wheeler, J. A. 2000, Exploring Black Holes: Introduction to General Relativity (Reading, MA: Addison Wesley)
't Hooft, G. 2001, Introduction to General Relativity (Princeton, NJ: Rinton)
Thorne, K. S., & Blandford, R. D. 2017, Modern Classical Physics: Optics, Fluids, Plasmas, Elasticity, Relativity, and Statistical Physics (Princeton, NJ: Princeton Univ. Press)
Will, C. M., & Yunes, N. 2020, Is Einstein Still Right?: Black Holes, Gravitational Waves, and the Quest to Verify Einstein's Greatest Creation (Oxford: Oxford Univ. Press)
Wald, R. M. 1984, General Relativity (Chicago, IL: Univ. Chicago Press)
Weinberg, S. 1972, Gravitation and Cosmology: Principles and Applications of the General Theory of Relativity (New York: Wiley)

Chapter 2

Sources of Gravitational Radiation

2.1 Compact Binaries: General Considerations

We now turn our attention to binary systems. These obviously have a large and varying quadrupole moment and have the additional advantage that we actually know that gravitational radiation is emitted from them in the expected quantities (based on direct detections and, earlier, observations of double neutron star binaries in our Galaxy). The characteristics of the gravitational waves from binaries, and what we could learn from them, depend on the nature of the objects in those binaries. We will therefore start with some general concepts and then discuss individual types of binaries.

First, let's get an idea of the frequency range available for a given type of binary. The frequency of a gravitational wave from a binary system is proportional to its orbital frequency, and the latter, by Kepler's third law, is $\sqrt{Gm/a^3}$, where we recall that m is the total mass of the binary and a is the semimajor axis of the orbit. There is obviously no practical lower frequency limit (just increase the semimajor axis as much as you want), but there is a strict upper limit. The two objects in the binary clearly won't produce a signal higher than the frequency at which they touch. If we consider two objects of mass $M/2$ and radius $R/2$, then the orbital frequency when they touch is $\sim\sqrt{GM/R^3}$. Notice that this is the same orbital frequency as that of a test particle (of infinitesimal size) in orbit just above the surface of an object of mass M and radius R.

But we can generalize the above statements about the maximum frequency! Noting that $M/R^3 \sim \rho$, roughly the density of an object of mass M and radius R, we can say that the maximum frequency is then $f_{\rm max} \sim (G\rho)^{1/2}$. Amazingly, this limit applies more generally than to just orbital frequencies. For example, a gravitationally bound object can't rotate faster than $f_{\rm max}$, because, if it did, it would fly apart! To see this, consider a test particle on the surface of such an object. This particle would move faster than the orbital speed if $f > f_{\rm max}$, and this is why $f_{\rm max}$, in this context, is called the Keplerian mass-shedding limit. In addition, you can convince

doi:10.1088/2514-3433/ac2140ch2

yourself that the frequency of a sound wave that involves most of the mass of the object can't be greater than $\sim(G\rho)^{1/2}$. Therefore, this is a general upper bound on dynamical frequencies, at least for signals that last for several cycles or more.

Major Payne: Aha! The authors' lack of rigor has caught up with them. If we apply this claim to the Sun then we find that we predict that sound waves have a frequency no greater than about 1/(3 hr). But in fact, pressure modes are seen with frequencies as high as 1/(200 s). This shows that this hand-waving approach has given us a completely wrong answer.

Captain Obvious: Nice try, but note the important caveat: the authors are talking about a sound wave that involves most of the mass of the object because otherwise any resulting gravitational waves will be extremely weak. Local conditions (e.g., sharp gradients or high-order modes) can produce higher frequencies, it's true. But when most of the mass is involved for at least several cycles, the statement is valid. This is indeed why the estimate works well for the maximum rotation frequency and the maximum orbital frequency.

This tells us a lot about the maximum frequencies of binaries composed of different types of stars. For example, binaries involving main-sequence stars can't have frequencies greater than $\sim 10^{-3}$–10^{-6} Hz; this is simply because the mass of a main-sequence star is $\sim(10^{-1}$–$10^{2})M_\odot$, while its radius is typically $\sim 10^6$ km, so $(G\rho)^{1/2} \sim (10^{-3}$–$10^{-6})$ Hz. Using similar arguments, binaries involving white dwarfs, which have masses of $\sim(10^{-1}$–$10^{0})M_\odot$ and radii of $\sim 10^3$–10^4 km, can't have frequencies greater than $\sim$(few $\times$ 10^{-2}–1) Hz, also depending on mass. For binaries containing neutron stars, however, the upper limit is much higher, at $\sim$(1000–2000) Hz, because neutron star masses are about $\sim 1M_\odot$, while their radii are about $\sim$10 km. For black hole binaries, the upper-frequency limit depends inversely on the total mass because, once the black holes touch, they form a remnant with mass roughly equal to the total mass of the binary m and horizon size roughly equal to the sum of the horizons of the individual components in the binary (because the horizon size of a black hole is proportional to its mass for fixed spin); then, the mass of this remnant divided by its volume is roughly $M/R^3 \sim m/m^3 \sim m^{-2}$. In particular, for black holes, the maximum frequency is on the order of few $\times$ $10^4(M_\odot/M)$ Hz at the event horizon, but in reality, the orbit becomes unstable and the objects plunge into each other at lower frequencies, at roughly their innermost stable circular orbit or ISCO (see Section 3.2 for more details). This tells us immediately that for ground-based gravitational-wave detectors, which have frequency ranges on the order of $\sim(10^1$–$10^3)$ Hz, we can detect neutron stars and stellar-mass black holes but not main-sequence stars, white dwarfs, or supermassive black holes.

Now suppose that the binary is well separated, so that the components can be treated as test particles, and we only need consider the lowest-order contributions to gravitational radiation.

Major Payne: Wait a second! Stars are not test particles, so treating them as if they have infinitesimal extent is not rigorously correct. Thus, if we want to be careful we need to consider what effects might be introduced for noninfinitesimal stars and decide, in a given situation, whether those effects can be ignored.

Captain Obvious: Ok, FINE! Let's see how large (or small) finite-size effects are.

The main finite-size effect is the tides that they raise on each other, which can be captured, to leading order, by the quadrupole deformation generated on star 1 due to star 2 (and vice versa). How big is such a quadrupole moment? Well, physically, the reason star 1 deforms due to star 2 is that different parts of star 1 feel different accelerations toward star 2. For example, the part of star 1 farthest away from star 2 is attracted less than the part that is closest. The quadrupole induced on body 1 must then be proportional to the gradient of the acceleration caused by body 2, or equivalently, by the second gradient of the gravitational potential due to body 2, i.e., $Q_{ij} \propto \partial_{ij}\phi_2$, which gives us $|Q_{ij}| \propto m_2/r_{12}^3$ because $\phi_2 = m_2/r_{12}$, where we recall that m_2 is the mass of body 2 and r_{12} is the orbital separation.

The problem with this is that the units are all wrong! We know, by dimensional analysis, that the quadrupole must scale as MR^2 (for a body of mass M and radius R), so we need to multiply by a factor that is like a length scale to the fifth power. In this case, that scale is the radius of star 1, because if the star has zero extent, then the effect must vanish. Including this factor, we then get $|Q_{ij}| \propto R_1^5(m_2/r_{12}^3)$. The constant of proportionality, which we will call k_1 here, is the so-called (electric, quadrupole) Love number of star 1, so that then

$$Q_1 = k_1 \frac{m_2 R_1^5}{r_{12}^3} = k_1 m_1 R_1^2 \left(\frac{m_2 m_1^2}{r_{12}^3}\right) C_1^{-3}, \tag{2.1}$$

where $C_1 = m_1/R_1$ is the compactness of star 1 (and we are using geometric units, of course). The Love number depends on the star's internal structure, as I'm sure the authors will explain later, maybe in Section 7.2.4 and Appendix C.3.

Now, the contribution due to the quadrupole moment on the (reduced) energy of the binary scales as $E_b^{Q_1}/m \sim Q_1/r_1^3$. You know this must be the case both by dimensional analysis and because you know how to expand the gravitational field in multipole moments. The reduced energy is then

$$\frac{E_b^{Q_1}}{m} \sim \frac{Q_1}{r_1^3} \sim \frac{m}{r_{12}}\left[k_1\left(\frac{m}{r_{12}}\right)^5 C_1^{-5}\right], \tag{2.2}$$

where in the second expression we have approximated $r_1 \sim r_{12}$ and $m_1 \sim m_2 \sim m$. We recognize this as a modification of $\mathcal{O}(v^{10}/c^{10})$ relative to the (reduced) Newtonian gravitational binding energy $E_b^{\rm Newt}/m = \mu/(2r_{12})$ because by the virial theorem[1] $\mathcal{O}(m/r_{12}) = \mathcal{O}(v^2/c^2)$.

[1] This is a highly useful theorem that states that in a system with moment of inertia I, kinetic energy T, and potential energy $V \propto r_{12}^n$, where r_{12} is the separation between particles, $\frac{1}{2}d^2I/dt^2 = 2T - nV$, which equals $2T + V$ for electrostatic particles or Newtonian gravity (because for these $n = -1$). For a system with long-term bounded motion, so that $(1/2)d^2I/dt^2$ averages to zero, $2\langle T\rangle + \langle V\rangle = 0$ for Newtonian gravity or electrostatic potentials, where the angle brackets indicate the long-term time average. There is also a generalization of the virial theorem, called the Layzer–Irvine equation, which applies to an expanding universe. For a quasi-circular binary then, using that $\langle T\rangle = (1/2)\mu v^2$ and $\langle V\rangle = \mu m/r_{12}$, the virial theorem then tells us that $v^2 = m/r_{12}$.

In perturbation theory, this is called a fifth post-Newtonian order effect (because the counting goes with the power of m/r_{12}), and it is thus extremely small. How small? Well, for a widely separated binary, the orbital velocities are much smaller than a tenth of the speed of light, so therefore the finite-size correction is suppressed by a factor of 10^{-10}. And before you ask, the answer is no. The Love number is $\mathcal{O}(0.1 - 1)$ for neutron stars and the compactness $C \sim 0.2$, so the factor of $k_1 C_1^{-5}$ cannot compensate for the factor of $(v/c)^{10}$, and thus, finite-size corrections are still much less than unity if $v/c \ll 0.1$. So, you see! It was a small effect after all, so it's alright to treat the stars as test particles.

Focusing then on a binary system composed of test particles and temporarily restricting our attention to circular binaries, how will their frequency and amplitude evolve with time? We argued in the previous section (see Equation (1.6)) that the amplitude at a distance $r \gg r_{12}$ from a binary with masses (m_1, m_2) and orbital separation r_{12}, is $h \sim (\mu/r)(m/r_{12})$, where we recall that $m \equiv m_1 + m_2$ is the total mass and $\mu \equiv m_1 m_2/m$ is the reduced mass. We can rewrite the amplitude using $f \sim (m/r_{12}^3)^{1/2}$, to read

$$h \sim \frac{\mu}{r}(mf)^{2/3} \sim \frac{\mathcal{M}}{r}(\mathcal{M}f)^{2/3}, \tag{2.3}$$

where $\mathcal{M}$ is the "chirp mass," defined by $\mathcal{M} = \eta^{3/5} m$, with $\eta = \mu/m$ the so-called symmetric mass ratio. The chirp mass is named that because it is this combination of μ and m that determines how fast the binary sweeps, or "chirps" (yes, like a bird), through a frequency band. When the physical constants are put back in, and we restore all the factors of 2 and π (to make the Gods of Perfectionism happy), the dimensionless gravitational-wave strain amplitude (i.e., the fractional amount by which a separation changes as a wave goes by) for the plus polarization, as measured a distance r from a circular binary is

$$h_+ = -\frac{G}{c^2}\frac{4\mathcal{M}}{r}\left(\frac{G}{c^3}\pi\mathcal{M}f_{\rm GW}\right)^{2/3}, \tag{2.4}$$

where $f_{\rm GW}$ is the gravitational-wave frequency, which as we shall see later is twice the orbital frequency ($2f = f_{\rm GW}$), and where we have assumed we observed the binary face on (i.e., the binary's inclination angle is zero). This expression is valid only to leading order in a weak-field expansion of the Einstein equations, but this suffices here for our purposes. Note that this is the same scaling we found above!

Dr. I. M. Wrong: The authors have no idea what they're doing! In fact, they can't even agree on how to define the mass ratio q in a binary. Various papers have different conventions. Some authors define q to be the ratio m_2/m_1 of the lower-mass to the higher-mass component of the binary, so that $0 \leqslant q \leqslant 1$. Some authors define q to be the reciprocal of this ratio, so that $1 \leqslant q \leqslant \infty$. But the gravitational-wave community seems to largely prefer to use the symmetric mass ratio η (sometimes also called ν, a symbol that looks a lot like the velocity v). If papers can't agree on something as simple as this, we can't trust them with real calculations!

Captain Obvious: This can indeed be confusing, but as long as we're self-consistent in our calculations, our conclusions are fine. There is a good reason why the gravitational-wave community prefers η. Recall from the text that the symmetric mass ratio is defined via $\eta = \mu/m = m_1 m_2/(m_1 + m_2)^2$. We can then also write $\eta = q/(1 + q)^2$, which is true whether $q \equiv m_2/m_1$ or m_1/m_2. So you see, this is one reason why the gravitational-wave folk like to use η! There is no longer any confusion about whether a "large" mass ratio means two very similar masses (if $q \equiv m_2/m_1$) or two very different masses (if $q \equiv m_1/m_2$).

A second reason is that in the series expansions often encountered in analytic papers on gravitational waveforms, η often appears directly. Because the maximum value of η is 1/4, for an equal-mass binary, this means that series expansions on η have the prospect of converging quickly.

The primacy of η also tells us why it is so difficult to get both masses of a comparable-mass binary from gravitational-wave observations alone. As we will explore more in Chapter 3, if many cycles of the binary coalescence are detected, then the chirp mass can be determined with high precision. But the next-order effects are much weaker and depend on η. Because η has a maximum for an equal-mass binary, it doesn't vary much with the mass ratio until the stars have very different masses from each other: for example, if $m_1/m_2 = 1$ then $\eta = 0.25$, but if $m_1/m_2 = 2$, then $\eta = 2/9 = 0.222\ldots$. Thus if the masses are comparable to each other, as they always are in a double neutron star system, η must be measured extremely precisely to break the degeneracy and yield reliable individual masses.

Let us now play the Fermi estimate game with the luminosity. Using the arguments we discussed earlier (see, e.g., Equation (1.12) and the discussion of the luminosity at the beginning of Chapter 1), the gravitational-wave luminosity must scale as

$$\mathcal{L} \sim (4\pi r^2) f^2 h^2 \sim (\mathcal{M} f)^{10/3} \sim \eta^2 \left(\frac{m}{r_{12}}\right)^5 . \tag{2.5}$$

The total energy of a circular binary of radius r_{12} is $E = -E_{\rm bin} \sim -\mu M/(2r_{12})$, so we then have the energy balance law:

$$\frac{dE}{dt} = -\mathcal{L}, \tag{2.6}$$

which then implies that the orbital separation r_{12} cannot possibly remain constant. Gravitational waves are draining energy from the binary and thus forcing the orbit to decay![2]

[2] As the Major pointed out in the previous chapter, you'll see lots of papers referring to dE/dt as the "energy flux" in the relativity community, but in reality it is a luminosity or power. For astronomers and astrophysicists, dE/dt is not a flux because it is energy per unit time. The flux of something is that something per area per time (e.g., the number flux of some particles or the energy flux).

Major Payne: Wait, we can't make a blanket statement like that! There are plenty of circumstances in which there are other processes that can affect the orbit with a magnitude that is comparable to or sometimes much larger than the effect of gravitational radiation. For example, the drag on Earth's orbit due to the aberration of the radiation from the Sun, tiny though it is, is much more important than the decay of the orbit due to gravitational radiation. Another example important to gravitational-wave astronomy is that supermassive black holes in roughly circular orbits with periods of years or more could have their inspiral more strongly affected by gravitational interactions with stars or by gas drag than by gravitational radiation. Thus, here we have to specify that the authors are implicitly assuming that there are no competing effects, and in a given situation, we need to see how good or bad that approximation is.

The energy balance law tells us that the source of the energy that gravitational waves carry is precisely the energy of the binary system. Thus, the energy must become more negative (or equivalently the gravitational binding energy must be increased) precisely as prescribed by the gravitational-wave luminosity. We can use this balance law to find a differential equation for the evolution of the orbital separation:

$$\begin{aligned} \frac{dE}{dt} \sim \frac{\eta}{2}\left(\frac{m}{r_{12}}\right)^2 \frac{dr_{12}}{dt} = -\mathcal{L} \sim -\eta^2\left(\frac{m}{r_{12}}\right)^5 \\ \frac{dr_{12}}{dt} \sim -\eta\left(\frac{m}{r_{12}}\right)^3. \end{aligned} \tag{2.7}$$

This equation tells us that the more extreme the mass ratio, the slower the inspiral, and thus, the longer it would take for the binary to merge. It also tells us that although the orbital separation changes due to gravitational-wave emission, the effect is truly tiny, because using the virial theorem, $dr_{12}/dt = \mathcal{O}(v^6/c^6)$. Just for kicks, Earth goes around the Sun at an orbital velocity of just about 30 km s^{-1}, so then $|dr_{12}/dt| < 10^{-24}$!

What if the binary orbit is eccentric? The formulae are then more complicated because the binary components will move faster when they are close to the pericenter and slower when they are at the apocenter. For this reason, we must average over the orbit to get sensible expressions for the rate of change of the orbital elements, such as the semimajor axis a and the orbital eccentricity e. This was done first (to lowest order in an expansion about small velocities and weak fields) by Philip Peters and Jon Mathews, by calculating the luminosity $\mathcal{L}$ and the torque $\mathcal{T}$ in gravitational waves at the lowest (quadrupolar) order. From these, they then determined the change in the orbital elements that would occur if the binary completed a full Keplerian ellipse in its orbit. That is, they assumed that, to lowest order, they could have the binary move as if it experiences only Newtonian gravity, and integrated the losses along that path.

Before quoting the results, we can understand one qualitative aspect of the radiation when the orbits are elliptical. From our derivation for circular orbits, we know that the loss of energy must force the semimajor axis to decrease and the elliptical orbit to decay. But from that derivation, we also see that the radiation is emitted much more strongly when the separation is small, because $\mathcal{L} \sim r_{12}^{-5}$. Consider what this would mean for a very eccentric orbit $(1 - e) \ll 1$. Most of the radiation would be emitted at pericenter, so this would have the character of an impulsive force. With such a force, the orbit will return to where the impulse was imparted.[3] But because energy is being lost, the binary cannot get quite as far away (to the same apocenter) as it did in its previous pass. This then means that the eccentricity must decrease.

Captain Obvious: Of course! This is clear, right? Another way to think about it is in terms of the radiated energy and angular momentum. For an elliptic orbit, the orbital energy depends on the semimajor axis but not on the eccentricity via $E \sim -\mu m/(2a)$. Therefore, if the energy is forced to change due to the emission of gravitational waves, it must do so by adjusting the semimajor axis.

Gravitational waves produced by elliptic orbits, however, also carry away angular momentum, so the gravitational-wave torque $\mathcal{T}$ must be balanced by a change in the angular momentum of the orbit $L_{\rm orb} \sim \eta M(Ma)^{1/2}\sqrt{1 - e^2}$. How can the orbital angular momentum change? Well, $L_{\rm orb}$ is already changing because the semimajor axis is changing, but this is not enough to balance $\mathcal{T}$, which then forces the eccentricity to also change. As in the case for the semimajor axis, the eccentricity decreases, but this time it is because gravitational waves also carry angular momentum away.

Major Payne: But the problem here is much more complicated! The gravitational waves are not really emitted in an instantaneous burst, and even if they were, the pattern of the radiation is very complicated. In the 1970s, Michael Turner and Clifford Will studied the emission of gravitational waves in eccentric orbits, and they found that the angular distribution of the gravitational-wave burst energy radiated, $dE/d\Omega$, is mostly orthogonal to the direction of propagation, with some beaming in the direction of propagation for large eccentricities. The resulting "antenna" pattern resembles that of electromagnetic bremsstrahlung radiation, which is why sometimes this type of emission is called gravitational bremsstrahlung. In practice, all of these effects have to be included carefully to get all the numerical factors right in the formulae for the luminosity and torque, especially when dealing with highly elliptical orbits.

The Peters formulae bear this out. If the orbit has semimajor axis a and eccentricity e, then their lowest-order rates of change are

$$\left\langle \frac{da}{dt} \right\rangle = -\frac{64}{5}\frac{G^3\mu M^2}{c^5 a^3(1 - e^2)^{7/2}}\left(1 + \frac{73}{24}e^2 + \frac{37}{96}e^4\right) \tag{2.8}$$

and

[3] Why? Because after the impulse, in bound motion (i.e., the two objects won't fly away from each other), the constants of motion are consistent with the location of the impulse. Therefore, the system can return to the same pericenter, with the same speed as it had at the impulse.

$$\left\langle \frac{de}{dt} \right\rangle = -\frac{304}{15} e \frac{G^3 \mu M^2}{c^5 a^4 (1 - e^2)^{5/2}} \left(1 + \frac{121}{304} e^2\right), \tag{2.9}$$

where the angle brackets indicate an average over an orbit. One can show (for example, through direct evaluation) that these rates imply that the quantity[4]

$$a\, e^{-12/19} (1 - e^2) \left(1 + \frac{121}{304} e^2\right)^{-870/2299} \tag{2.10}$$

is constant throughout the inspiral, at least, to lowest order in the small velocity and weak-field expansion used to derive Equations (2.8) and (2.9). Because $0 \leqslant e \leqslant 1$, the final factor varies between 0.88 and 1, which means that we aren't that wrong in saying that $a\, e^{-12/19}(1 - e^2)$ is approximately conserved.

In physics and math, it is commonly helpful to look at limits. In the limit $e \to 1$, the quantity $e^{-12/19} \approx 1$, and thus, $a(1 - e^2)$ is approximately conserved. We note that the pericenter distance $r_p = a(1 - e)$, and thus, if $e \approx 1$, then $1 + e \approx 2$ and $a(1 - e^2) = r_p(1 + e) \approx 2r_p$; this means that when $e \to 1$ the pericenter distance is approximately conserved, as we derived with our rough argument about the effect of gravitational radiation being impulsive for highly eccentric orbits, and as shown in Figure 2.1. In the opposite limit, $e \to 0$, the last two factors of Equation (2.10) tend to 1, so that $a\, e^{-12/19}$ is approximately conserved. We can write this as $e \propto a^{19/12}$, which means that as the semimajor axis shrinks, the eccentricity decreases, as shown in Figure 2.1. When we note that the orbital frequency in Newtonian gravity scales as $f \sim a^{-3/2}$ by Kepler's third law, and thus $a \sim f^{-2/3}$, we see that $e \propto f^{-19/18}$. Thus, in the small-eccentricity limit, the eccentricity scales roughly as the inverse of the frequency. Together, the high-eccentricity and low-eccentricity limits tell us that if we observe a binary at a frequency much higher than the orbital frequency at the pericenter of the initial orbit, we expect the eccentricity to be much smaller than unity. That is, an elliptic orbit circularizes due to the emission of gravitational waves! This leads to a welcome simplification of the task of constructing waveform templates, which we will describe in Chapter 3.

Did we have evidence that these formulae actually work before the direct detection of gravitational waves using ground-based gravitational-wave detectors? Yes! Nature has been kind enough to provide us with the perfect test sources: binary neutron stars. Several such systems are known, all of which have binary separations orders of magnitude greater than the size of a neutron star, so the lowest-order expansions in weak fields and small velocities should work. Indeed, the *da/dt* predictions have been verified to better than 0.1% in a few cases through measurements of the decay of the orbital period, but the *de/dt* predictions will be much tougher to verify. The reason for the difference is that *de/dt* has to be measured by

[4] You may have notice the funny fraction that appears in the exponent of the expression in Equation (2.10). This is not a typo, but rather a mean trick of the gods that must be laughing at us while we try to keep all these fractions straight. It is nonetheless remarkable that, at this order in the expansion about weak fields and slow velocities, there exists a closed-form solution for the evolution of the semimajor axis with the eccentricity.

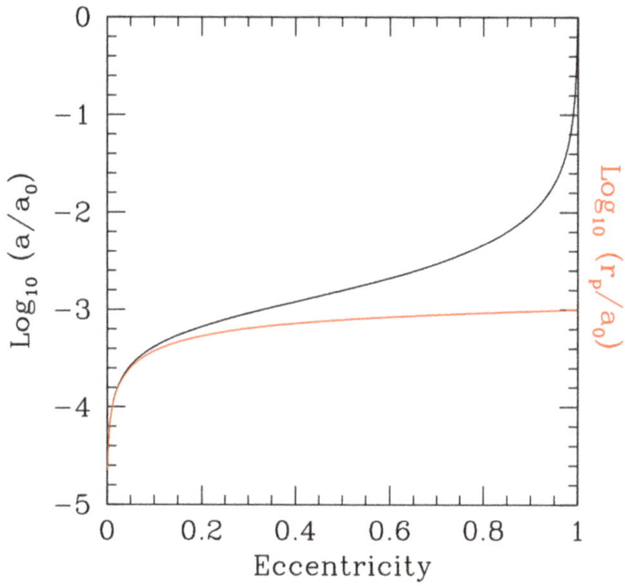

Figure 2.1. Semimajor axis versus eccentricity using the Peters equations and the constant of motion in Equation (2.10). The binary starts with a semimajor axis a_0 and eccentricity $e_0 = 0.999$. The figure shows the $\log_{10}$ of the semimajor axis divided by a_0 (black line) and the $\log_{10}$ of the pericenter $r_p = a(1 - e)$ divided by a_0 (red line). As advertised, the pericenter distance is close to constant until the eccentricity decreases significantly from 1.

determining the eccentricity orbit by orbit, while da/dt has a direct manifestation in the orbital period, and thus the total orbital phase of the binary. In particular, over the time in which a changes only slightly, and thus da/dt is roughly constant, f changes linearly with time; thus, because the phase difference from a constant-frequency orbit is the time integral of the frequency difference, the phase difference accumulates quadratically with time. These systems provide a spectacular verification of general relativity, deservedly leading to the 1993 Nobel Prize in physics going to Hulse and Taylor, who discovered the first such binary with radio observations.

Thus, even prior to the direct detection of gravitational waves, the community was quite confident that, at least in weak gravity, gravitational radiation exists as advertised. What happens in strong and dynamical gravity when binaries plunge and collide? That is, at least in part, the topic of this book, and we will cover it in much more detail in later chapters.

Before doing so, however, let us continue on our broad exploration of the coalescence of compact binaries, focusing on black holes for the moment. Generically, we divide the whole coalescence process into three stages that of course overlap with each other. The first stage is the inspiral, which follows the binary from large separations to when the binary has reached a stage of dynamical instability. Dynamical instability occurs roughly when the binary is separated by several m (or several Gm/c^2 if we put the constants back in), which happens to coincide approximately with the ISCO of a test particle in motion around a black hole of mass equal to the total mass of the system. Once the dynamical instability sets in, the orbital motion becomes a plunge, and this happens on a dynamical timescale (i.e., of order the time needed for free fall). Of course, a proper plunge only occurs in the extreme mass-ratio limit, whereas for comparable-mass binaries, the transition from inspiral to merger is much smoother. As the (apparent) horizons disturb each other and finally touch, the spacetime becomes extremely complicated and must be treated numerically. This is called the merger phase.

Ultimately, of course, the "no hair" theorems guarantee that, when the merger product is a black hole devoid of any external influences (and under some sensible mathematical restrictions, including the validity of general relativity), the system must settle into a Kerr spacetime. It does this by radiating away its bumpiness as a set of quasi-normal modes. The lowest order, and longest lived, of the modes is the $l = 2$, $m = 2$ mode. When the nonlinear gravitational waves produced shortly after merger die down, and all that is left are quasi-normal modes, the system has entered the period of ringdown. This period can be treated both numerically, by investigating the late stage of full numerical relativity simulations, as well as semianalytically through black hole perturbation theory. The result is that the frequency $f_{\rm qnr}$ of the gravitational radiation for a given mode (where "qnr" stands for "quasi-normal ringdown"), as well as its quality factor $Q \equiv \pi f_{\rm qnr} \mathfrak{T}$ (where $\mathfrak{T}$ is the characteristic duration of the mode), which measures roughly how many cycles the ringing lasts, depends on the final black hole mass $M_{\rm fin}$ and the spin $\chi_{\rm fin} \equiv cJ_{\rm fin}/GM_{\rm fin}^2$ of the final black hole (sometimes also called $\hat{a}_{\rm fin}$), as well as the mode numbers. For the typically dominant $l = 2$, $m = 2$ mode, fitting formulae valid to $\sim$5% are

$$\begin{aligned} f_{\rm qnr} &\approx [1 - 0.63(1 - \chi_{\rm fin})^{0.3}](2\pi M_{\rm fin})^{-1} \\ Q &\approx 2(1 - \chi_{\rm fin})^{-0.45}. \end{aligned} \tag{2.11}$$

Thus, more rapidly spinning remnants have higher frequencies and last for more cycles. For example, for $\chi_{\rm fin} = 0$ and $M_{\rm fin} = 10\ M_\odot$, $f_{\rm qnr} \approx 1200$ Hz, whereas for $\chi_{\rm fin} \approx 1$ and $M_{\rm fin} = 10\ M_\odot$, $f_{\rm qnr} \approx 3240$ Hz. This could allow identification of the spin and mass based on the character of the ringdown.

Dr. I. M. Wrong: What's with the "quasi" nonsense? Any system can be broken up into normal modes, as you'd know if you actually understood the physics.

Captain Obvious: I'm sure the authors will explain all of the above in more detail in Chapter 3, but okay, let me explain this bit here my way. Let's start by first recalling ordinary normal modes from classical mechanics. Think of a wire, clamped at both ends, which is then plucked. We know that the boundary conditions admit a variety of separate modes, e.g., one with no nodes between the ends, one with one node, one with two nodes, and so on. Because these are normal modes, we also know that any dissipationless oscillation of the wire can be decomposed into a linear superposition of the normal modes. In the linear approximation, those modes all oscillate separately (although Major Payne insists that I note that mode–mode coupling is possible once you leave the linear regime). Decomposition of a system into its normal modes is an extraordinarily powerful analysis method in many situations (and is sometimes the topic of PhD-qualifying examinations!).

Something like this exists for the oscillations of stars, including compact objects such as neutron stars and black holes. However, the necessary decay of such modes due to dissipation via gravitational radiation means that, unlike with normal modes, one cannot for example represent the actual oscillations of a merger remnant settling into a Kerr state as a linear superposition of these modes (because at linear order, normal modes do not decay). Thus, these are quasi-normal, rather than normal, modes. In this sense, black holes ring down more like a bell that is struck or like a violin string that is plucked, whose vibrations and oscillations decay due to material stresses, air resistance, and other dissipative forces.

We can make rough estimates of the energy released in each phase as a function of the reduced mass μ and total mass m of the system. As the inspiral phase goes from infinite (spatial) separation to the ISCO, the energy released is roughly μ times the specific gravitational binding energy at the ISCO, so $E_{\rm inspiral} \sim \mu m/(2r_{12}^{\rm ISCO}) \sim \mu$, where we recall that r_{12} is the orbital separation and (as you will learn in general relativity) $r_{12}^{\rm ISCO} = (1-9)m$, depending on the magnitude and direction of the black hole spin. What about the merger and ringdown phases? We know that the strain amplitude is $h \sim (\mu/r)(m/r_{12})$, where we recall that r is the distance to the observer. For the merger and ringdown phases, $r_{12} \sim m$, so $h \sim \mu/r$. We also know that the luminosity is $\mathcal{L} \sim r^2h^2f^2 \sim \mu^2f^2$, and if the phase lasts a characteristic time $\mathfrak{T}$, then the total energy released is $E \sim \mu^2f^2\mathfrak{T}$. But because the light-crossing time is m, the characteristic frequency is $f \sim 1/m$ and the characteristic time is $\mathfrak{T} \sim m$, so we have finally $E \sim \mu^2/m$, or a factor $\sim\mu/m$ times the energy released in the inspiral. For an equal-mass coalescence, the amount of energy released during inspiral is then comparable to that emitted during merger and ringdown. But for highly unequal-mass binaries ($\mu \ll m$), the inspiral produces much greater total energy than the merger or ringdown. One way to picture this is to note that the smaller $\eta \equiv \mu/m$ is for a fixed total mass m, the weaker the gravitational radiation is per unit time. The inspiral takes longer for smaller η, which helps compensate for the weakness of the radiation, but the merger and plunge occur on a dynamical timescale of order m regardless of how small η is. This is one reason why analyses of extreme mass-ratio inspirals (with $\eta \ll 1$) have ignored the merger and ringdown phases (which we will discuss in more detail in Chapter 3). Of course, all of this discussion ignores the fact that detectors are typically operational in a given frequency band, and the merger frequency scales inversely with the total mass of the system. So, even if less energy is emitted in the merger and ringdown, for massive enough systems, the inspiral cannot be detected by ground-based instruments at all, and thus, all that matters is the post-inspiral phase.

The coalescence of neutron stars is somewhat different. The initial inspiral phase is the same, of course. But as the neutron stars get closer to each other, they become tidally distorted. As we mentioned in Chapter 1, this is because any one star in a binary will feel a tidal acceleration due to its companion, and this will cause the matter inside the star to move and redistribute itself, inducing a quadrupole (and higher-order mutipole) deformation, both of its shape and of its gravitational field. The tidal Love number (or usually just the Love number for short; see Chapter 7 for an extensive discussion) is precisely a measure of how much the gravitational field is multipolarly deformed due to an external tidal perturbation. This doesn't happen for black holes for subtle reasons.

Major Payne: Wait, what? Surely the authors are not implying that the horizon of black holes (i.e., the geometry of the induced two-dimensional surface that defines the apparent horizon) is not deformed in a merger. Indeed, careful and very high-order black hole perturbation theory calculations carried out by Eric Poisson and his group have shown analytically that the horizon is indeed deformed. Numerical relativity simulations have also shown this, through visualizations of tidal tendexes.

Captain Obvious: Indeed, for black holes things are different, and the tidal Love number is zero, in spite of their shape becoming deformed right before merger. As Eric Poisson puts it, intuitively this is because black holes do not have any matter in their interior, except at the singularity itself. They are vacuum solutions to the Einstein equations after all! Therefore, when in the presence of an external tidal acceleration, there is no matter inside the black hole to move and redistribute itself, and thus to generate a quadrupolar or higher multipolar deformation to the gravitational field. This is why the tidal Love number must be zero for black holes.

These statements, however, do not mean that the horizon of a black hole does not deform. Indeed, there are other Love numbers, conspicuously called shape Love numbers, that control how much of a deformation the surface of the body (for a black hole, its horizon) acquires when in the presence of a tidal force. For black holes, these shape Love numbers are indeed nonzero, as expected from high-order perturbation theory calculations and numerical relativity simulations.

If you want a picture to understand why a black hole isn't tidally deformed even though the shape of its horizon changes due to the nearby presence of a gravitating object, think about it this way. Very heuristically, you can think of a nonrotating black hole as an object whose mass is all concentrated at its center, sort of like a point mass in Newtonian gravity. Because of this, the mass can't get redistributed and the tiny region where this mass is located can't be deformed. For this reason, the gravitational field that this mass produces cannot be deformed when doing a multipolar expansion far from the object. However, such an object's surface can be deformed when you think of it as an equipotential surface. We know that, although equipotentials are spheres, if the mass is by itself in Newtonian gravity, they are tidally distorted by the presence of another object. Therefore, you find yourself in a situation in which the mass distribution of the object (and therefore the gravitational field it produces far away) is not multipolarly deformed, yet its surface is tidally deformed. In other words, the geometry of the horizon of the black hole is tidally deformed, although its gravitational field far from the horizon is not. Of course, like any heuristics in this book, this explanation should be taken as a metaphor and not as a rigorous picture.

Eventually, of course, the neutron stars crash into each other's surface, and, after the collision, there are several possible fates depending on the mass of the remnant. In decreasing the order of total mass, these fates are the following: (1) immediate collapse into a black hole, (2) temporary formation of a differentially rotating remnant with a significant accretion disk that, when it approaches solid-body rotation, collapses (this is called a hypermassive neutron star), (3) formation of a solidly rotating object that is stable at its initial spin rate, but collapses when its rotation slows down enough (this is called a supramassive neutron star), and finally (4) formation of an object that is stable even at zero rotation. During the merger and accretion from a disk, quite a bit of matter can be ejected, some of which can escape or return to the remnant and then be accreted. There are several paths to electromagnetic emission during and after the merger, as we will discuss in Section 7.2.6.

For mixed binaries, composed of a neutron star and a black hole, much of the coalescence resembles that of a binary black hole. The main difference here is that, depending on the black hole mass, the neutron star may either be tidally disrupted or be swallowed whole. Perhaps somewhat counterintuitively, the former occurs when the black hole is light enough and not when it is heavy. The tidal acceleration, given by contractions of the Riemann tensor with the 4-velocity of the observer, scales as $M\,R/r^3$, for a body of radius R, separated from another body of mass M by a distance r. Now, when their separation is comparable to the horizon of the black hole, $r \sim M$, the tidal acceleration at the horizon scales as R/M^2, and it is therefore larger for lighter black holes. Numerical simulations show that for a neutron star to be tidally disrupted by a black hole, the latter must have a mass smaller than about 10 solar masses. If tidal disruption occurs, a large electromagnetic signal is also expected, and people are now working out the details including whether there is a clear signature of a black hole–neutron star merger versus a binary neutron star merger.

Captain Obvious: Surely we can Fermi estimate the black hole mass at which disruption occurs, right? In detail, of course, this depends somewhat on the equation of state (EOS) of the dense matter in the neutron star, the spin parameter of the black hole, and the orientation of the orbit. However, as a Fermi estimate, we can compare the Newtonian tidal force on the neutron star from the black hole, with the neutron star's self-gravity, at the event horizon. Say that the black hole has a mass $M_{\rm BH}$ and that the neutron star has a mass of $M_{\rm NS}$ and a radius of $R_{\rm NS}$. If the separation between the center of the black hole and the center of the neutron star is $r \gg R_{\rm NS}$, then when $GM_{\rm NS}^2/R_{\rm NS}^2 \approx 2GM_{\rm BH}M_{\rm NS}R_{\rm NS}/r^3$ we expect disruption. Putting in $r = 2GM_{\rm BH}/c^2$ for the event horizon radius gives $M_{\rm BH} \approx 10\ M_\odot$ for $M_{\rm NS} = 1.4\ M_\odot$ and $R_{\rm NS} = 12$ km. This is not too far off from the value of $M_{\rm BH} \approx 7\ M_\odot$ found doing full numerical relativity simulations with spinning black holes and neutron stars.

Let us conclude by discussing one last interesting effect that emerges from the higher-order studies of binary inspirals: gravitational waves not only carry away energy and angular momentum, but they also carry away net linear momentum. This has the effect of forcing the center of mass of the system to move in an ever-widening spiral and bob up and down if the black holes are spinning in the right orientation. We can understand this as follows, following a nice heuristic idea by Alan Wiseman. In an unequal-mass binary, the lower-mass object moves faster. As the orbital speed increases, the gravitational radiation from each object becomes more beamed, with the lower-mass object producing more beaming because it moves faster. Therefore, at any given instant, there is a net kick against the direction of motion of the lower-mass object. If the binary were forced to move in a perfect circle, the center of mass of the system would simply go in a circle as well. However, because in reality the orbit is a tight and diminishing spiral, the velocity of the center of mass, the time-dependent recoil, becomes larger and larger with time and the center of mass moves in an expanding spiral; eventually, the objects merge, at which point the recoil stops changing with time, it acquires its final value and the entire system moves with a constant velocity. Note that by symmetry, equal-mass nonspinning black holes can never produce a linear momentum kick and that if the mass ratio is gigantic, the fractional energy release is small and therefore so is the kick. For nonspinning black holes in a quasi-circular orbit, the optimal ratio for a kick is about 2.6.

Major Payne: Maybe this gets the right sense of the answer, but it is mathematically sloppy! The problem is that when we think of radiation being beamed, it is natural to have a mental picture in which the radiation has a short wavelength and comes out tightly, as if a gravitational-wave flashlight were pointed toward the direction of motion of an orbiting mass. One of the many problems with this picture is that, in fact, the wavelength of a gravitational wave is larger than the system! To see this, think about a circular system, for which the gravitational-wave frequency is of order the orbital frequency. But gravitational waves travel at the speed of light and the orbit can't travel faster than the speed of light. Thus, a gravitational-wave period times the speed of light, which gives us the wavelength, is larger than the size of the orbit (and it can be much larger if the orbital speed is much less than the speed of light). Thus, we cannot think of gravitational waves as localized to one binary partner or the other; they come from a full, nonlinear, disturbance of the spacetime, properly defined very far away from the system.

This process is potentially important astrophysically because if the final merged remnant of a black hole inspiral is moving very rapidly, it could be kicked out of its host stellar system, with possibly interesting implications for supermassive black holes and hierarchical merging. There have therefore been a number of calculations of the expected kick, and it has turned out that these are very challenging. The primary reason is that most of the action is near the end, when the black holes are close to each other and slow-motion/weak-field approximations to the orbit are inaccurate. Indeed, analytic calculations suggest that the kick due to inspiral (from infinity to right before plunge) is minimal, but the final plunge could produce interesting speeds. Tremendous progress in numerical relativity has shown that kick velocities can be as high as 5000 km s^{-1} when the binary components are spinning at the right rate and in the right orientation. For nonspinning components, the maximum kick velocity is much smaller, less than 200 km s^{-1} if the system is in a quasi-circular orbit.

2.2 Nonbinary Sources

Once we move away from binaries, we enter, for the most part, unknown territory. All nonbinary sources are of unknown strength, which is another way of saying that if they are detected, we can learn a lot of astrophysics. We now discuss continuous, burst, and stochastic sources, following the classification we introduced in Section 1.2.

2.2.1 Continuous Sources

The first class of uncertain sources that we will discuss is continuous sources. This class consists of all sources that emit gravitational waves at essentially a constant frequency. One example is rigidly rotating stars with some nonaxisymmetric lump, like a "mountain," on their surface. Another example is a very widely separated binary that evolves in frequency only slightly during the period of observation. Sure, the frequency will not remain exactly constant, but it will also not change nearly as drastically as it changes for coalescing binary sources.

The approximate constancy in the frequency of continuous sources has important implications for their observability. A binary increases its frequency a lot as it loses energy, meaning that searching for an unknown binary requires potentially involved data analysis. In contrast, a spinning source can in principle emit gravitational waves at approximately a single frequency for a long time, so the signal builds up in a narrow frequency bin. As a result, particularly for high frequencies observable with ground-based detectors, continuous-wave sources are interesting because they can in principle be seen even at relatively low amplitudes.

What amplitude can we expect? From Chapter 1 we know that if the (reduced, trace-free) quadrupole moment is I_{ij} (see Equation (1.2)), then the gravitational-wave metric perturbation is

$$h_{ij} \sim \frac{G}{c^4}\frac{1}{r}\frac{\partial^2 I_{ij}}{\partial t^2}. \tag{2.12}$$

For binaries we argued that $|I_{ij}| \sim \mu a^2$, where we recall that μ was the reduced mass and a the semimajor axis, and we also had a relation between a, the orbital angular frequency $\omega^2 \sim \partial^2/\partial t^2$, and the total mass m using Kepler's third law. However, for a spinning source, these relations do not have to hold. For a gravitationally bound source (e.g., a neutron star), its spin or rotational angular frequency Ω cannot be greater than the Keplerian angular velocity, but it can certainly be less. In addition, unlike for binaries, not the entire quadrupole moment is involved in gravitational-wave generation (indeed, if the spinning source is axisymmetric, no gravitational radiation is emitted at all!).

Let us say then that some fraction ϵ of the quadrupole moment is nonaxisymmetric (generically, this could be, e.g., a lump or a wave), such that $h_{ij} \sim (G/c^4)(1/r)\Omega^2\epsilon I_{ij}$. The luminosity is then

$$\mathcal{L} \sim r^2|h_{ij}|^2 f^2 = \frac{32}{5}\frac{G}{c^5}\epsilon^2 I^2 \Omega^6, \tag{2.13}$$

where we have put in the correct numerical factors for rotation around the minor axis of an ellipsoid. The quantity I is the moment of inertia around the minor axis as the principal axis components of the reduced and traceless quadrupole moment and the moment of inertia tensors are the same. The quantity ϵ is the ellipticity in the equatorial plane, defined by $\epsilon = (a - b)/(ab)^{1/2}$, where the principal axes of the ellipsoid are $a \geqslant b > c$.

Major Payne: Wait! There is a critical and unstated assumption in the above, which is that the whole object is rotating as a rigid body. This might be reasonable for something like a mountain, but internal waves can produce differential rotation. This can matter a lot. For example, consider the classical problem of a homogeneous, incompressible, self-gravitating fluid of a given mass and a given angular momentum. At low angular momentum, the equilibrium shape is an oblate spheroid, but at higher angular momenta the lowest-energy configuration is a triaxial ellipsoid. But there is a family of such solutions, ranging from rigid rotation (the Jacobi ellipsoids) to ellipsoids that maintain a fixed orientation but have fluid that circulates internally (the Dedekind ellipsoids). The former case has a varying mass quadrupole and can then emit significant gravitational radiation. In the latter case, the mass quadrupole is constant, and thus any gravitational radiation has to come from a changing mass current, which is much weaker. As an example of this, it occasionally occurs to someone (starting with Chandrasekhar in the 1970s) that rapid rotation during a core-collapse supernova could produce a triaxial ellipsoid that would then be a prolific source of gravitational radiation. But it has been shown that because gravitational radiation conserves circulation, the ellipsoid would be driven in the direction of Dedekind ellipsoids so that the gravitational signal would likely be very weak.

The luminosity has an extremely strong dependence on Ω. The rotational energy of such an object would be roughly $E_{\rm rot} = \frac{1}{2}I\Omega^2$, where again here I stands for the moment of inertia, so if the part of the star generating the gravitational waves (e.g., a lump) is coupled to the rest of the star, then by the energy balance law we must have

$$\frac{dE_{\rm rot}}{dt} = I\Omega\dot{\Omega} = -\frac{32}{5}\frac{G}{c^5}\epsilon^2 I^2 \Omega^6 \quad \rightarrow \quad \dot{\Omega} = -\frac{32}{5}\frac{G}{c^5}\epsilon^2 I \Omega^5. \tag{2.14}$$

For pulsars, we can relate this to the dimensionless period derivative $\dot{P} = -2\pi\dot{\Omega}/\Omega^2$ to find

$$\dot{P} = \frac{64\pi}{5}\frac{G}{c^5}\epsilon^2 I \Omega^3. \tag{2.15}$$

The gravitational waves emitted by mountains on neutron stars force them to spin down.

And what is even better is that pulsars have been observed to spin down! Observations suggest that for young pulsars $\dot{P} \sim 10^{-13}$, while for the most stable of the millisecond pulsars $\dot{P} \sim 10^{-21}$–10^{-22}. Let us pretend, for a second, that the only reason that pulsars spin down is because they have lumps on their surface that force them to emit gravitational waves. If so, we can then estimate how large of a lump these neutron stars must have for the gravitational-wave-driven spindown to match observations. For a typical neutron star moment of inertia $I \approx 10^{45}$ g cm^2 and a young pulsar like the Crab with $\Omega \approx 200$ rad s^{-1} and $\dot{P} \approx 10^{-13}$, this implies $\epsilon \sim 3 \times 10^{-4}$. By the same argument, a millisecond pulsar with $\Omega \approx 2000$ rad s^{-1} and $\dot{P} \approx 10^{-21}$ has $\epsilon < 10^{-9}$.

Is this large or small? Earth's ellipticity is about 3×10^{-3}, but Earth's surface gravity is $\sim 10^{11}$ times less than that of a neutron star, which means that Earth can have much larger mountains relative to its size than can a neutron star. The ellipticity of neutron stars is unknown, but it is expected to be much smaller than Earth's. Therefore, the observed spindown is probably caused by other effects, notably magnetic braking (more on this in Chapter 5). As a result, the lack of observation of gravitational waves from a continuous source can be used to place an upper limit on the ellipticity, and therefore, on the magnitude of the surficial deformation.

If the ellipticities are so small, what strain amplitudes should we expect? When the correct factors are put in, we find that the characteristic strain amplitude from a pulsar of rotational frequency Ω at a distance r is

$$h_c = \frac{4G\,I\,\epsilon\,\Omega^2}{c^4\quad r} \leqslant \frac{1}{r}\left[\frac{5G}{2c^3}\frac{I|\dot{\Omega}|}{\Omega}\right]^{1/2}, \tag{2.16}$$

where in the second expression we have inserted ϵ as a function of Ω assuming that all of the spindown rate $\dot{\Omega}$ is due to gravitational-wave braking. Putting in some numbers, we get

$$h_c \approx 4 \times 10^{-28}\left(\frac{I}{10^{45}\text{g cm}^2}\right)\left(\frac{\epsilon}{10^{-4}}\right)\left(\frac{1\text{ sec}}{P}\right)^2\left(\frac{1\text{ kpc}}{r}\right). \tag{2.17}$$

For the Crab pulsar, $P = 0.03$ s, $r = 2$ kpc, and $\epsilon < 3 \times 10^{-4}$, so the maximum amplitude is $h_c \approx 6 \times 10^{-25}$. For a millisecond pulsar with $P = 0.003$ s, $r = 1$ kpc, and $\epsilon < 10^{-9}$, the maximum amplitude is $h_c \approx 4 \times 10^{-28}$.

Dr. I. M. Wrong: These amplitudes are extremely small! There is no way you can detect them, which shows you again how these physicists are exaggerating their claims ...

Captain Obvious: These amplitudes seem extremely small, but the coherence of their signal (and the fact that the frequency is known from radio observations) means that searches can go extremely deep. Suppose that you have a detector with a strain density[5]

[5] Time for some definitions (more about this in Section 4.1). Say that you have a detector which, because of noise, has its fractional length change from its equilibrium length by $n(t)$ as a function of time; because $n(t)$ is a fractional length, it is dimensionless. It is often useful to take the Fourier transform of $n(t)$, i.e., $\tilde{n}(f) = \int_{-\infty}^{+\infty} dt\, n(t)e^{-i2\pi ft}$, to represent the detector noise as a function of frequency. If the noise is stationary and Gaussian, then it is fully characterized by a power spectral density $S_n(f)$, where $\langle \tilde{n}(f)\tilde{n}^*(f')\rangle = \frac{1}{2}\delta(f - f')S_n(f)$, so $S_n(f)$ has units of 1/frequency. Detector strain density is the square root of $S_n(f)$. You can do similar things with the signal, e.g., $h(t)$, the dimensionless strain amplitude, is the fractional length change introduced by gravitational waves as a function of time.

$S_n^{1/2}$, which is usually measured in units of Hz $^{-1/2}$. Say now that you are looking at a periodic source with some (constant) frequency, which has a dimensionless strain amplitude h_c. If you detect the signal for a time T, then the signal (which is actually periodic) is spread out over a frequency range $1/T$ (think about taking a Fourier transform and computing the power!). The power at the peak is therefore roughly $h_c^2 T$, so the strain density of the signal at the peak is roughly $h_c T^{1/2}$. Because gravitational-wave detectors measure the strain amplitude directly (not the power), the signal to noise is roughly $h_c T^{1/2}/S_n^{1/2}$. If we need a signal-to-noise ratio ρ to claim a detection, this means that $\rho = h_c T^{1/2}/S_n^{1/2}$, from which we can solve for the time needed for that detection, namely $T = (\rho S_n^{1/2}/h_c)^2$. For example, at design sensitivity, advanced LIGO is expected to have a 1σ strain density at 60 Hz (roughly the frequency of the Crab signal, which would be at twice the rotation frequency) of $S_n^{1/2} \sim 6 \times 10^{-24}$ Hz $^{-1/2}$. Therefore, in principle, a coherent signal at the Crab maximum could be detected confidently in a time $T \sim [10 \times 6 \times 10^{-24}\ \mathrm{Hz}^{-1/2}/(6 \times 10^{-25})]^2 \approx 10^4$ s, or around three hours. For a very stable millisecond pulsar, though, the required integration time would be more than 10^{10} s, which is prohibitively large.

The full sensitivity of advanced LIGO has not quite been reached yet, but the strongest currently reported bounds on the ellipticities are about $\epsilon \lesssim 10^{-8}$ to 95% confidence. That's pretty good; it means that those pulsars don't deviate from axisymmetry by more than $\sim 10^{-2}$ cm over their $\sim 10^6$ cm radius!

2.2.2 Burst Sources

The third type of individual source of gravitational waves that we will consider are bursts. This class consists of all sources that emit "short" gravitational waves relative to the observation time, and which, well, look like a flare, a twitch, or a "spasm" of the measured strain.

The prototypical example of a burst source is a core-collapse supernova. Of course, if a collapse were to be perfectly spherically symmetric then it would emit zero gravitational radiation. However, a slight amount of net angular momentum will produce gravitational radiation during the collapse (recall that an axisymmetric source will emit gravitational radiation if it expands or contracts). Even without rotation, simulations show that the turbulence and instabilities that are produced during a core collapse will produce gravitational radiation that lasts for some tenths of a second and could be seen using second- or third-generation ground-based detectors if they are in our Galaxy.

Another example of a burst source is a magnetar giant flare. Magnetars are neutron stars with extra-strong magnetic fields (which is really saying something for neutron stars; here we're talking about surface field strengths in the $B \sim (10^{14}\text{–}10^{16})$ G range). The magnetic field tries to reorient itself to reduce the energy, and in doing so, it drags through the crust. The crust resists, but at some breaking point, energy on the order of $\sim 10^{45}\text{–}10^{47}$ ergs can be released suddenly (only three such giant flares have been seen, in contrast to hundreds of weaker events). In X-ray observations of giant flares, clear quasi-periodic oscillations at many frequencies, from tens to hundreds of Hertz, have been detected. Their origin is not certain, but to the degree that they involve coherent mass motion, they will also produce gravitational radiation (although it is difficult to forecast their detectability).

A third example of a burst source is a pulsar glitch. An isolated pulsar spins down over the long term. However, numerous pulsars have been observed to "glitch," meaning that their rotation frequency increases rapidly over a period that is probably less than minutes. The leading model for this behavior involves the sudden coupling of the normal crust of the neutron star with a neutron superfluid within the crust of the neutron star. This changes the mass currents within the star, and therefore it will also produce gravitational radiation. However, it takes heroic optimism to believe that these will be detected in gravitational waves by even third-generation detectors, unless of course the source was next door.

Yet another example of a burst source is a highly eccentric binary[6]. Consider one such eccentric binary that, near pericenter, emits gravitational waves in the sensitivity band of a detector, but that for most of its orbit emits gravitational waves at too low a frequency to detect. The pericenter passages therefore appear as short-duration bursts of gravitational waves. The bursts are spaced by the orbital period of the binary, and their frequency is inversely proportional to the amount of time the binary spends near pericenter. This characteristic timescale can be Fermi estimated as the pericenter distance divided by the pericenter velocity, so $\mathfrak{T} \sim r_p^{3/2}[m(1 + e)]^{-1/2}$. Initially then, the bursts should look roughly like rectangular tiles in a frequency-time spectrogram, with temporal width $\mathfrak{T}$ and frequency width $1/\mathfrak{T}$. However, as the inspiral proceeds, the spacing between these tiles should shrink and their shapes should change, as the eccentricity and the orbital period decrease, until the signal becomes that of an inspiral. Detection of such sources would benefit from a coherent strategy that makes use of these analytic expectations.

High-mass black-hole coalescences can also be thought of as burst sources, if all you get to see with your detector is the merger. Imagine that a binary has a total mass that is so large that the inspiral portion of the coalescence cannot be detected, yet still small enough that the merger or ringdown can be seen by a given detector. Then, the detectable portion of the signal lasts for only a small number of cycles in the detector, and thus, it looks like a burst of radiation. We see then that what a burst is and what it is not is not just a property of the source, but also a property of the instrument we use to detect the gravitational radiation it emits.

Bursts of gravitational radiation are therefore produced by sources that are quiescent at first, then experience a short period of very violent or very fast accelerations of matter, and then return to quiescence. As you may have expected, these sources are extremely hard to model, so sometimes they fall in the "other" category. For binaries or continuous sources, we can construct, in a parameterized way, a waveform family that represents the waves from the source. As we will see in Chapter 4, this means that we can optimize the detection of the source. But for burst sources, this is not the case.

Even though bursts are hard to model, we do know they exist in nature. Consider again the gravitational waves expected from a core-collapse supernova. At least at

[6] Yes, this is the nonbinary subsection of this chapter, but here we're thinking about just a small part of the binary orbit, so it's a bit different!

this time, although the rough frequency of the waves is believed to be hundreds of Hertz and numerical models show a degree of convergence, the waveform is not known well enough that parameterized models can be set up with confidence. Similarly, we could suppose that there is a currently unimagined type of source with, therefore, an unknown waveform, whose features resemble that of a burst.

In Chapter 4 we will discuss in more detail how such sources can be detected if they are strong enough. The key difference between a burst of gravitational waves and simple noise is that if the recorded burst is really a gravitational wave, then distinct detectors will record a correlated signal, whereas most types of noise will be uncorrelated between detectors except by chance. Thus, the burst detector pipelines for gravitational-wave instruments act as a catch-all. Indeed, the first gravitational-wave event (GW150914, which was the coalescence between two heavy, stellar-origin black holes) was detected first using the burst pipeline (which is very fast to run) and only later by the binary search analysis (which was not on at the time of the burst and had no difficulty detecting the event when it came online).

> **Captain Obvious:** We can make a simple order-of-magnitude estimate of the strength of a burst of gravitational waves using one of our previous results. The luminosity in gravitational waves is $\mathcal{L} \sim r^2 h^2 f^2$ for a source of frequency f, dimensionless strain amplitude h, and distance r. Thus, if the source emits a total energy E over a time t, then for r and f fixed, we see that $h \propto t^{-1/2}$. It turns out (as the authors discuss in Chapter 4) that the strain sensitivity of detectors also scales as $h \propto t^{-1/2}$, which means that, in a rough sense, the detectability of a burst does not depend strongly on the duration of the event, just the fluence (which is the total energy per area integrated over the event), as long as the duration is much shorter than the time over which the detector's properties change significantly.

Of course, if a particular category of events is understood better, or if enough such events occur that waveform parameterization is possible, then searches using that template will improve the sensitivity to those events, as we shall see in Chapter 4.

2.2.3 Stochastic Backgrounds

Our first three categories of sources have all been about individual objects or events. In contrast, stochastic gravitational waves are by definition the combination of many individually unresolved sources. These could be individual sources that are simply too faint to detect (e.g., double compact object coalescences that are too far away to yield confident detections) or, excitingly, could involve early-universe processes. In any case, the collection of many sources into a stochastic background means that such a background is usually considered to have significant breadth in frequency.

To get a handle on these issues, we need to think in terms of broad bands of frequency with many sources, rather than the signal produced by an individual source. For this purpose, it is useful to discuss the relation between the gravitational-wave spectral density from a class of sources, and the total energy released in those sources over their lifetimes.

Before moving into the details, following a careful treatment by Sterl Phinney, we can state the general idea very simply. Suppose we have a closed box, and sources in the box that emit some total average energy $\langle E \rangle$ during their lifetimes. If the number density of the sources is n, then the energy density in the box is $n\langle E \rangle$. If the energy emitted between frequencies f and $f + df$ for a single source is $(dE/df)df$, and if there is no redshifting, then the total energy density between f and $f + df$ is $n(dE/df)df$. The trick then is to relate this quantity to an integral over cosmological redshift of the gravitational-wave energy emitted by sources at different redshifts.

Before we can do that, however, the expansion of the universe changes these expressions, as we will now explore. Let the frequency of a gravitational wave in the rest frame of a source be f_r, and the observed (redshifted) frequency be $f = f_r/(1 + z)$, where z is the cosmological redshift (as you will see when you take general relativity). Let the energy emitted by a single source between f_r and $f_r + df_r$ be $(dE/df_r)df_r$, which is the total energy emitted over the lifetime of the source, in all directions, and is measured in the rest frame of the source. Let us also define $\Omega_{\rm GW}(f)$ as follows. Suppose that $\left(d\rho_{\rm GW}(f)c^2/df\right)df$ is the present-day energy density in gravitational waves[7] of frequencies between f and $f + df$. Let $\rho_c c^2 = 3H_0^2 c^2/(8\pi G)$ be the mass–energy density needed to close the universe (here $H_0 \approx 70$ km s^{-1} Mpc^{-1} is the Hubble constant; more about H_0 in Chapter 6). Then

$$\Omega_{\rm GW}(f) \equiv \frac{1}{\rho_c c^2}\frac{d\rho_{\rm GW}(f)c^2}{d\ln f}. \tag{2.18}$$

That is, $\Omega_{\rm GW}(f)$ is the ratio of the present-day energy density in gravitational waves in a logarithmic interval around f to the critical energy density.

The total current energy density in gravitational waves $\rho_{\rm GW}(f)c^2$ can then be computed by integrating $\left(d\rho_{\rm GW}(f)c^2/df\right)df$, which gives us

$$\rho_{\rm GW}(f)c^2 = \int_0^\infty \rho_c c^2 \Omega_{\rm GW}(f)\frac{df}{f} = \int_0^\infty \frac{\pi}{4}\frac{c^2}{G}f^2 h_c^2(f)\frac{df}{f}. \tag{2.19}$$

Here, the second equality (including the strange factor of $\pi/4$) comes from Equation (1.11) (without the average over multiple wavelengths), and $h_c^2(f) \equiv (h_+^2(f) + h_\times^2(f))/2$ is the characteristic gravitational-wave amplitude over a logarithmic frequency interval around f. This characteristic gravitational-wave amplitude is related to the one-sided (meaning frequencies from 0 to ∞ rather than $-\infty$ to ∞) spectral power density[8] $S_{h,1}(f)$ by $h_c^2(f) = fS_{h,1}(f)$.

[7] We know the notation here can be a bit confusing, but $\rho_{\rm GW}(f)$ is not the same as the $\rho_{\rm GW}$ of Chapter 1. Back then, we defined $\rho_{\rm GW}$ by integrating over multiple wavelengths, while here we are interested in the density between frequencies f and $f + df$. Therefore, do not compute $\rho_{\rm GW}(f)$ by taking the derivative of $\rho_{\rm GW}$ in Equation (1.11) to conclude that $d\rho_{\rm GW}(f)c^2/d\ln f = fd((\pi/4)(c^2/G)f^2 h_{c,0}^2)/df = 2(\pi/4)(c^2/G)f^2 h_{c,0}^2$.

[8] We will talk more about spectral densities in Chapter 4, but for now, just think of the power spectral density as a measure of the mean-squared fluctuations in gravitational-wave background, i.e., $\langle h_{ij}^2 \rangle = \int S_h(f)df$, as the Captain mentioned when talking about continuous sources.

We want to relate this to the gravitational-wave energy radiated throughout the history of the universe, including the effects of redshifts. Suppose that the number of sources between z and $z + dz$ per unit comoving volume is $N(z)dz$. The phrase "comoving volume" refers to the volume that the region in question occupies now and follows the expansion of the universe; for example, at a redshift $z = 1$ the scale factor of the universe was $1/(1 + z) = 1/2$ what it is now, so the region encompassed by a given comoving volume had one-eighth of the volume it does now (but would contain the same number of galaxies, assuming that the galaxies all followed the expansion of the universe).

The present-day gravitational-wave energy density can then be expressed with the following integrals over the frequency and redshift. Here the gravitational-wave energy, as measured at the sources, between source-frame frequencies f_r and $f_r + df_r$ is $(d\rho_{\mathrm{GW}}/df_r)df_r$, so if this was emitted at redshift z, the energy would be redshifted by a factor $(1 + z)^{-1}$. Integrating over all frequencies and redshifts we get

$$\begin{aligned}\rho_{\mathrm{GW}}(f) &= \int_0^\infty \int_0^\infty N(z)(1+z)^{-1}\, d\rho_{\mathrm{GW}}\, dz,\\ &= \int_0^\infty \int_0^\infty N(z)(1+z)^{-1} f_r (d\rho_{\mathrm{GW}}/df_r)\,(df_r/f_r)\, dz,\\ &= \int_0^\infty \int_0^\infty N(z)(1+z)^{-1} f_r (d\rho_{\mathrm{GW}}/df_r)\, dz\,(df/f),\end{aligned} \tag{2.20}$$

where in the third line we use $df_r/f_r = d[f(1+z)]/f(1+z) = df/f$ for the observed frequency f.

We now have two expressions (Equations (2.19) and (2.20)) for the same quantity $\rho_{\mathrm{GW}}(f)$ and can equate them at each frequency:

$$\rho_c c^2 \Omega_{\mathrm{GW}}(f) = \frac{\pi}{4}\frac{c^2}{G} f^2 h_c^2(f) = \int_0^\infty N(z)(1+z)^{-1} f_r (d\rho_{\mathrm{GW}}/df_r) dz. \tag{2.21}$$

This equation relates the energy density to the current number density of event remnants, and the energy they released over their lifetimes, and it is extremely general. It is independent of cosmology. It is also unaffected by beaming as long as the beams are randomly oriented. If there are multiple types of sources, we just need to add their contributions. This equation therefore allows us to compute the background due to any given class of source, at any given frequency. It turns out that the current-day energy density at a given frequency is not even all that sensitive to $N(z)$, assuming that the sources are not concentrated too much at one given redshift.

To proceed further, we would now need to specify some information about what is causing the gravitational-wave background, so that this can be inserted in the right-hand side of Equation (2.21). A background due to processes in the early universe (say, before the production of the cosmic microwave background) would be very exciting because it would contain information that is unavailable otherwise. In principle, one could see gravitational waves from very early in the universe, because gravitons have a very small interaction cross section. We need to state clearly, however, that, even by the standards of gravitational-wave astronomy, these processes are all highly speculative. One consequence of this is that although it would be extremely exciting to detect a background of early-universe gravitational radiation, a nondetection would not be surprising.

We will examine some specific possibilities in Chapter 6, but for now let's just establish a generic scaling using our Fermi techniques. What can $\Omega_{\rm GW}(f)$ depend on? We know that $\Omega_{\rm GW}(f)$ itself is dimensionless. From its definition, it has a factor of H_0^{-2} (from $1/\rho_c$) and must also be proportional to $S_{h,1}$. It could be proportional to some power of f as well. You might think that a factor of G could also be involved, but because $\Omega_{\rm GW}(f)$ is dimensionless and all the other factors have units of s or s^{-1}, there cannot be any powers of G. The only combination that works is

$$\Omega_{\rm GW}(f) \propto H_0^{-2} f^3 S_{h,1}(f). \tag{2.22}$$

The result of this is that unless in some frequency range $\Omega_{GW}(f)$ increases with f more steeply than $\Omega_{GW}(f) \propto f^3$, the spectral density decreases with increasing frequency. As a result, for most realistic sources of background, it will be easier to detect the background at lower frequencies.

2.3 Exercises

1. This problem shows the limits of order-of-magnitude calculations in some cases. Let's say you'd like to estimate the recoil speed of a merged black hole remnant, due to linear momentum carried away by gravitational radiation. To simplify things, suppose we have two nonrotating black holes of masses m_1 and m_2 that collide head on, so there is no spin at any point. A theorem from black hole thermodynamics says that the square of the "irreducible" mass of the final black hole cannot be less than the sum of the squares of the irreducible masses of the initial black holes. For nonrotating black holes, this becomes

$$m_{\rm final}^2 \geqslant m_1^2 + m_2^2. \tag{2.23}$$

Like the increase in entropy, this is an inequality, but for our order-of-magnitude estimate, we will assume $m_{\rm final}^2 = m_1^2 + m_2^2$. With that assumption, compute the final speed of the remnant (as a fraction of the speed of light, and as a function of m_1 and m_2), assuming that all the radiated energy is carried away in a single direction. For comparison, the maximum kick occurs when $m_1/m_2 \approx 2.76$ and it is $\sim$175 km s^{-1}.

2. Major Payne claimed that drag on Earth due to the Sun's radiation is stronger than the effective drag from the gravitational radiation emitted by Earth's orbit. The drag due to radiation, which is called Poynting–Robertson drag, reduces Earth's orbital energy at a rate (in the limit of Newtonian motion around a point source of luminosity)

$$\mathcal{L}_{P-R} = \mathcal{L}_s \frac{r_\oplus^2 \Omega_{\rm orb}^2}{4c^2}, \tag{2.24}$$

where $r_\oplus$ is Earth's radius, $\mathcal{L}_s$ is the Sun's luminosity, and $\Omega_{\rm orb}$ is the angular velocity of Earth. Given this:

(a) To within a factor of 10, compute the ratio of $\mathcal{L}_{P-R}$ to $\mathcal{L}_{\rm GW,\oplus}$ (the gravitational-wave luminosity from Earth's orbit).

(b) Suppose that two 1.4 $M_\odot$ neutron stars orbit each other in a circle. Each star is cooling off and therefore emits radiation; suppose that the luminosity of the radiation is 10^{30} erg s^{-1}, which might be typical of an isolated neutron star of moderate age. Compute the orbital radius, and the corresponding orbital frequency, such that the orbital energy loss due to Poynting–Robertson drag is equal to that due to gravitational radiation.

3. Using Newtonian gravity as an approximation, determine the largest mass of a nonrotating black hole such that the tidal acceleration from the hole, at the event horizon, exceeds the surface gravitational acceleration of (a) the Sun, (b) a white dwarf with mass 0.6 $M_\odot$ and radius 10^9 cm, and (c) a neutron star with mass 1.4 $M_\odot$ and radius 12 km. Assume that the star/white dwarf/neutron star is dropped into the hole radially, with no rotation.

4. Derive the constant of motion associated with the inspiral according to the Peters equation. Hint: define $y \equiv e^2$ to get da/dt and dy/dt, then look for a constant in the form $C = af(y)$.

5. Consider the ringdown produced by two 10 $M_\odot$ black holes. Suppose that the ringdown lasts for two cycles and emits a total of 1% of the mass–energy of the final black hole. Assuming a nonrotating black hole ($j \equiv cJ/Gm^2 = 0$), what would be the frequency of the radiation and how long would it last? The frequency is in the range of human hearing (although, of course, not audible!), and sound amplitude is measured in decibels, where 0 dB has an intensity of 10^{-9} erg cm^{-2} s^{-1}. If the binary black hole merger occurs at the distance of the Virgo Cluster (about 50 million light years), compare the intensity of the ringdown at Earth with the intensity of the loudest scream ever registered (129 dB, by Jill Drake of the UK).

6. Dr. I. M. Wrong has come to you with a brilliant idea: LISA (the Laser Interferometer Space Antenna, which has an expected launch date of 2034) will be the ideal instrument to detect satellites around extrasolar planets. In particular, he envisions a $m = 6 \times 10^{26}$ g satellite (about 10% of Earth's mass, bigger than any satellite in the solar system) orbiting with an orbital frequency of $f_{\rm orb} = 5 \times 10^{-5}$ Hz around a planet with mass $M = 2 \times 10^{31}$ g, about 10 times Jupiter's mass. At gravitational-wave frequencies $f_{\rm GW} < 10^{-3}$ Hz, LISA's spectral density sensitivity at a signal to noise of 1 is $10^{-19}(10^{-3}\ {\rm Hz}/f_{\rm GW})^2$ Hz$^{-1/2}$. Assuming an observing time of 10^8 s, evaluate the detection prospects if the system is at a distance of 10 pc (about 3×10^{19} cm).

7. We claim in this chapter that the frequency of a sound wave that involves most of a gravitationally bound object can't exceed $\sim(G\rho)^{1/2}$. But is that

correct? Consider a spherical star of mass M and radius R. Recalling that the sound speed is $c_s = (dp/d\rho)^{1/2}$, where p is the pressure and ρ is the (baryon) density, do a simple order-of-magnitude calculation to determine the sound speed and thus the crossing time of a sound wave across the star. The inverse of that time is of order the frequency. What are some objects that could potentially rotate or pulsate at a frequency much greater than $\sim(G\rho)^{1/2}$, and why?

8. Suppose that you are performing radio observations of a double pulsar system, in which both neutron stars are visible as pulsars. We'd like to determine, qualitatively, how overdetermined the system is. That is, we'd like to know how many aspects of the system can be measured, versus how many parameters there are. This is a deliberately vague question to get you thinking about the process of measurement. If more quantities can be measured than there are parameters, then the system is overdetermined and the underlying theory can be tested.

9. Dr. I. M. Wrong has rushed to fill a much-needed gap in the literature. As a starting point, because of their crusts, Earth-like planets are not perfect oblate spheroids. In fact, it is estimated that $\epsilon \approx 3 \times 10^{-5}$ for the Earth. Dr. Wrong therefore proposes that LISA would be the perfect satellite to detect the rotation of Earth-like planets in exoplanet systems, especially close ones like the planet believed to orbit Proxima Centauri. The European Space Agency has asked you to look into this possibility. You are to assume that Proxima Centauri (distance 1.3 pc) has an Earth clone with the same mass, rotation rate, and ϵ as Earth. Assume that LISA's sensitivity for a 10^8 s observation at a half-day period (half, because gravitational waves for circular motion have twice the frequency of the motion) is $h = 10^{-20}$. Report on the prospects for detection.

10. Another system that could be considered a burst source is a highly eccentric binary because such a binary will emit almost all of its radiation in a burst near pericenter. Suppose that you have two 10 $M_\odot$ black holes in a highly eccentric orbit around each other with an orbital frequency at pericenter of 50 Hz. Treat the approach at pericenter as $\sim$1 circular orbit at that frequency. How close would such a system have to be to produce a strain amplitude of $\sim 5 \times 10^{-22}$, which is the minimum needed for a confident advanced LIGO detection of one pericenter passage of such a source?

11. The first publication that claimed a possible detection of gravitational waves was by Joe Weber in 1969. His experiments involved aluminum bars with resonance frequencies of about 1660 Hz. The minimum strain that could be detected was $h \approx 10^{-16}$, and a few candidate events were seen per day, which extrapolates to $\sim$1000 per year. Assuming that the events came from the center of our Galaxy (distance about 8 kpc), that they lasted about 1 s each, and that their flux was roughly isotropic, compute the implied

mass–energy per year emitted in gravitational waves from our Galaxy. Comment on the implications.

12. Darth Sidious has taken an interest in the generation of gravitational waves. He plans to use the Death Star to destroy the planet Alderaan in such a way as to make the resulting gravitational waves visible throughout the universe to detectors that (coincidentally?) have roughly the sensitivity of advanced LIGO. He has brought you in to consult on this plan. Can he do it? Be careful how you deliver your answer: the Emperor does not like bad news.

13. Suppose that your stochastic background consists of a large number of circular binaries that extend to frequencies much larger than that of interest to you (so that the binary coalescences are not cut off by mergers or affected by tides). Then, for a very large population of binaries that spans the observable frequencies:
    (a) Calculate the dependence of the strain amplitude h on the gravitational-wave frequency $f_{\rm GW}$.
    (b) Calculate the number of binaries simultaneously resident in a small frequency interval between $f_{\rm GW}$ and $f_{\rm GW} + df_{\rm GW}$ ($df_{\rm GW} \ll f_{\rm GW}$). In particular, suppose that there are 10^8 double white dwarf binaries in our Galaxy that will merge within 10^{10} years, so that in steady state there are 10^{-2} double white dwarf mergers per year. Treating the white dwarf binaries as equal-mass binaries with total mass $\sim 1\ M_\odot$, estimate the frequency above which there will be fewer than one binary per 10^{-8} Hz wide bin.

Useful Books

Auger, G., & Plagnol, E. 2017, An Overview of Gravitational Waves: Theory, Sources and Detection (Singapore: World Scientific)

Carroll, S. M. 2019, Spacetime and Geometry: An Introduction to General Relativity (Cambridge: Cambridge Univ. Press)

Frolov, V., & Novikov, I. D. 1998, Black Hole Physics: Basic Concepts and New Developments (Berlin: Springer)

Lightman, A. P., Press, W. H., & Teukolsky, S. A. 1975, Problem Book in Relativity and Gravitation (Princeton, NJ: Princeton Univ. Press)

Maggiore, M. 2018, Gravitational Waves: Volume 2: Astrophysics and Cosmology (Oxford: Oxford Univ. Press)

Misner, C. W., Thorne, K. S., & Wheeler, J. A. 1973, Gravitation (Princeton Univ. Press: Princeton, NJ)

Thorne, K. S., Price, R. H., & Macdonald, D. A. 2010, Black Holes: The Membrane Paradigm (New Haven, CT: Yale Univ. Press)

Chapter 3

Gravitational-wave Modeling of Binaries

As we will explore in more detail in the next chapter, the expected weakness of many gravitational-wave sources means that we need to use optimal detection techniques. In turn, this requires that we understand enough about possible sources to use tailored algorithms. In this chapter, we discuss the modeling of binaries, which are the only individual sources yet detected and for which there is a clear set of waveform templates that can be derived (albeit with much effort!) from general relativity. In the next chapter, we will discuss how these templates are applied to data. We defer to the next chapter a discussion of the detection of burst and stochastic sources because these do not have waveform families in the same sense as binaries.

3.1 Approximations Rule!

General relativity is hard. It's very hard. You may think it's easy, because, after all, you may have seen in undergraduate general relativity that Einstein's theory is encompassed in just one equation: $G_{\mu\nu} = 8\pi T_{\mu\nu}$ (up to quantum corrections, supposedly). But this simple tensor equation, relating the curvature of spacetime (contained in the Einstein tensor $G_{\mu\nu}$) to the matter content of spacetime (contained in the stress–energy tensor $T_{\mu\nu}$) hides 10 coupled partial and nonlinear differential equations. This is quite different from other theories we are used to, such as Newtonian gravitation or Maxwellian electrodynamics. In both cases, these theories are also described by coupled partial differential equations, albeit fewer than 10, but the key thing is that those equations are linear.

What does linearity mean in this context? It just means that the principle of (linear) superposition holds, and thus, that the sum of two solutions to the theory is also a solution to the theory. This is perhaps more clearly understood in terms of waves. Imagine you have two electromagnetic waves of a certain frequency ω_1 and ω_2 that are solutions to the Maxwell equations:

doi:10.1088/2514-3433/ac2140ch3

$$\vec{E}_1 = E_1 \sin[\omega_1(t - r)]\hat{r}, \qquad \vec{E}_2 = E_2 \sin[\omega_2(t - r)]\hat{r}. \tag{3.1}$$

Then, the superposition $\vec{E}_3 = \vec{E}_1 + \vec{E}_2$ also satisfies the Maxwell equations. In your gut, you already know this must be true from your experiences with water waves (fluid mechanics); if you've ever been to a beach, you've probably seen two waves in the ocean coming together to form a bigger, more intimidating wave. So wave plus wave equals meaner wave.

But nature is nonlinear. How do we know? Because we've seen and measured how pendula move. Yes, we know, you think the pendulum equation is linear because you probably think pendula obey the simple harmonic oscillator (ordinary differential) equation

$$\frac{d^2\theta}{dt^2} + \frac{g}{L}\theta = 0, \tag{3.2}$$

where θ is the angle of oscillation, $g = 9.8$ m s^{-2} is the local acceleration due to gravity on Earth (which, by the way, is not a constant on Earth, and depends on altitude, latitude, etc.), and L is the length of the pendulum. But as you may also know, this equation is only an approximation (valid to linear order in small oscillations) to the actual pendulum equation of Newtonian mechanics:

$$\frac{d^2\theta}{dt^2} + \frac{g}{L}\sin\theta = 0. \tag{3.3}$$

The trigonometric function makes all the difference, and this is assuming the pendulum is "simple," meaning not attached to other pendula, which can create really interesting and sometimes chaotic nonlinear motion. Nonlinear partial differential equations are truly everywhere, from the motion of fluids with viscosity, as described by the Navier–Stokes equations, to the behavior of the particles subject to the strong force, as described by the equations of quantum chromodynamics. In many ways, it is the nonlinear nature of the universe that makes life really exciting, really fun, but really hard at the same time.

Life is hard when you do not have an exact solution to the nonlinear partial differential equations that describe the physical system you are trying to model. Let's go back to the simple example above. In the case of the simple harmonic oscillator, we know the solution in terms of sines and cosines, but for the case of the true pendulum equation, the solution is only known through an elliptic integral, which, in turn, is only known through tabulations of numerical solutions.[1] Of course, there are linear theories in which certain problems also do not have exact, closed-form solutions, as in the case of Newtonian gravity. We all know that the two-point-mass problem in Newtonian gravity has a closed-form solution (Keplerian orbits), but once you allow for more than two celestial bodies (even just three), then a generic and exact closed-form solution is not known (the handful of closed-form solutions known to the three-body problem are not astrophysically realistic). Unfortunately, if

[1] The deep reader may object that sines and cosines are also only known through tables constructed from numerical solutions. Today, however, these trigonometric functions are so familiar to us that we treat them as basic functions, just like we would polynomials.

we know anything for sure it is that the universe has more than just two bodies, as evidenced by the solar system, and there is no general, closed-form, mathematical description of their motion even in Newtonian gravity.

Needless to say, exact solutions in Einstein's theory of general relativity are few and far between. In fact, we only have exact solutions in the most simplified and symmetric situations we can consider. For example, we know of exact solutions outside of isolated, nonspinning (spherically symmetric), and static (unchanging in time) bodies (the Schwarzschild solution), and even for isolated, spinning (axisymmetric), and stationary black holes (the Kerr solution). But for nonisolated, or nonaxisymmetric, or nonstationary spacetimes, there are no exact solutions. For our purposes, the main condition we would like to relax is that of isolation, and even if we just consider one other body, there is still no known exact solution. Let us say that again: there is no exact solution for the general two-body problem in general relativity, which in Newtonian gravity would be described by Keplerian orbits. In some sense, this is because in Einstein's theory the "two-body" problem becomes a "one-spacetime" problem, as our colleagues sometimes say. What this means is that we cannot separate the motion of the bodies as induced by the curvature of spacetime, from the generation of curvature by the bodies themselves. We are then forced to consider approximate solutions.

Approximations are a fact of life. Most of us are already familiar with what physicists sometimes call "analytic approximations." The word "analytic" here just means that the solutions obtained can be expressed in closed form and in terms of known functions, such as polynomials and trigonometric functions. Analytic approximations typically start by approximating the differential equations that describe the physical system we are trying to model.

> **Major Payne:** Technically, an analytic function is one that possesses a convergent Taylor expansion locally about any point in its domain. And so you would expect analytic solutions to be composed of analytic functions, but that's not necessarily the case here. So the use of the word "analytic" here is a bit sloppy, but I'll allow it.

For example, we may be interested in the small oscillations of a pendulum, in which case we can approximate Equation (3.3) with Equation (3.2), because $\sin\theta \sim \theta[1 + \mathcal{O}(\theta^2)]$ for $\theta \ll 1$. For the two-body problem, when the objects have comparable masses, we could approximate the Einstein equations by their expansion assuming the gravitational field is weak, the sources are weakly gravitating, and they are slowly moving. Alternatively, we may wish to consider the two-body problem when one of the objects has a mass much smaller than the other, in which case we can expand the Einstein equations in the mass ratio of the system. The former approximation is called post-Newtonian (or PN for short), while the latter system is called an extreme mass-ratio inspiral (or EMRI for short).

You may hope that approximations are not actually necessary, because today we have so much computing power at our fingertips, that we could just throw the entire problem at a computer and let it solve it for us. Numerical solutions are extremely powerful, but they are also approximations to the true solution, in the sense that computers cannot keep an infinite number of digits in their computations. Most computer codes are limited to 16 digits of precision (a.k.a. "double" precision), which means that every time a numerical computation is performed, then the result is only accurate to that level of precision at best. Moreover, numerical solutions may also suffer from numerical instabilities, depending on the numerical formulation of the problem, and error in the extraction of physical observables, which in relativity are defined at (spatial or null) infinity (a place that is not typically in the numerical grids used). In this (very formal mathematical) sense, numerical solutions are also approximate, although their accuracy can be extremely high if all of its sources of error are controlled.

A slight complication is that numerical solutions can be obtained either from an exact set or from an approximate set of differential equations. That is, we could drop in the full Einstein equations (properly decomposed) into a supercomputer and ask it to solve them. Alternatively, we could take the Einstein equations and expand them in some approximation scheme, such as the PN approximation or in the small mass-ratio approximation, to then solve the resulting approximate equations numerically. The latter approach is typically called a semianalytic solution because it combines analytic methods with numerical ones. Full numerical solutions will not agree with semianalytic solutions exactly because they solve different differential equations.

Why would you ever want to solve numerically an approximate set of differential equations? Because sometimes solving the full set of Einstein equations is too computationally expensive, and you may need to obtain millions or billions of solutions for your particular application, such as when comparing to data. Hopefully, it is clear at this point that the method one uses to solve the differential system describing the physical model of interest depends sensitively on the application one has in mind for this solution. If one needs a few very precise and short solutions, then full numerical evolutions may be the way to go. If one needs billions of very long solutions, then analytic or semianalytic methods may be more appropriate. As we will see in later chapters, all of these methods (analytic, semianalytic, and fully numerical) are essential for the accurate construction of models for the gravitational waves emitted in the coalescence of compact objects.

Before concluding this generic discussion, let us say a few words about the accuracy, or equivalently, the level of error incurred by these different methods to obtain solutions. Analytic solutions are found through an approximation scheme that typically employs an expansion in a small parameter. Because all calculations are done analytically, one has control of the order to which these expansions are kept, which means one knows the order of the terms ignored. This order allows you to make an estimate of the error incurred by the analytic approximation. For example, if we try to describe the motion of a pendulum through Equation (3.2), we know we are making a relative error of $\mathcal{O}(\theta^2)$ in the differential equation, because $\sin\theta = \theta[1 + \mathcal{O}(\theta^2)]$. Therefore, the solution

$$\theta = C_1 \cos\omega t + C_2 \sin\omega t, \tag{3.4}$$

with $\omega^2 = g/L$ and $C_{1,2}$ determined by initial conditions, is only accurate to $\mathcal{O}(C_{1,2}^2)$. Thus, given any set of initial conditions, we know how to estimate the error in the approximate solution.

This, of course, is a very simplified example, because errors can (and typically do) also grow secularly with time in analytic approximations, accumulating and eventually rendering the approximate analytic solution invalid. To see this more clearly, let us add one more term in our expansion of the sine function, i.e., $\sin\theta = \theta - \theta^3/6 + \mathcal{O}(\theta^5)$. Equation (3.3) then becomes Duffing's equation, namely

$$\frac{d^2\theta}{dt^2} + \frac{g}{L}\theta\left(1 - \frac{\epsilon^2}{6}\theta^2\right) = 0, \tag{3.5}$$

where we have inserted the bookkeeping parameter ϵ to remind ourselves that $\theta \ll 1$. If we tried to solve this equation as a series expansion in ϵ, we would find

$$\theta = C_1 \cos\omega t - C_1^3\frac{\epsilon^2}{6}\left(\frac{1}{32}\cos 3\omega t - \frac{1}{32}\cos\omega t - \frac{3}{8}\omega t\sin\omega t\right) + \mathcal{O}(\epsilon^2), \tag{3.6}$$

where we have assumed the initial conditions $\theta(0) = C_1 \ll 1$ and $\theta'(0) = 0$ for concreteness. Clearly, this solution becomes highly inaccurate when t becomes large enough, i.e., when $t \sim 6/(\omega C_1^2)$, setting the bookkeeping parameter to unity.

When secular errors occur in analytic approximations, one is either forced to use a more sophisticated mathematical technique (such as multiple-scale analysis), or a resummation technique. We are not going to explain in detail what multiple-scale analysis is,[2] but in essence, one changes the perturbative ansatz to a function of two timescales (a slow one and a fast one), thus also allowing the frequencies in the solution to admit a perturbative expansion. When performing multiple-scale analysis, one finds the solution

$$\theta = C_1 \cos\left[\omega t\left(1 - \frac{1}{16}\epsilon^2\right)\right] + \mathcal{O}(\epsilon^2). \tag{3.7}$$

Inserting this function into Equation (3.5) one sees the equation is not satisfied at $\mathcal{O}(\epsilon^2)$, but this is because we have only gone to leading order in multiple-scale analysis here. What this solution does do, however, is eliminate the secularly growing mode, i.e., the term proportional to (ωt) in Equation (3.6). One can easily verify that the "exact" (i.e., numerical) solution to Equation (3.5) is indeed very well approximated by Equation (3.7).

But now a miracle happens. Recall we said that when secularly growing terms appear in approximate solutions one can either use fancy math or employ a resummation technique. What is the latter? Resummation colloquially means a procedure through which one repackages a function at the cost of introducing higher-order terms. Applying this to Equation (3.6), one obtains Equation (3.7) automagically.

[2] For that, we refer the reader to the book by Carl Bender and Steven Orzsag (see the bibliography at the end of this chapter).

Captain Obvious: I think it'd be useful to give an example here. Imagine we had a function $f(x) = 1 + x + \mathcal{O}(x^2)$. We may resum this function by writing it as $f(x) = (1 - x)^{-1} + \mathcal{O}(x^2)$, which is formally identical to the series expansion written above, up to uncontrolled remainders. It turns out we can do exactly the same thing to the perturbative solution of Equation (3.6). If we focus on the ϵ^0 term and the term proportional to $\epsilon\omega t$, one finds one can resum it exactly as given in Equation (3.7). In this sense, the multiple-scale solution is nothing but the resummed version of the perturbative solution to Duffing's equation!

Major Payne: The Captain's description is too simplified. The power series $f(x)$ cannot just be resummed as $(1 - x)^{-1}$, but in fact, there is an infinite number of functions that, when Taylor-expanded about $x \ll 1$, lead to the power series, such as $(1 + a\,x)^{1/a}$ for any real constant a. The more terms you calculate in the Taylor expansion, the more sure you can be that you have the right resummation. For example, if we knew that $f(x) = 1 + x + x^2 + x^3 + \mathcal{O}(x^4)$, then the resummation $(1 - x)^{-1} + \mathcal{O}(x^4)$ would still work, but $(1 + a\,x)^{1/a}$ would not unless $a = -1$.

When dealing with numerical solutions, the error incurred cannot be computed in the same way as when dealing with analytic approximate solutions. Rather, in this case, several numerical evolutions are typically calculated, with various choices of discretization. This is because numerical algorithms typically become more accurate as the discretization step is reduced (within a given window controlled by the absolute accuracy of the data type). Thus, in this approach, the most accurate numerical solution (the one with the smallest discretization step within the allowed window) can serve as an estimate of the error incurred. Another way to estimate the error, of course, is to use the numerical solution to calculate the degree to which certain physical constraints are satisfied with time. In Maxwell's electrodynamics, for example, one can discretize and solve the differential equations that describe the evolution of an electromagnetic wave, and then insert this solution into the constraint equations of Maxwell's theory to check the degree to which they are satisfied in time.

In the semianalytic approach, one must contend with both sources of error described above. On the one hand, one will incur an error due to approximating the original differential system, and this can be estimated analytically as before. On the other hand, one will also incur an error due to discretizing this approximate differential system when solving it numerically, and this can be estimated numerically as above. Both of these sources of error must be checked and controlled with time to ensure that the semianalytic solutions remain accurate.

3.2 Compact Binaries

Let us now focus on binaries composed of compact objects, such as neutron stars and black holes. In general, their coalescence can be divided into three (not so very well-defined) parts: (i) the inspiral, (ii) the plunge and merger, and (iii) the ringdown (see the left-top panel of Figure 1.2, and in more detail, Figure 3.1). The inspiral phase is typically defined as part of the coalescence during which the compact objects are widely separated, i.e., $r_{12} \gg \mathfrak{R}_{1,2}$, where $\mathfrak{R}_{1,2}$ is the characteristic radius of the compact objects. For a neutron star, this is simply its radius, while for a black hole, it is its mass (times G/c^2 to get the units right). When objects are so widely separated, their characteristic velocity is much smaller than the speed of light. This is clearly so for quasi-circular inspirals, because by Kepler's laws we have that $v \sim (m/r_{12})^{1/2}$, where we recall that v is the relative orbital velocity, and m is the total mass. For eccentric orbits, this is also true because the requirement of large separations translates to requiring that the semimajor axis $a \gg \mathfrak{R}_{1,2}$ and indeed also the pericenter distance $r_p = a(1 - e) \gg \mathfrak{R}_{1,2}$ for orbital eccentricity e.

However, the emission of gravitational waves means that the motion of two bodies in general relativity is not conservative. This implies that quantities we would have thought of as "constants of the motion," such as the energy and angular momentum of the orbit, are not actually constant. True, their temporal variation is tiny if the bodies have a large separation, as we saw in Chapter 2, but nonetheless, it is still not zero. This means that the orbit will decay, with the circle or ellipse that the bodies trace becoming smaller and smaller with time, while possibly precessing more and more violently. One can think of this process as that of many ellipses osculating into each other, or alternatively, as a single ellipse whose orbital elements are slowly varying functions of time. Either way, the inspiral regime is the realm of PN theory and semianalytic methods for EMRIs.

Eventually, the objects get so close to each other that this osculating approximation becomes highly inaccurate. This is because the strength of the gravitational waves emitted scales inversely with the orbital separation, as we saw in Chapter 2. Therefore, when the separation becomes comparable to the characteristic radius $\mathfrak{R}_{1,2}$, the rate at which gravitational waves remove energy from the orbit becomes large. This, in turn, implies that the orbit changes more and more rapidly, until eventually, the two bodies plunge into each other and merge. We therefore define the plunge-and-merger phase as that which occurs when the bodies are at an orbital separation $r_{12} \sim \mathfrak{R}_{1,2}$.

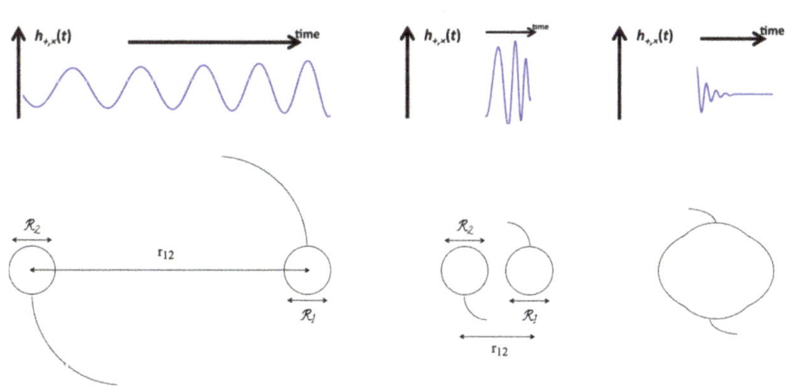

Figure 3.1. Sketch of the three phases of coalescence: inspiral (left), plunge and merger (middle), and postmerger/ringdown (right).

To understand better where the plunge phase begins, recall that in general relativity the effective gravitational acceleration is larger than it would be in Newtonian gravity (by greater amounts when the binary separation is small; this is sometimes called the "pit in the (effective) potential"). One consequence is that for a test particle in a circular orbit around a black hole, there is an orbital separation that minimizes the specific angular momentum of the system. This is in strong contrast with Newtonian gravity, for which the specific angular momentum of circular orbits decreases monotonically with decreasing radius. The existence of a minimum in the specific angular momentum implies that inside that radius, circular orbits are unstable, and hence this determines the ISCO that we mentioned in Chapter 2.

Captain Obvious: I think it helps to provide a bit of context here. The circumferential radius (i.e., the one you get by dividing the circumference of a circle by 2π) of the ISCO for a test particle in orbit around a nonrotating black hole of mass M is $6GM/c^2$, which is 3× the Schwarzschild radius and is thus ~9 km for an $M = 1\ M_\odot$ black hole. The ISCO radius for equatorial orbits moving in the same direction as a maximally rotating black hole is GM/c^2, and for equatorial orbits moving in the opposite direction as a maximally rotating black hole is $9GM/c^2$. When the authors talk about the ISCO for a binary system as defining the transition between inspiral and plunge, what they mean is the ISCO of the effective problem, i.e., the ISCO of a test particle of mass $\mu = m_1 m_2/(m_1 + m_2)$ in orbit around a black hole with mass $M = m_1 + m_2$.

For a test particle moving under the influence of gravity alone, the ISCO is a precisely determined radius, and it is where the plunge phase begins. But finite losses of angular momentum (e.g., from magnetic effects or gravitational radiation) blur the line. For comparable-mass binaries, one can compute the separation and angular momentum at the ISCO to different PN orders, and although the various approaches agree qualitatively (e.g., Taylor-like PN expansions or Padé resummation in the equivalent one-body approaches), the corrections from, e.g., 2PN to 3PN order are disturbingly large. For such comparable masses, numerical simulations show that, rather than there being a distinct plunge phase, the orbits have the character of progressively more open spirals as the masses approach each other. This is in part why it's so hard (and perhaps not perfectly well defined) to define the transition between inspiral and merger. For the sake of concreteness, one typically picks the ISCO of the effective problem, as the Captain mentioned above.

Soon after the bodies touch, the "remnant"' of the collision is a highly distorted object, whose final fate will depend on its type and its mass, as we mentioned in Chapter 2. As a refresher, recall that if the two compact objects that collided are black holes, then the remnant will always be a black hole with a mass somewhat less than the sum of the masses of the initial two black holes (because gravitational radiation carries away mass–energy) and an area that is larger than the sum of the areas of the original black holes (this is the "'area law," which came out of several nifty papers by researchers such as James Bardeen, Jacob Bekenstein, Brandon Carter and Stephen Hawking).

Dr. I. M. Wrong: Wait a minute! Clearly this area "law" cannot be right, because if I take two stars and accelerate them toward each other in a trajectory such that their surfaces just graze, I can unbind material and produce three objects from the collision. Two of the resulting three objects will have an area smaller than the two objects before they collided. So here is an example of how the area law doesn't work. Yet another example of how your fancy math leads to unphysical results!

Captain Obvious: Your intuition is right when it comes to "fluffy" material bodies, like main-sequence stars, but our everyday intuition fails in physics in many circumstances, and dealing with black holes is one of them. Black holes are special and, although their exterior gravitational field may look like that of a star from far away, they are drastically different from up close.

For example, imagine you had a spaceship and you flew close to a supermassive black hole. Remember that tidal forces scale inversely with the black hole mass, so in principle, you could do this without your spaceship getting tidally disrupted. Now, imagine you put a spacesuit on (with the appropriate jetpack of course) and got really close to the black hole, like an arms length from its horizon. What do you think would happen if you "stuck your arm" in and then you tried to pull it out? Well, because nothing can escape from inside the horizon of a black hole (unless it can travel faster than light), once your arm is in, it would stay in, and it would drag the rest of you in as well! Well, I suppose you could cut your arm off so that you don't fall in, but let's not get gory here.

Something similar happens in the grazing collision you described. As soon as the black holes get close enough to each other, the shape of their horizons deforms, stretching toward each other as if they were trying to touch. A short time later, a common horizon forms around both objects, and at that instant, a new black hole is formed. The mass of the newly formed black hole is about the sum of the masses of the individual black holes (and before Major Payne chimes in, yes, a little bit of energy is lost to gravitational waves in this process, and so the mass of the new object is a bit less than the sum of the individual masses). The surface area of a black hole goes as the square of its radius, so the square of its mass. And therefore, for a merger of two comparable-mass black holes, the area of the newly formed object is about twice the sum of the areas of the individual black holes, and in particular, definitely larger than either of the individual areas.

If one of the compact objects is a neutron star and the other is a black hole, then the neutron star will either be absorbed by the black hole without disrupting, or it will be tidally stripped and disrupted prior to the collision. If the two compact objects are neutron stars, then the remnant can either be another neutron star or a black hole, depending on how much mass was ejected during the merger event. In any case, the plunge-and-merger regime is highly dynamical and nonlinear, and thus, this is the realm of numerical relativity.

After the remnant has formed, it will settle down through the emission of gravitational waves, or, in the case of the neutron star remnant, also through the emission of electromagnetic waves, high-energy charged particles, and neutrinos. Recall from Chapter 2 that if the remnant is a massive neutron star, supported by rapid rigid rotation (a supramassive neutron star) or by rapid differential rotation over an extended disk (a hypermassive neutron star), then ellipticity or oscillations can lead to the emission of gravitational waves with certain characteristic frequencies. If the remnant is a black hole, or if the massive neutron star collapses to a black hole after spinning down, the remnant will ring down through quasinormal gravitational-wave modes.

As with the inspiral–plunge transition, the merger–ringdown transition is not defined precisely; loosely, we can say that the ringdown stage is when the oscillation amplitudes are small enough to be treated as independent linear modes. This requires that there be exponential damping of the modes, which is expected to occur after a few light-crossing times. Given that the size of a black hole is proportional to its mass, one typically says that the ringdown starts at $t \sim t_{\rm merger} + {\rm few} \times M_{\rm remnant}$, where $t_{\rm merger}$ is the time when the merger occurred, and $M_{\rm remnant}$ is the mass of the remnant.

3.2.1 Comparable-mass Binaries

The description of the inspiral, plunge-and-merger, and ringdown stages of coalescence now allows us to describe in more detail how to model the gravitational waves emitted in each stage. Here we provide an overview, with some key physical results that will make your life easier if you decide to do a deep dive into any one of these topics in the future.

To set up our discussion, we recall from Chapter 2 that at the lowest order, Philip Peters and Jon Mathews treated the evolution of two point masses in orbit around each other, due to gravitational radiation. As we indicated previously, they essentially imagined that the point masses move in a Keplerian ellipse with a given semimajor axis and eccentricity, over a single orbit, and computed the energy and angular momentum lost during that single orbit. The energy and angular momentum of the new orbit correspond to a new semimajor axis and eccentricity. If the change per orbit of the energy and angular momentum is very small, we can approximate the evolution as a gradual change in the semimajor axis and eccentricity.

The real orbit will not be a Keplerian ellipse, even before we take into account gravitational radiation. Instead it will precess, and for a given semimajor axis and eccentricity, the "pit in the potential" means that the masses move faster than they would in Newtonian gravity. In addition, because the masses are not actually points, there will be tidal effects to consider, and because the objects themselves have some rotation (only in a negligible set of cases will they not be rotating at all), there are additional effects, such as spin–orbit and spin–spin coupling as well as frame-dragging.

As we will discuss in more detail in Chapter 4, a rough measure for the importance of an effect is whether, over the period of observation, its neglect would lead to a significant "mismatch" with the data. This can happen if the effect is weak but the number of cycles is large, or if the number of cycles is small but the effect is strong. For instance, harkening back to the pendulum example, if the initial angle θ_0 from the vertical is large, then the oscillation period is different enough from the low-amplitude limit that, relative to the linear approximation, the pendulum will rapidly get out of phase. If $\theta_0 \ll 1$, then the period is very close to the low-amplitude limit, but after enough oscillation periods, the pendulum will still get out of phase with the linear solution.

As a result, if a given source has many gravitational-wave cycles in the band of a given detector, then very subtle effects must be considered. For example, double neutron star binaries can have thousands of detectable cycles in ground-based detectors, depending on the starting frequency. A full numerical run for so many cycles is infeasible, which means that it is important to develop analytical or semianalytical techniques that can capture the most important effects over such a long time, both in terms of the motion and in terms of the resulting gravitational radiation. We now describe an approach, PN theory, which in its various forms has proven to be invaluable in such analysis.

3.2.1.1 Inspiral and Post-Newtonian Theory

PN methods allow us to, in effect, use series solutions to model both the orbital motion and the gravitational waves emitted by comparable-mass compact binaries. In this approach, we expand and solve the Einstein equations assuming weak fields (i.e., in powers of G) produced by slowly moving sources (i.e., in inverse powers of c). The method can be envisioned as an iterative technique, starting with the Newtonian equations of motion. Their solution tells us how these bodies move in a flat spacetime. The (linearized) Einstein equations then tell us how these bodies curve the spacetime and induce waves in spacetime. Both of these effects, in turn, correct the trajectories of the bodies, and on and on the merry-go-round goes.

Major Payne: None of that makes sense without at least a few equations! Many of these equations will be familiar from a first course in general relativity, but let's present them here again (they are good for the soul!).

Because we assume weak fields produced by slowly moving sources, we can decompose the metric tensor in terms of a (flat Minkowski) background $\eta_{\alpha\beta}$ and metric perturbation $h_{\alpha\beta}$. The latter is actually defined via the relation $\sqrt{-g}\,g^{\alpha\beta} = \eta^{\alpha\beta} - h^{\alpha\beta}$, where g is the determinant of the metric, which reduces to $g_{\alpha\beta} = \eta_{\alpha\beta} + h_{\alpha\beta} + \mathcal{O}(h^2)$ to linear order only. Inserting this into the Einstein equations leads to their relaxed form, namely

$$\Box_\eta h^{\alpha\beta} = -\frac{16\pi G}{c^4}\tau^{\alpha\beta}, \tag{3.8}$$

where $\Box_\eta$ is the D'Alembertian operator associated with Minkowski spacetime and $\tau^{\alpha\beta}$ is an effective stress–energy tensor constructed from the stress–energy tensor of the sources and terms quadratic in the metric perturbation. In deriving the relaxed Einstein equations, we have employed the harmonic gauge condition, namely $\partial_\alpha h^{\alpha\beta} = 0$, which implies that the coordinates used are harmonic, i.e., they satisfy $\Box_g x^\alpha = 0$.

The Bianchi identities, $\nabla_\mu G^{\mu\nu} = 0$, ensure that the divergence of the above equation vanishes identically, meaning that

$$\partial_\alpha \tau^{\alpha\beta} = 0, \tag{3.9}$$

which reduces to the conservation of the stress–energy tensor of matter to leading order in the metric perturbation. Equation (3.8), which tells us the spacetime corrections due to moving sources, and Equation (3.9), which tells us how sources move in spacetime, are then what we need to solve iteratively. Of course, that's easier said than done, because there is a lot of art and math in solving these equations iteratively.

In a sense, much of PN theory reduces to solving the wave equation with ever more complicated sources. The sources, however, are complicated enough that one typically divides the spacetime into different zones, inside which the integrands can be expanded slightly differently. These three zones are the far or wave zone, the near zone, and the inner zones of the two bodies (see Figure 3.2). At a simply descriptive level, the far zone is where $r \gtrsim \lambda_{\rm GW}$ and the near zone is where $r \lesssim \lambda_{\rm GW}$, where we recall that $\lambda_{\rm GW} \sim r_{12}/v$ is the characteristic wavelength of the gravitational waves emitted by the binary system, with r_{12} the orbital separation and v the relative orbital velocity. Fundamentally, the reason that the distinction matters is that in the far zone, all fields depend on the retarded time $\tau_r \equiv t - r/c$, where t is the time coordinate and r is the field-point distance as measured from the center of mass of the system; in the near zone, the field-point distance r is so small that we can expand retarded time as $\tau_r \approx t$.

In the near zone, because we are working with slowly moving sources, time derivatives are smaller than spatial derivatives, i.e., any near-zone field ϕ will have as an argument $(\omega t, r)$ for some characteristic frequency ω, and then, $\partial_t\phi/\partial_r\phi \propto r\omega/c \ll 1$, because by Kepler's law, $\omega \propto r_{12}^{-3/2}$ for a binary system, and $r_{12} \gg Gm/c^2$ for slowly moving sources. This, in turn, implies that in addition to expanding the field equations in weak gravitational fields (which turns out to be equivalent to performing expansions in powers of G), we can also expand them in small velocities, or alternatively in inverse powers of the speed of light c^{-1}. This double expansion in G and $1/c$ is formally called a PN expansion. Of course, because we are expanding in powers of G, the near zone excludes the inner zones of the two bodies, i.e., the region close enough to either of them that a weak-field expansion is no longer valid (see Figure 3.2).

In a bit of a twist, however, it turns out that PN theory can be used not just to describe weak-field sources, but also to describe black holes by treating them as point masses, i.e., through a distributional Dirac delta function stress–energy tensor. This approximation works provided one never evaluates the solution too close to the gravitating source. This additional restriction reduces to $r \gg m_{1,2}$ for black hole binaries, which defines the inner zone (see Figure 3.2). At first sight, one may wonder why this is so. The answer is that as long as the bodies remain far from each other, the way they move in spacetime and the gravitational waves they emit do not care about the precise geometric structure close to the bodies. This is a direct consequence of the strong equivalence principle that you probably studied in your general relativity introduction and which, of course, need not hold in modified theories of gravity. The gravitational field that a body feels due to its companion, and which controls how the

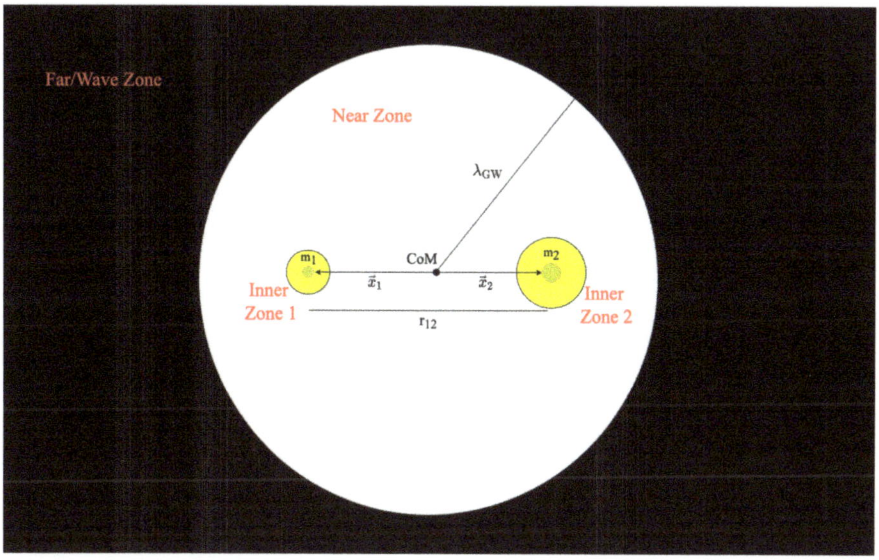

Figure 3.2. "Fried egg"' diagram of the different zones used in post-Newtonian theory. The black region (the "griddle") represents the far/wave zone, the white region (the "egg white") the near zone, and the yellow regions (the "'yolks") the inner zones.

body moves through spacetime, is still quite weak at the location of this body, far from its companion. The same, of course, cannot be said about compact binaries that are close to each other, and are thus exiting the inspiral regime.

In the near zone, the metric tensor allows one to find the PN equations of motion for our sources. This can be derived in many ways, such as directly from Equation (3.9), or from the geodesic equations associated with the PN metric. We are not going to present the derivation here because it would take us into too long of a tangent (indeed, there are entire books dedicated just to the PN approximation!). Instead, let us focus here on the final result for a binary system of test particles, for which all approaches lead to the same equations of motion, namely

$$\begin{aligned} u_1^i = &- \frac{Gm_2}{r_{12}^2} n_{12}^i \\ &+ \frac{1}{c^2}\left\{ n_{12}^i \left[\frac{5G^2 m_2 m_1}{r_{12}^3} + \frac{4G^2 m_2^2}{r_{12}^3} + \frac{Gm_2}{r_{12}^2}\left(\frac{3}{2}(n_{12}^j v_{2j})^2 - v_1^2 + 4v_{1j}v_2^j - 2v_2^2 \right) \right] \right. \\ &\left. + v^i \frac{Gm_2}{r_{12}^2}\left[4(n_{12j}v_1^j) - 3(n_{12j}v_2^j) \right] \right\} + \mathcal{O}(c^{-4}), \end{aligned} \tag{3.10}$$

where, as indicated in Figure 3.2, x_1^i and x_2^i are 3-vectors indicating the positions of the bodies as measured from the center of mass, with magnitudes $r_1 = |x_1^i|$ and $r_2 = |x_2^i|$, $v_1^i = \partial_t x_1^i$ and $v_2^i = \partial_t x_2^i$ are the velocities of the bodies; $r_{12} = |x_2^i - x_1^i|$ is the orbital separation; and $n_{12}^i = n_1^i - n_2^i = x_1^i/r_1 - x_2^i/r_2$ is a unit separation vector. There is a similar expression for a_2^i with the rule $1 \leftrightarrow 2$. As we can see from this expression, the orbital motion here is identical to what we would expect in Newtonian gravity, but with corrections that enter at $\mathcal{O}(c^{-2})$. Notice of course that we have truncated the expression here at $\mathcal{O}(c^{-2})$, because there are terms of $\mathcal{O}(c^{-4})$ and higher that we have not written down. The terms inside curly braces are said to be of first PN (1PN) order, because they are of $\mathcal{O}(1/c^2)$ relative to the leading-order term, which in turn is said to be "Newtonian."

Dr. I. M. Wrong: Aha! Once more, the authors' sloppiness betrays them. I see repeated indices in the equation of motion, and this doesn't make sense. Sure, this is relativity, so the authors' must be using the Einstein notation convention and forgot to tell us. But the Einstein summation convention is just shorthand for inserting the metric tensor and then summing over the corresponding components such as $n_{12\,i}v_1^i = g_{ij}n_{12}^i v_1^j = \sum_{i,j=1}^3 g_{ij}n_{12}^i v_1^j$. But what metric should I use to carry out this summation? The authors' lack of care is just unbelievable!

Captain Obvious: Hold your horses! The authors may be skipping through a few details here, but the equation above is actually correct! You are right that repeated indices imply the Einstein summation convention. This simply means that repeated indices are to be summed over, so, for example, $n_{12\,i}v_1^i = n_{12\,x}v_1^x + n_{12\,y}v_1^y + n_{12\,z}v_1^z$ in a Cartesian coordinate system, because the vectors (and one-forms) are purely spatial. Now to evaluate this expression, you are right that one usually converts the one-form into a vector through the metric, so in our example $n_{12\,i}v_1^i = g_{ij}n_{12}^i v_1^i = g_{xx}n_{12}^x v_1^x + g_{yy}n_{12}^y v_1^y + g_{zz}n_{12}^z v_1^z$. But recall that the metric has been expanded into the Minkowski metric plus a metric perturbation, $g_{\mu\nu} = \eta_{\mu\nu} + h_{\mu\nu}$, and the expression where these contractions appear in Equation (3.10) are all of $\mathcal{O}(1/c^2)$. Therefore, you can replace the metric here by the flat spacetime metric at the cost of introducing only higher-order terms in G/c^2, which lump together with the $\mathcal{O}(1/c^4)$ terms they didn't write down, and thus can be ignored.

Once we have the acceleration, we can also derive a very useful formula: the relativistic extension of Kepler's third law. This can be achieved easily for circular orbits by noting that $r_{12}\omega^2 = \langle \hat{n}_i a_{12}^i \rangle$, where $a_{12}^i = a_1^i - a_2^i$ is the relative acceleration, $\hat{n}^i$ is a spatial unit vector orthogonal to the circular orbit (i.e., a radially pointing vector), and the angled brackets stand for averaging over one orbit. Using Equation (3.10) in the center-of-mass frame, one then finds

$$\omega^2 = \frac{v^2}{r_{12}^2} = \frac{Gm}{r_{12}^3}\left[1 - (3 - \eta)\frac{Gm}{c^2 r_{12}} + \mathcal{O}(c^{-4})\right], \tag{3.11}$$

where the first equality is a definition of the angular velocity and the second equality comes from the relative acceleration, recalling that the symmetric mass ratio $\eta = m_1 m_2/m^2$. This is an important expression because it relates one of the characteristic frequencies of the problem (the orbital frequency) to one of the characteristic distances (the orbital separation) and to one of the characteristic velocities (the orbital velocity). And because the orbital frequency is related to the gravitational-wave frequency by an integer factor (a factor of 2 in the case of circular orbits), we can then calculate the orbital separation of a binary when it enters the sensitivity band of detectors. For example, to Newtonian order, $\pi f_{\rm GW} = (m/r_{12}^3)^{1/2}$, $r_{12} = m(\pi m f_{\rm GW})^{-2/3}$, and $v = (m/r_{12})^{1/2} = (\pi m f_{\rm GW})^{1/3}$, so at a gravitational-wave frequency of $f_{\rm GW} = 10$ Hz, a neutron star binary with total mass $m = 2.8 M_\odot$ is at a separation of about 722 km (if we use the 1PN accurate expressions) or 726 km (if we use the Newtonian expressions), and has a velocity of $v = 0.0763c$ (if we use the 1PN accurate expressions) or $0.0760c$ (if we use the Newtonian expressions). We see then that as long as the separations are large and the velocities are small, then the PN terms are small corrections of the Newtonian results.

The motion as prescribed to this PN order, however, is purely conservative because gravitational waves first enter the orbital evolution at $\mathcal{O}(c^{-5})$. Gravitational radiation, like electromagnetic radiation, carries energy and momentum, and therefore, it is essential to include the "radiation-reaction force." Ignoring radiation reaction induces secular errors, i.e., errors that grow with time and don't average out. For example, because gravitational waves remove energy from the binary system, the orbital separation r_{12} changes with time, whereas it is constant in Newtonian gravity. The dissipative effect of radiation reaction can be incorporated into a quasi-circular orbit through the energy balance law we saw back in Chapter 2, $dE/dt = -\mathcal{L}$, where we recall that E is the binary's orbital energy and $\mathcal{L}$ is the gravitational-wave luminosity. The energy can be derived from the equations of motion, for example, through the calculation of a Lagrangian or a Hamiltonian, to find

$$\begin{aligned} E = {} & \frac{1}{2} m_1 v_1^2 + \frac{1}{2} m_2 v_2^2 - \frac{G m_1 m_2}{r_{12}} + \frac{1}{c^2}\Bigg[\frac{G^2 m_1 m_2 m}{2 r_{12}^2} + \frac{3}{8}(m_1 v_1^4 + m_2 v_2^4) \\ & + \frac{G m_1 m_2}{2 r_{12}}\Big(-(n_{12,i} v_1^i)(n_{12,j} v_2^j) + 3 v_1^2 + 3 v_2^2 - 7(v_{1,i} v_2^i)\Big)\Bigg] + \mathcal{O}(c^{-4}), \end{aligned} \tag{3.12}$$

before going to the center-of-mass frame, where $m = m_1 + m_2$ is the total mass. One can now take a time derivative of this quantity and set it equal to (the negative of) the gravitational-wave luminosity to find an expression for the rate of change of otherwise constant quantities, such as the orbital separation r_{12} as we did in Equation (2.7). But to get the right answer at 1PN order, we also need $\mathcal{L}$ to this order.

In the far zone, the gravitational-wave luminosity can be computed through the gravitational-wave pseudotensor and solutions for the gravitational-wave spacetime perturbation in the far zone. We saw a bit of this already in Chapter 1, where Major Payne explained how to compute the GW energy density (see Equation (1.8)).

The gravitational-wave luminosity is nothing but $\mathcal{L} = c\rho_{\rm GW}$, and so, expanding the metric in (far-zone) multipole moments, one has

$$\begin{aligned}\mathcal{L} = c\,\rho_{\rm GW} &= \frac{c}{32\pi G}\left\langle \dot{h}_{ij}\dot{h}^{ij}\right\rangle,\\ &= \frac{G}{5c^5}\dddot{\mathcal{I}}_{\langle ij\rangle}\dddot{\mathcal{I}}^{\langle ij\rangle} + \frac{G}{c^7}\left[\frac{1}{189}\ddddot{\mathcal{I}}_{\langle ijk\rangle}\ddddot{\mathcal{I}}^{\langle ijk\rangle} + \frac{16}{45}\dddot{\mathcal{J}}_{\langle ij\rangle}\dddot{\mathcal{J}}^{\langle ij\rangle}\right] + \mathcal{O}(c^{-9}),\end{aligned} \tag{3.13}$$

where the angled brackets in the indices stand for the symmetric and trace-free part operation, $\mathcal{I}_{ij}$ is the radiative-mass quadrupole moment, $\mathcal{I}_{ijk}$ is the radiative-mass octupole moment, and $\mathcal{J}_{ij}$ is the radiative current quadrupole moment. For a quasi-circular binary of point particles, these moments are the same as the near-zone multipole moments of the source to lowest PN order, and thus $\mathcal{I}^{ij} = m_1 x_1^i x_1^j + 1 \to 2 + \mathcal{O}(1/c^2)$, $\mathcal{I}^{ijk} = m_1 x_1^i x_1^j x_1^k + 1 \to 2 + \mathcal{O}(1/c^2)$, and $\mathcal{J}^{ij} = m_1 x_1^i \epsilon^j{}_{kl} x_1^k v_1^l + 1 \to 2 + \mathcal{O}(1/c^3)$, so then

$$\mathcal{L} = \frac{32}{5}\eta^2\left(\frac{Gm}{c^2 r_{12}}\right)^5\left[1 + \left(-\frac{2927}{336} - \frac{5}{4}\eta\right)\frac{Gm}{c^2 r_{12}} + \mathcal{O}(c^{-4})\right], \tag{3.14}$$

where once more recall that $\eta = m_1 m_2/m^2$ is the symmetric mass ratio. Notice now a curiosity of the PN conventions: in the above expression, the term in parentheses inside the square brackets is a 1PN correction relative to the leading-order term (because it is of relative $\mathcal{O}(1/c^2)$ smaller than unity), but the leading-order term itself (the so-called Newtonian piece of the luminosity) is proportional to G^5/c^{10}! Indeed, in Newtonian theory, there are no gravitational waves or gravitational-wave luminosity, yet the leading-order term is still called "Newtonian."

Captain Obvious: PN approximations are often labeled in terms of an (at first somewhat cryptic) order-counting scheme, which the authors have touched on in passing. Because this is so important, and because it has caused more than one headache to nonexperts, let's explain it further here.

Given any approximation, the first nonvanishing term, called the "leading-order" term, is always said to be of "Newtonian" order. Yes, we know, this is silly. If we are talking about gravitational waves, saying that the leading-order expression is of Newtonian order makes little sense, because there are no gravitational waves in Newtonian gravity. Alas, we are stuck with this nomenclature, so we'd better get used to it. As you can imagine then, the first-order correction to this leading-order term is said to be a first post-Newtonian (1PN) correction. In general relativity, these are of $\mathcal{O}(G, c^{-2})$ smaller than leading order, which means they are suppressed by factors of $\mathcal{O}(M/r_{12})$ or $\mathcal{O}(v^2)$.

When computing quantities associated with the conservative dynamics of the binary, like its gravitational binding energy, PN corrections are typically even in order, so they are proportional to $\mathcal{O}[(M/r_{12})^N]$ or $\mathcal{O}(v^{2N})$, and so we say these terms are of NPN order. If you think about it, this has to be the case because conservative dynamics must be even under time reversal by definition. That is, it doesn't matter whether you play the movie forwards or in reverse when you consider, for example, a conserved circular orbit. Corrections that are odd in velocity would be odd under time reversal, so they cannot be allowed.

Of course, there are exceptions to the "even" rule described above. If one considers a term that is, for example, linear in the spin of the objects, then under time reversal, this term would flip signs, because the spin would flip sign. Therefore, to be invariant under time reversal, such a linear-in-spin term would have to be proportional to an odd power of velocities. Indeed, the first spin correction to the conservative dynamics of a binary is linear in spin and multiplied by three powers of orbital velocities, thus making it a 1.5 PN order term. If we ignore spins, though, the "even" rule explained above is valid.

For quantities associated with the dissipative dynamics of the binary, such as the rate of change of the binding energy due to gravitational-wave emission, PN corrections can be odd or even in order. For example, there can be corrections of $\mathcal{O}(v^3)$, which would be of 1.5PN order, and corrections of $\mathcal{O}(v^5)$, which would be 2.5PN order. These odd terms are very important because some of them are associated with nonlinear effects related to the scattering of a gravitational wave off of the background spacetime geometry (a so-called "tail effect") or the emission of gravitational waves due to energy contained in gravitational waves themselves (a so-called "memory effect"). But of course, there can also be even-order dissipative corrections at a higher PN order: for example, at second order in dissipative corrections, you can have modifications that enter at odd PN orders squared, which are then even PN order. People therefore talk about "tail of the tail" effects, which enter at 3PN order (as the square of a 1.5PN order term).

To bring this discussion home, let's go back to the authors' discussion of Equation (3.14), where they wrote down the gravitational-wave luminosity in terms of the total mass, the symmetric mass ratio, and the orbital separation. As the authors pointed out, the leading-order term, the $(32/5)\eta^2[Gm/(c^2 r_{12})]^5$, is the Newtonian term. In absolute order, this term scales as $1/c^{10}$ so you may be tempted to call it a 5PN order term, but don't do that! PN counting is always applied to terms relative to the leading-order one (the so-called "controlling factor" in asymptotic mathematics), and when people try to use it to denote absolute orders, that's when confusion begins to creep in. With the Newtonian term understood, we can then say that the second term inside the square brackets of Equation (3.14) is a 1PN correction, because it is of $\mathcal{O}(1/c^2)$ relative to the leading-order term.

With the luminosity at hand, we can now return to our discussion of the radiation reaction or "self-acceleration." This quantity can be thought of as the "force" that makes the binary inspiral, which of course must be due to the loss of gravitational waves. For a Fermi estimate, it suffices to relate this energy loss to the self-force via $|\mathcal{L}| = |f^i_{\rm RR} v_{12,i}|$, and from this one finds that

$$| a^i_{\rm RR} | \sim \frac{|f^i_{\rm RR}|}{\mu} \sim \frac{32}{5}\frac{\eta}{m} v^9[1 + \mathcal{O}(c^{-2})], \tag{3.15}$$

where one uses the virial theorem $v^2 \sim m/r_{12}$. How does this compare to the Newtonian acceleration? Well, this acceleration goes as $a^i_{\rm Newt} \sim 1/r_{12}^2 \sim v^4$, and therefore $|a^i_{\rm RR}|/|a^i_{\rm Newt}| \sim \mathcal{O}(c^{-5})$. We see then, once more, that radiation reaction due to emission of gravitational waves enters at 2.5 PN order in general relativity. Observe also from this estimate that as the mass ratio of the binary goes to zero (i.e., in the extreme mass-ratio limit), the gravitational-wave luminosity, and thus, the self-acceleration, vanishes, leading to noninspiraling trajectories.

Now that we have the orbital energy, Kepler's third law, and the luminosity, we can use the energy balance law to find the rate of change of the orbital separation as a function of time to 1PN order and can derive other quantities from it. One quantity of particular importance to gravitational-wave astrophysics is the rate at which the orbital frequency f changes as a function of time. Before calculating this in detail, let's estimate df/dt using our Fermi machinery. We know that $E \propto \eta v^2$ and $\mathcal{L} \propto \eta^2 v^{10}$, so then $df/dt \propto (dE/df)^{-1}\mathcal{L} \sim (f/E)\mathcal{L} \sim (f/v^2)v^{10} \sim v^{11}$, where in the last approximation we used that $f \sim v^3$ by Kepler's third law. We then know the scaling of the chirping rate with frequency, $df/dt \propto (2\pi m f)^{11/3}$, but the units are all wrong, because the left-hand side is in Hz^2 and the right-hand side is dimensionless (once you put back factors of c). The only constant scale in the problem is the total mass of the binary, so we then have that $df/dt \sim m^{-2}(2\pi m f)^{11/3}$. We can now check whether this Fermi estimate is correct by combining Equations (3.12) and (3.14) to find $df/dt = (dE/df)^{-1}(-\mathcal{L})$, and thus in the center of mass frame

$$\frac{df}{dt} = \frac{48\eta}{5\pi}\left(\frac{c^6}{G^2m^2}\right)\left(\frac{2\pi G m f}{c^3}\right)^{11/3}\left[1 - \left(\frac{743}{336} + \frac{924}{336}\eta\right)\left(\frac{2\pi G m f}{c^3}\right)^{2/3} + \mathcal{O}(c^{-4})\right]. \quad (3.16)$$

Notice that our Fermi estimate for df/dt is up to a factor of order unity identical to the Newtonian part of Equation (3.16). This equation is important because it tells us the chirping rate of gravitational waves, i.e., the rate at which the gravitational-wave frequency increases with time. This is because for quasi-circular binaries, the gravitational-wave frequency is twice[3] the orbital frequency, $f_{\rm GW} = 2f$. The chirping rate is important because it controls the temporal evolution of the gravitational-wave phase, which is precisely the quantity to which interferometric detectors are most sensitive. As you can thus imagine, the chirping rate plays a critical role in the construction of waveform models with which to filter the data, as we will see in Chapter 4.

For now, however, let us content ourselves with a short little Fermi calculation that uses the chirping frequency. A quantity that is often of interest in gravitational-wave astrophysics is the amount of time a signal is "in band." By this, we mean the amount of time it takes to sweep from some initial frequency f_0 to merger (if it merges in band) or exit the sensitivity band. We can estimate this quantity by solving Equation (3.16) to find

$$\delta t \sim \frac{m}{4\pi}\left[\left(\pi m f_{\rm GW,low}\right)^{-8/3} - \left(\pi m f_{\rm GW,high}\right)^{-8/3}\right], \quad (3.17)$$

where we have used that for a quasi-circular binary $2f = f_{\rm GW}$ with $f_{\rm GW}$ the gravitational-wave frequency, and where we have ignored the numerical prefactor in Equation (3.16), which is of order unity for equal-mass binaries. We see then that the binary

[3] In fact, for a binary with any orbital eccentricity, the gravitational waveform can be expressed as sums of harmonics of the orbital phase, where the latter is just the time integral of the orbital frequency.

spends a tremendous amount of time at low frequencies, and then, as it inspirals to higher frequencies, it spends less and less time. We also see that if $f_{\rm GW,low} \ll f_{\rm GW,high}$, then to a very good approximation, $\delta t \sim m/(4\pi)(\pi m f_{\rm GW,low})^{-8/3}$. This is the case, for example, for a binary neutron star detectable with ground-based instruments. For example, if for a detector $f_{\rm GW,low} = 10$ Hz, then because $f_{\rm GW,high} \gtrsim 10^3$ Hz, $\delta t \sim 10^3$ s, or about 16 minutes. What this approximation does not convey is that for binaries with extreme mass ratios, the chirping rate is suppressed (i.e., Equation (3.16) is proportional to η), so such a binary will spend a much longer time in any frequency decade than would a comparable-mass binary of the same total mass.

We are almost done now. The far-zone calculation gave us expressions for the gravitational-wave luminosity, while the near-zone calculations gave us expressions for the equations of motion of the source and Kepler's third law. The only ingredient left to discuss is the gravitational-wave itself. This quantity is only properly defined very far from the source, so in the far zone, where it can be expanded in terms of radiative multipole moments,

$$h_{ij} = \frac{2G}{c^4 D_L}\left[\ddot{\mathcal{I}}_{\langle ij\rangle} + \frac{1}{3c}\dddot{\mathcal{I}}_{\langle ijk\rangle}N^k - \frac{4}{3c}\epsilon^l{}_{k(i}\ddot{\mathcal{J}}_{\langle j)k\rangle}N^l + \mathcal{O}(c^{-2})\right], \tag{3.18}$$

where N^i is a vector from the center of mass to the observation point and D_L is the luminosity distance. Because the above expressions are to be evaluated in the far zone, one must remember that the radiative multipole moments are functions of retarded time, and not just of regular time. Projecting out the + and × modes, then to Newtonian order we have

$$h_+ = -\frac{2G\eta m}{c^2 D_L}\left(\frac{G}{c^3}\pi m f_{\rm GW}\right)^{2/3}[(1+\cos^2\iota)\cos 2\phi + \mathcal{O}(c^{-1})] \tag{3.19}$$

$$h_\times = -\frac{4G\eta m}{c^2 D_L}\left(\frac{G}{c^3}\pi m f_{\rm GW}\right)^{2/3}[\cos\iota\sin 2\phi + \mathcal{O}(c^{-1})], \tag{3.20}$$

where ι is the inclination angle (defined as the angle between the orbital angular momentum and the line of sight) and ϕ is the orbital phase. Notice that the above expression has the functional form we foreshadowed back in Equation (2.4). With this last ingredient, one can now in principle calculate gravitational waves emitted by a binary system. "All" one has to do is solve for the orbital motion of the binary (i.e., the position of the objects as measured from the center of mass as a function of time, $x_1(t)$ and $x_2(t)$) using the PN equation of motion enhanced with gravitational-wave dissipation, and then insert these trajectories into the radiative multipole moments to compute the gravitational-wave perturbation.

Ok, so there you have it: the PN approximation of inspiraling binaries in a nutshell. Before moving on to the merger phase, however, let us discuss a few topics related to the PN approximation that we have left out. First, to what order have all of these terms been computed? As of the writing of this book, the conservative dynamics (e.g., the equations of motion, the gravitational binding energy, etc.) have been fully, rigorously established up to 4PN order for non-spinning binaries, while the dissipative dynamics (e.g., the rate at which energy and angular momentum are carried away) have been established to 3.5PN order. The algebra is daunting and serious technical challenges exist that make it difficult to determine unambiguously the coefficients in each succeeding set of terms. Fortunately, tidal effects only enter at the 5PN order, which one can justify by realizing that tidal couples have an $(\mathfrak{R}/r)^6$ energy dependence (where $\mathfrak{R}$ is the characteristic size of the object), or five powers of $1/r$ smaller than the Newtonian potential (see the discussion around Equation (2.2) in Chapter 2). Therefore, for many purposes, tidal effects can be ignored when $r \gg \mathfrak{R}$. However, tidal effects are crucial when we model the gravitational waves emitted in the late inspiral, plunge and merger of neutron star binaries, and in turn, they play a crucial role in estimating their radii and thus in constraining the equation of state of dense matter, as we will see in Chapter 7.

Another interesting fact is that the dynamics of compact binaries can be described with precision within the PN approximation, even when the binary is not in a quasi-circular orbit or when it's spinning. In the quasi-circular limit, all expressions simplify greatly, because one can then neglect an important and otherwise generic effect: precession. When binaries are in eccentric orbits, their Keplerian ellipses precess within the orbital plane, an effect probably known to you from earlier studies of the precession of the perihelion of Mercury. Unlike in that case, which considers weakly gravitating bodies such as the Sun and a planet, moving extremely slowly relative to the speed of light, in-plane precession due to eccentricity can be much more extreme for binary black holes or neutron stars in their late inspiral. Similarly, when compact objects spin about some axis, their spin angular momentum will couple with the orbital angular momentum of the binary, producing out-of-plane precession if the rotational and orbital axes are not aligned. Such an effect is particularly important for binaries containing black holes, as in this case the magnitude of the spin angular momentum need not be small. The effect of either type of precession is to make many harmonics of the gravitational-wave signal important, and these harmonics introduce a modulation of the amplitude that is correlated with the precession timescale. We will not present here equations or waveforms for such eccentric and spin–precessing models, but just know that they can be constructed within the PN approximation.

A fascinating observation is related to the convergence properties of the PN approximation. The PN approach is useful but tricky because succeeding terms are not always much smaller than the terms before them and often they come in with alternating signs! For example, the Newtonian acceleration is overwhelmingly dominant for an extremely wide range of separations (out to spatial infinity, in fact), but the range in which the 1PN term is necessary but the 2PN term is negligible is small. And this becomes even more true for the higher-order terms, if the masses are not comparable to each other, and if the binary is not in a quasi-circular orbit. This is a telltale sign that the PN theory may be an asymptotic or a divergent series, although a rigorous proof has not yet been found.

Dr. I. M. Wrong: Wait a second! Divergent series! What? If your series is divergent, then the next term in the series is larger than the previous term, and any truncation will therefore be necessarily inaccurate. This proves that the PN approximation is worthless and people should just not use it.

Major Payne: That statement couldn't possibly be farther from the truth. Asymptotic series and divergent series appear all the time in physics, and we have been able to use them very effectively to predict the solution to differential equations, and thus, the evolution of physical systems. You may have encountered many examples of asymptotic series in the past without actually knowing it, such as when you sum Feynman diagrams in quantum field theory (if you haven't seen that yet, you may see it later on in graduate school), or even when people design the wing of a plane using boundary layer theory.

There are entire books dedicated to asymptotic expansions, so I will not go into a long tangent about them. The two key points to remember are the following. One: asymptotic expansions need not be convergent, and in fact, they can be divergent, but that's ok. And two: the accuracy of an asymptotic expansion typically increases as more terms are added to the series for a while, but then, after a given term, adding more terms will push the approximation farther and farther away from the correct answer. Therefore, if one truncates the series at a sufficiently low order, then the resulting sum can approximate the true answer very effectively (and in many cases, even more effectively than a regular Taylor series!).

Let's do a quick example to see how this comes about. Let's say that you want to evaluate the error function for a large argument, where

$$\mathrm{erf}(x) = \frac{2}{\sqrt{\pi}} \int_0^x e^{-t^2} dt, \tag{3.21}$$

which appears frequently in statistics when considering normal distributions. The prefactor of $2/\sqrt{\pi}$ is so that the error function goes to one as $x \to \infty$. One can find a Taylor expansion for this function by series-expanding the integrand about small t and integrating term by term to find

$$\mathrm{erf}(x) = \frac{2}{\sqrt{\pi}} \sum_{n=0}^{\infty} \frac{(-1)^n}{(2n+1)n!} x^{2n+1}. \tag{3.22}$$

Although this series is convergent (the ratio of the $(n+1)$th term to the nth term is smaller than one for large-enough n and fixed x), it is a horrible approximation for large x; I mean, you need 31 terms to get erf(3) right to one part in 10^5! It turns out that you can also find an asymptotic expansion for this function, namely

$$\mathrm{erf}(x) \sim 1 - \frac{e^{-x^2}}{\sqrt{\pi}} \sum_{n=0}^{\infty} (-1)^n \frac{(2n+1)!!}{2^n} \frac{1}{x^{2n+1}} \qquad x \to \infty. \tag{3.23}$$

I'm not going to go over the details here (because the math gets a bit heavy), but essentially you can derive this answer by repeated integration by parts. This series diverges for any value of x because the $(n+1)$th term in the series grows faster than the nth term for large-enough n. However, this asymptotic series can provide a phenomenal approximation to the right (read: numerically-calculated) answer; for example, you need only two terms to get erf(3) right to one part in 10^5! Note, however, that if you add more and more terms in this series, eventually your error will increase and lead to a very poor approximation, because the asymptotic series does formally diverge after all. So there you have it: divergent series can "converge" to the right answer faster than convergent series!

One can often make good progress by using only the "Newtonian," lowest-order terms in PN expressions when describing binaries. For instance, one can use the Newtonian piece of the equation of motion and the Newtonian piece of the balance law to describe an "ok" inspiral. This, of course, will not be nearly accurate enough for gravitational-wave modeling and parameter estimation. But it can still be useful in other applications concerning astrophysics.

A final comment should be made about attempts to improve the accuracy of the PN approximation. Indeed, various clever attempts have been made to recast the expansions into forms that approach the right answer faster than a Taylor series. For example, one path is to pursue equivalent one-body spacetimes in which an effective test particle moves, and to then graft on the effects of gravitational radiation losses. One can also use Padé resummation, in which the different PN terms in a series are recast as ratios of polynomials, in the hopes that this can more naturally model the presence of the ISCO or the light ring in black hole spacetimes. There are other, more sophisticated resummation techniques that can be employed, but it is not obvious, a priori, that in practice this confers special advantages. By another "miracle" of general relativity, however, a subset of these resummations techniques do seem to help and provide more accurate PN approximations, relative to full numerical relativity evolutions.

3.2.1.2 Merger

The merger phase is, in many ways, much messier than the inspiral phase because characteristic timescales that separate in the inspiral no longer do so. What are these timescales? For generic binaries, there are typically three dominant ones: the orbital timescale (tied to the orbital period), the precession timescale (tied to the time it takes the binary to complete a full precession cycle), and the radiation-reaction timescale (tied to the amount of time required for otherwise constant quantities to change appreciably). From Newtonian mechanics, we know that the orbital timescale is given just by $T_{\rm orb} = 2\pi r_{12}^{3/2}/(Gm)^{1/2} \sim Gm/v^3$ by Kepler's third law. The precession timescale can be estimated roughly to be $T_{\rm prec} = \vec{S}_{1,2}/\dot{\vec{S}}_{1,2} \sim m/v^5$ from spin–orbit precession; we could have defined the precession timescale in terms of the orbital angular momentum vector, but the answer would be the same. The radiation-reaction timescale can be estimated via $T_{\rm rr} = v/\dot{v} \sim m/(\eta v^8)$. The ratios $T_{\rm orb}/T_{\rm prec} \sim v^2$ and $T_{\rm prec}/T_{\rm rr} \sim v^3$ are much less than unity only when $v \ll 1$. Close to plunge and merger, however, the orbital speed can become close to 1/2, at which point these fundamental timescales do not separate nicely. This implies that PN methods break down and one must rely on full numerical relativity simulations.

Numerical relativity has advanced tremendously in the last several decades. We have learned how to split the spacetime into one dimension of time and three of space and how to decompose the Einstein equations in such a way that they are well suited for coding on a supercomputer (i.e., what we call a strongly hyperbolic formulation). This is by no means trivial because the obvious decomposition (that leads to a weakly hyperbolic formulation) produces a system of equations that will become unstable when evolved for long enough on a computer. We have also learned how to prescribe initial data that satisfy the Einstein equations on a $t = \text{const}$ "hypersurface" of spacetime. This is crucial because, just like in electrodynamics, the Einstein equations contain constraints that must be satisfied everywhere in spacetime, and especially on the initial hypersurface. Prescribing constraint-satisfying initial data is hard because a PN ansatz (or some other simpler ansatz) will not satisfy the Einstein equations exactly; numerical improvements of this ansatz will solve the constraint equations on that initial hypersurface, but they will also introduce modifications to the initial ansatz that will generically lead to physical systems that are different from what one wanted to model in the first place, e.g., through the introduction of unwanted eccentricity. We have also figured out how to construct smart choices of our coordinate systems that change in time and have fantastic properties. These coordinates can "damp" away numerical inaccuracies, they allow us to "move" the singularities in the numerical grid so that they never hit a grid point (and thus are never evaluated, because if they were, computers would return a nan[4]), and they ensure that the grid points don't all bunch up at the event horizon or close to the singularities as we evolve forward in time. We have finally also figured out how to extract physical information from the numerical simulations by considering the behavior of our solutions far away from the sources, where spacetime is almost flat, and using certain curvature quantities that encode the gravitational-wave perturbations.

As you can imagine, all of this is numerically very challenging, and so a highly accurate numerical simulation that is decently long (say 10 orbits) can take days to weeks on computer clusters. And this is provided that the binary is "simple," meaning just black holes (so that we don't have to worry about matter and electromagnetic fields), with no spins (so that we don't have to worry about spin–orbit precession), in quasi-circular orbits (so that we don't have to worry about eccentric precession), and of comparable masses (so that we don't have to worry about EMRIs). If you make the orbit eccentric or if you increase the mass ratio, then the fundamental timescales increase significantly. For example, the radiation-reaction timescale is proportional to the inverse of the symmetric mass ratio, $1/\eta$, so when $m_2 \ll m_1$ it takes a very, very long time for the orbit to decay; intuitively, this makes sense because the strength of the gravitational waves emitted scale with the mass ratio, so the smaller η is, the weaker the waves are, and the smaller the rate at which they remove energy from the orbit. Moreover, if the mass ratio is extreme, then numerical codes have to resolve very disparate scales (scales of $\mathcal{O}(m_1)$ and scales of $\mathcal{O}(m_2)$, as well as scales of many times $\mathcal{O}(\lambda_{\rm GW})$ in the gravitational-wave extraction region, far from the source). All of this taxes the computational resources so significantly that numerical simulations are mostly (but not entirely) restricted to comparable-mass ratios, quasi-circular orbits, and spin magnitudes that are not too

[4] Not the delicious Indian bread naan, but rather the acronym "not a number."

close to their extreme values of $|\vec{S}_{1,2}| = m_{1,2}^2$. These simulations, however, are crucial to establishing a baseline with which to construct hybrid models able to model the inspiral, merger, and ringdown, as we will see in Section 3.2.1.4. Moreover, recent work has begun to consider using only numerical relativity simulations to extract very high-mass gravitational-wave signals, because advanced detectors can only see the merger and ringdown phases of such events.

3.2.1.3 Ringdown

In contrast with the highly dynamic and nonlinear merger portion of a binary coalescence, the ringdown phase can be treated with analytic and semianalytic techniques from black hole perturbation theory. As we noted in Chapter 2, lots of objects have characteristic frequencies of oscillations, from wires to stars—why not black holes? Indeed, the theory of black hole ringdown (from the end stages of a perturbed black hole to the final state as a stationary and axisymmetric Kerr spacetime) has been worked out to the extent that future high-signal observations of ringdown are expected to test general relativity; indeed, crude tests of the no-hair theorem have already been conducted on some of the strongest early LIGO/Virgo events.

In general, we can look for the characteristic frequencies of a system and their decay by introducing a perturbation around an equilibrium state and then determining the evolution of those perturbations. Because a black hole is just spacetime (in contrast to a star, which also has matter), it is logical to decompose the spacetime into a background $\bar{g}_{\mu\nu}$ plus a spacetime perturbation $h_{\mu\nu}$. Unlike in the PN case, however, the background cannot be a Minkowski spacetime, but rather it must be the Kerr spacetime when considering perturbations of a spinning black hole, or the Schwarzschild spacetime for perturbations of a nonrotating black hole. We then have $g_{\mu\nu} = \bar{g}_{\mu\nu} + h_{\mu\nu}$, where $h_{\mu\nu}$ is a metric perturbation assumed to be much smaller than the background.

But how shall we parameterize these perturbations? We can continue to follow the playbook of characteristic frequencies by realizing that any function can be written as a sum of coefficients times basis functions, if those basis functions are complete. Of course, for a given problem some basis functions are better than others. For example, if you are interested in the transverse oscillations of a wire that is clamped at both ends, a Fourier decomposition is logical; a Taylor expansion would not be particularly useful. Essentially, you would like to be able to decompose an arbitrary perturbation into normal modes that, to linear order, evolve independently of other normal modes. Then, each normal mode has a spatial and temporal dependence that can be evaluated.

When we try this approach with the ringdown of black hole perturbations, we run into trouble. Our first try would be to note that the Schwarzschild spacetime is spherically symmetric (so it must be independent of the azimuthal angle ϕ) and stationary, so perhaps a good decomposition would be something like

$$h_{\mu\nu} = \sum_{\ell} e^{-i\omega_{\rm QNM}t} R_\ell(r) P_\ell(\cos\theta)\delta_{\mu\nu}, \tag{3.24}$$

where QNM means "quasinormal mode," $\omega_{\rm QNM}$ is some (possibly complex) quasinormal ringdown frequency, and $\delta_{\mu\nu}$ is the four-dimensional identity matrix. Our specific goal with this ansatz is to separate the radial and temporal behavior of the Einstein equations from the angular behavior. If you try this, however, you will fail miserably! Even though an ansatz like that of Equation (3.24) works well to separate the (scalar) wave equation, it just does not work for the Einstein equations. We clearly need something more general.

For a nonrotating black hole, it turns out that one can find a coordinate system (a so-called "Regge–Wheeler gauge") in which we can write the metric perturbation as

$$h_{\mu\nu} = \sum_{\ell m} \begin{pmatrix} H_0^{\ell m} Y^{\ell m}(\theta, \phi) & H_1^{\ell m} Y^{\ell m}(\theta, \phi) & h_0^{\ell m} S_\theta^{\ell m}(\theta, \phi) & h_0^{\ell m} S_\phi^{\ell m}(\theta, \phi) \\ \cdot & H_2^{\ell m} Y^{\ell m}(\theta, \phi) & h_1^{\ell m} S_\theta^{\ell m}(\theta, \phi) & h_1^{\ell m} S_\phi^{\ell m}(\theta, \phi) \\ \cdot & \cdot & r^2 K^{\ell m} Y^{\ell m}(\theta, \phi) & 0 \\ \cdot & \cdot & \cdot & r^2 \sin^2\theta K^{\ell m} Y^{\ell m}(\theta, \phi) \end{pmatrix}, \quad (3.25)$$

where $(H_0^{\ell m}, H_1^{\ell m}, H_2^{\ell m}, h_0^{\ell m}, h_1^{\ell m}, K^{\ell m})$ are all functions of radius and time that we must determine, while $Y^{\ell m}$ are regular spherical harmonics and $(S_\theta^{\ell m}, S_\phi^{\ell m}) = (-\partial_\phi Y^{\ell m}/\sin\theta, \sin\theta \partial_\theta Y^{\ell m})$ are vector spherical harmonics. The functions $(H_0^{\ell m}, H_1^{\ell m}, H_2^{\ell m}, K^{\ell m})$ and $(h_0^{\ell m}, h_1^{\ell m})$ are said to be "polar" and "axial" because the metric components in which they appear pick up a factor of $(-1)^\ell$ and $(-1)^{\ell+1}$ respectively under a parity transformation ($\theta \to \pi - \theta$ and $\phi \to \pi + \phi$), just as the Major explained back in Chapter 1. And when one uses this parameterization of the metric perturbation, a miracle happens: the (t, r) sector of the Einstein equations separates from the angular sector!

What does that mean? It means that the Einstein equations become a set of coupled partial differential equations of only radius and time for the perturbation functions $(H_0^{\ell m}, H_1^{\ell m}, H_2^{\ell m}, h_0^{\ell m}, h_1^{\ell m}, K^{\ell m})$. But this is still a nasty coupled set of complicated partial differential equations, so what do we do now? Let us introduce the following master functions:

$$\Psi_{\rm RW}^{\ell m} = \frac{r}{1 + \lambda_\ell}\left[K^{\ell m} + \frac{f}{\Lambda_\ell}\left(f H_2^{\ell m} - r\partial_r K^{\ell m}\right)\right], \qquad \Psi_{\rm ZM}^{\ell m} = -\frac{f}{r} h_r^{\ell m}, \quad (3.26)$$

which we will call the Regge–Wheeler and the Zerilli–Moncrief functions, named after Tullio Regge, John Wheeler, Frank Zerilli, and Vincent Moncrief. In these equations, $f = 1 - 2M/r$, $\lambda_\ell = (\ell + 2)(\ell - 1)/2$, and $\Lambda_\ell = \lambda_\ell + 3M/r$. When one uses these master functions, then a second miracle occurs: the separated Einstein equations decouple!

This means that the nasty coupled set of partial differential equations becomes a single equation for the master functions. Defining $\Psi_{\rm RW/ZM}^{\ell m}(t, r) = e^{-i\omega_{\rm QNM} t}\tilde{\Psi}_{\rm RW/ZM}^{\ell m}(r)$ (or better yet, working in the Fourier domain), the perturbed Einstein equations decouple into

$$f^2 \partial_{rr}\tilde{\Psi}_{\rm RW/ZM}^{\ell m} + \frac{2M}{r^2} f \partial_r \tilde{\Psi}_{\rm RW/ZM}^{\ell m} + \left(\omega_{\rm QNM}^2 - V_{\rm RW/ZM}\right)\tilde{\Psi}_{\rm RW/ZM}^{\ell m} = 0, \quad (3.27)$$

with the Regge–Wheeler and Zerilli–Moncrief potentials

$$V_{\rm RW} = \frac{f}{r^2}\left[\ell(\ell+1) - \frac{6M}{r}\right], \tag{3.28}$$

$$V_{\rm ZM} = \frac{f}{r^2\Lambda_\ell}\left[2\lambda_\ell^2\left(1+\lambda_\ell+\frac{3M}{r}\right) + 18\frac{M^2}{r^2}\left(\lambda_\ell+\frac{M}{r}\right)\right]. \tag{3.29}$$

If you were able to solve these differential equations for the master functions, then you could reconstruct the metric perturbation (i.e., the gravitational-wave polarizations) via

$$h_+ - ih_\times = \frac{1}{2r}\sum_{\ell\geqslant 2,m}\sqrt{\frac{(\ell+2)!}{(\ell-2)!}}\left(\Psi_{\rm ZM}^{\ell m} + i\Psi_{\rm CPM}^{\ell m}\right)\,{}_{-2}Y^{\ell m}, \tag{3.30}$$

where we have defined the Cunningham–Price–Moncrief master function via $\Psi_{\rm CPM}^{\ell m} = 2\dot{\Psi}_{\rm RW}^{\ell m}$ and where ${}_{-2}Y^{\ell m}$ are called spin-weighted spherical harmonics (functions of the regular $Y^{\ell m}$ and their derivatives). And from these, you could even calculate the gravitational-wave luminosity

$$\mathcal{L} = \frac{1}{64\pi}\sum_{\ell\geqslant 2,m}\frac{(\ell+2)!}{(\ell-2)!}\left(\left|\dot{\Psi}_{\rm ZM}^{\ell m}\right|^2 + \left|\dot{\Psi}_{\rm CPM}^{\ell m}\right|^2\right), \tag{3.31}$$

as well as the rate at which waves carry away angular and linear momentum.

Ok, cool, so "all" we have to do is solve those Regge–Wheeler and Zerilli–Moncrief equations, and we are done … but how do we do that? It turns out that this equation is not bad at all. In fact, it looks a lot like the Schrödinger equation, doesn't it? If you do a coordinate transformation to tortoise coordinates $r \to r^*$ (defined via $dr/dr_* = f$), the first and second terms of Equation (3.27) become simply $(d^2/dr_*^2)\Psi_{\rm RW/ZM}^{\ell m}$, which then turns this equation into exactly a one-dimensional Schrödinger equation (albeit with a nasty potential). Solving this equation is then an issue of finding the frequencies $\omega_{\rm QNM}$ that yield a solution with the right boundary conditions at the Schwarzschild horizon and at spatial infinity. Because the horizon is a "one-way membrane," we must have that $\Psi_{\rm RW/ZM}^{\ell m} \sim e^{-i\omega_{\rm QNM}(t+r^*)}$ for r close to the horizon (i.e., an ingoing wave). Similarly, at spatial infinity, we must have $\Psi_{\rm RW/ZM}^{\ell m} \sim e^{-i\omega_{\rm QNM}(t-r^*)}$ for $r \to \infty$ (i.e., an outgoing wave).

It turns out that there is a discrete family of frequencies $\omega_{\rm QNM}$ that are labeled by the ℓ harmonic number and the n "overtone" number (as well as the m azimuthal number in the spinning case), and this family is complex and infinite. This means that the frequencies possess a real part and an imaginary part, the latter of which is related to the damping time of the mode. Because the family is infinite, we typically label the modes through an overtone number n, with $n = 0$ being the dominant (least damped, longest damping time) mode. This is cool, and perhaps somewhat familiar from your studies of the modes of a vibrating string, but the situation here is different, in part because we have dissipation. For example, if we consider a dissipationless vibrating wire, its oscillations can be represented, for all time, by a sum of normal modes. But if there is dissipation then the modes decay (usually exponentially), which means that if we project the system infinitely backward in time, then the modes would have infinite strength! This means that when ringdown is represented as a sum of quasinormal modes, there needs to be a starting time designated; and it turns out that at sufficiently late times there is also a power-law tail, discovered by Richard Price, which also cannot be represented as a sum of exponentially decaying oscillations. Fortunately, for much of the ringdown (roughly from the peak of the waveform to when the amplitude has decayed by 10 or more orders of magnitude), the waveform can be well represented by a sum of exponentially decaying quasinormal modes.

Can we at least estimate what these magical quasinormal frequencies are? After all, these are the numbers that matter the most for our waveforms, because they tell us the frequency at which the waves oscillate (through their real parts) and the rate at which they decay (through their imaginary parts). There are numerous ways to calculate the quasinormal frequencies, so let's try a few in ascending order of difficulty and rigor.

Our first attempt will be ridiculous. We will treat our nonrotating black hole as a Newtonian sphere of uniform density and ask how fast it oscillates. If matter is moving in and out as part of the oscillation, we'd expect that the angular frequency of the oscillation would just be the orbital angular frequency: $\omega_R = (GM/R^3)^{1/2}$, so when we put in $R = 2GM/c^2$, we get $\omega_R = 2^{-3/2}/M \approx 0.35/M$.

For the imaginary part, we need to work a bit harder. We can use Equation (2.13) for the luminosity,

$$\mathcal{L} \sim r^2 h^2 f^2 = \frac{32}{5}\frac{G}{c^5}\,\epsilon^2\, I^2\, \Omega^6, \tag{3.32}$$

where I is the moment of inertia along the smallest axis ($I = (2/5)MR^2$ for a uniform-density sphere) and $\epsilon \ll 1$ is the fractional deviation from sphericity. When we put in $R = 2GM/c^2$ and $\Omega = (GM/R^3)^{1/2}$, we get $\mathcal{L} = (1/500)(c^5/G)\epsilon^2$. Now, when we compute the gravitational potential energy of a uniform-density oblate spheroid with two equal axes and a third axis that is a factor of $(1 - \epsilon)$ times those axes, we find that the energy is $(4/75)(GM^2/R)\epsilon^2$ larger than the gravitational potential energy of a sphere of the same mass and density (again with $\epsilon \ll 1$). Thus, when we substitute $R = 2GM/c^2$, we find that the energy difference is $\Delta E = (2/75)Mc^2\epsilon^2$ (note that because a sphere minimizes the potential energy, the energy difference must scale as ϵ^2 rather than ϵ). The imaginary part of our quasinormal mode frequency is then the inverse of the characteristic time $T = \Delta E/\mathcal{L}$ to reduce the energy, so $\omega_I = 1/T \approx 0.075/M$ (where now $G \equiv c = 1$ again). Stunningly, these numbers turn out to be quite close to the correct values of $\Re\left(\omega_{\rm QNM}^{\ell=2}\right) = 0.37/M$ and $\Im\left(\omega_{\rm QNM}^{\ell=2}\right) = 0.09/M$ for the fundamental $\ell = 2$ mode!

But using a uniform-density Newtonian star as a stand in for a black hole is so unreasonable that it is useful to think about other more sophisticated approaches. Our next attempt will be a slightly more sophisticated Fermi estimate, in the sense that we will at least use the master equations, though still it will be hilariously unreasonable. Dusting off our Fermi tools, we replace $\partial_r \Psi^{\ell m}_{\rm RW/ZM} \to \Psi^{\ell m}_{\rm RW/ZM}/r$ in Equation (3.27). This gives us an equation for $\omega_{\rm QNM}$ in terms of radius, but what radius should we evaluate the quasinormal frequencies at? It makes sense to evaluate them where the potential has a maximum, which for the $\ell = 2$ mode is at $r_{\rm max} \approx 3.28M$ in the Regge–Wheeler case and $r_{\rm max} \approx 3.10M$ in the Zerilli–Moncrief case. Doing so, we find $\omega^{\ell=2}_{\rm QNM} \approx 0.34/M$, which is quite close to the correct value for the real part of the quasinormal frequencies of the fundamental $\ell = 2$ mode (which we recall is $\Re(\omega^{\ell=2}_{\rm QNM}) \approx 0.37/M$). But what is more, we find the same $\omega_{\rm QNM}$ using either the Regge–Wheeler or the Zerilli–Moncrief equations (with the respective potential and $r_{\rm max}$)! The fact that both polar and axial modes oscillate with the same frequency is called isospectrality and it holds exactly even when you solve the problem numerically (as opposed to just through a Fermi estimate).

Dr. I. M. Wrong: Wait a minute! Something smells seriously rotten. The authors spent a considerable amount of time convincing us that the vibrations of black holes are quasinormal, so the frequencies must be complex! The Fermi nonsense they did above gave them purely real frequencies, so this must be all wrong. I knew it! These physicists are just making things up...again.

Captain Obvious: You are right that the quasinormal frequencies must be complex. After all, the perturbations of a black hole must decay, so that it can return to its background equilibrium configuration. So what went wrong? Well, the authors were doing a simple Fermi estimate to show us that the modes are isospectral. They didn't intend to give us a precise answer for what the actual quasinormal frequencies are when, for example, you solve the problem numerically.

The Fermi problem gave us a hint of isospectrality, and it got the real part of the frequency alright, but it completely missed the imaginary part. It turns out, there is a slightly more sophisticated Fermi estimate we can make to do a bit better. Let's return to Equation (3.27), but now consider that these master functions represent gravitational waves, so a set of traveling gravitons. What happens if we try to treat the wave quantum mechanically? Let's try it out! Using the definition of the momentum operator $\hat{p} = -i\hbar(d/dr)$, let's replace the derivatives in Equation (3.27) with $\hat{p}$. Then, let's use the uncertainty principle to argue that a measurement of momentum and position for a graviton must satisfy $(\delta p)(\delta r) \geqslant \hbar/2$, so then the lowest energy level must have $\delta p = \hbar/(2\delta r)$. Replacing this in the $\hat{p}$ of Equation (3.27) and replacing δr with just r yields an expression for $\omega_{\rm QNM}$ that depends only on r and is complex!

There is a hint now that we may be in the right track, because a complex equation will typically render a complex result, which is what we want. But at what radius should we evaluate $\omega_{\rm QNM}$? Well, because we are considering the fundamental mode, this corresponds to the longest-lived mode, so it makes sense to evaluate $\omega_{\rm QNM}$ at the value of r that minimizes the energy. Taking the derivative of $\omega_{\rm QNM}(r)$ and setting it to zero, we find that, for the $\ell = 2$ mode, this value is $|r| \approx 3.29M$ in the Regge–Wheeler case and $|r| \approx 3.11M$ in the Zerilli–Moncrief case. Notice that these values are extremely close to the values that extremize the Regge–Wheeler and the Zerilli–Moncrief potentials, as found by the authors. Now, evaluating the complex expression we found for the quasinormal frequencies at these values of r, we find $\omega^{\ell=2}_{\rm QNM} \approx 0.39/M - i0.01/M$ again for both axial and polar modes (in spite of using different values of $|r|$ and different potentials). This new Fermi estimate preserved isospectrality, but now we also find that the quasinormal frequencies are complex, with a negative imaginary part. This is great, because it indicates that the modes are exponentially damped (and do not grow exponentially, which would have indicated the existence of an instability). Unfortunately, this is as far as we can go with Fermi estimates, because as you can see, the value of the imaginary part is still off by about an order of magnitude (recall that the authors told us that the correct answer is $\Im(\omega_{\rm QNM}^{\ell=2}) \approx -i0.09/M$). If we want a more accurate value for the quasinormal frequencies, we are going to have to solve the problem properly.

Ok, so Fermi estimates could only get us part of the way to an answer, so it is clear that we will have to think harder. One way to get a better approximate answer is to realize that the radii at which all of the action is happening is very close to $3M$. In fact, if we evaluate the radius at which the Regge–Wheeler and the Zerilli–Moncrief potentials are maximum for large ℓ, we find $r_{\max}/M = 3 + 1/\ell^2 - 1/\ell^3 + \mathcal{O}(1/\ell^4)$ for both potentials! Indeed, in the $\ell \to \infty$ limit, also known as the eikonal limit, $r_{\max} = 3M$. In general relativity, this radius has a special meaning in the context of nonspinning black holes: it is the only radius at which massless particles (like light or gravitational waves) can exist in a circular orbit. What is the characteristic frequency of such an orbit? The only frequency available to us is the orbital one. For a generic orbit of a null particle (called a null geodesic in relativity), the orbital frequency $\Omega \equiv d\phi/dt = (f/r^2)(L/E)$, where $f = 1 - 2M/r$, L is the orbital angular momentum, and E is the orbital energy. Now, circular orbits are special because they have $\dot{r} = 0$, and thus, they minimize the effective radial potential, like in Newtonian gravity. This turns out to happen at $r_c = 3M$, which is called the light ring or photon sphere, and when $L_c/E_c = r_c/f_c^{1/2}$ for a null geodesic. Using these conditions in the orbital frequency, we then find $\Omega_c = f_c^{1/2}/r_c = 1/(3\sqrt{3})M^{-1}$. Is this close to the quasinormal frequencies of a nonrotating black hole? Well, for the $\ell = 2$ mode it looks like $\Re(\omega_{\rm QNM}^{\ell=2}) \approx 2\Omega_c = 0.38/M$, so it makes sense to postulate $\Re(\omega_{\rm QNM}^{\ell}) \approx \ell\Omega_c$. Does this work? A numerical calculation reveals that $\Re(\omega_{\rm QNM}^{\ell=3}) = 0.60/M$ and our approximation is $3\Omega_c = 0.58/M$, while $\Re(\omega_{\rm QNM}^{\ell=4}) = 0.81/M$ and our approximation is $4\Omega_c = 0.77/M$. This geodesic approximation has given us excellent results! And even though the approximation has a hope of working only for large ℓ, we see that even when $\ell = 2$ it yields a very good estimate.

Can we also use the geodesic analogy to estimate the imaginary part of the quasinormal frequencies? It turns out that we can! As we mentioned above, null geodesics in general relativity can exist in a circular orbit only at the light ring, but these orbits are unstable. What does this mean? It means that a small perturbation will send them flying into the black hole or out to spatial infinity. We can quantify the degree of instability of a dynamical system through something called the Lyapunov exponent. Simply put, this exponent measures the rate at which trajectories that start very close to each other separate upon evolution. In our context, we can use the Lyapunov exponents associated with null geodesics at the light ring to quantify the rate at which two nearby geodesics (one that starts infinitesimally close to the light ring and one that stays at the light ring) will separate and spiral out to spatial infinity or in toward the black hole. A perturbation analysis of geodesics near the light ring shows that the (square of the) Lyapunov exponent is

$$\lambda^2 = \left[-\frac{r^2}{2f}\frac{d^2}{dr_*^2}\left(\frac{f}{r^2}\right)\right]_{r_c} = \left\{-\frac{r^2}{2f}\left[f^2\partial_{rr} + \frac{2M}{r^2}f\partial_r\right]\left(\frac{f}{r^2}\right)\right\}_{r_c}. \tag{3.33}$$

Evaluating this expression, we find that $\lambda = 1/(3\sqrt{3})M^{-1} \approx 0.19/M$, which we see is very close to twice the imaginary part of the quasinormal frequency! Indeed, in the geodesic approximation, we set $\Im(\omega_{\rm QNM}^{\ell}) \approx -i\,\lambda/2$ for the fundamental mode. Recalling that $\Im(\omega_{\rm QNM}^{\ell=2}) \approx -i0.09/M$ and noting that $-i\,\lambda/2 \approx -i0.096/M$, we see that the geodesic approximation is very good! Moreover, we note that the geodesic approximation implies the imaginary part of the quasinormal frequencies of the fundamental mode are independent of ℓ, and indeed, this is what one also finds numerically, e.g., $\Im(\omega_{\rm QNM}^{\ell=3}) \approx -i0.093/M$ and $\Im(\omega_{\rm QNM}^{\ell=4}) \approx -i0.094/M$. Once more, the geodesic approximation has given us excellent results!

Composing the real and imaginary parts of the quasinormal frequencies in the geodesic approximation, we have

$$M\omega_{\rm QNM,\,geod}^{\ell} = \ell\Omega_c - i\frac{1}{2}\lambda = \frac{\ell}{3\sqrt{3}} - i\frac{1}{6\sqrt{3}}, \tag{3.34}$$

which looks a lot like what one would obtain if one searched for $\omega_{\rm QNM,\,geod}^{\ell}$ in a $\ell \gg 1$ approximation, which we will refer to as the Wentzel–Kramers–Brillouin (WKB) approximation. You might have encountered the latter in quantum mechanics, for example, when solving the time-independent Schrödinger equation in the "semi-classical limit" (meaning in a power series in $\hbar$). By the way, this is physically why Captain Obvious' Fermi estimate based on quantum mechanics roughly worked! What went wrong in the Captain's treatment is that one must derive a new Bohr–Sommerfeld quantization rule (which can be obtained by solving the Regge–Wheeler and Zerilli–Moncrief master equation near the peak of the potential with WKB methods), instead of using the uncertainty principle. Carrying out such WKB analysis, one finds the quasinormal frequencies

$$\begin{aligned} M\omega_{\rm QNM,\,WKB}^{\ell} = &\frac{1}{3\sqrt{3}}\left[\ell + \frac{1}{2} - \frac{281}{216}\left(\frac{1}{\ell} - \frac{1}{\ell^2}\right) + \mathcal{O}(\ell^{-4})\right] \\ &- i\frac{1}{6\sqrt{3}}\left[1 - \frac{1591}{3888}\frac{1}{\ell^2} + \mathcal{O}(\ell^{-3})\right]. \end{aligned} \tag{3.35}$$

We see clearly how when $\ell \gg 1$, $\omega_{\rm QNM,\,WKB}^{\ell} \sim \omega_{\rm QNM,\,geod}^{\ell}$ to leading order in ℓ. Numerically, we also see that the WKB expansion is quite accurate even when $\ell = 2$, getting of course better for larger ℓ. More concretely, we have $M\omega_{\rm QNM,\,WKB}^{\ell=2} = 0.42 - i0.086$, $M\omega_{\rm QNM,\,WKB}^{\ell=3} = 0.62 - i0.092$, and $M\omega_{\rm QNM,\,WKB}^{\ell=4} = 0.82 - i0.094$, which is quite close to the actual numerical answer!

Major Payne: AHH!!! I can't take it anymore. This lack of precision in the language and in the mathematics is what creates confusion and leads to students making mistakes. In real life, you want to compute the quasinormal modes accurately because you are likely to use them to construct models of gravitational waves and compare those to data. That means we cannot rely on "simple" asymptotic expansions (or use Fermi estimates). You have to solve the problem exactly, and because a closed-form expression does not exist mathematically, the best we can do is go for a numerical solution.

In mathematics, differential equations of the type of Equation (3.27) are well suited to the Green's function method. Recall that Green's function is just a function that satisfies the original differential equation, but with any sources replaced with a Dirac delta function. Once you know Green's function, then the solution is simply given by the integral of the product of Green's function with any sources of the differential equation. It turns out there is a general prescription to find Green's functions if you have two linearly independent solutions to the homogeneous problem. All you do is construct the product of these linearly independent solutions and divide by their Wronskian, where the latter is defined via $W \equiv (\partial_r f_1) f_2 - f_1(\partial_r f_2)$ for $f_1(r)$ and $f_2(r)$ two independent solutions. It then follows that the poles of Green's function occur exactly where the Wronskian vanishes, and this is because when this occurs, the two solutions are not linearly independent any longer.

How do we apply this to Equation (3.27)? We want a solution that satisfies two boundary conditions (one at the horizon and one at spatial infinity). We can then construct two linearly independent solutions by starting one numerical integration at spatial infinity (or equivalently far enough away from the black hole) and integrating in, and by starting another integration at (or close to), the horizon and integrating out, in both cases with the same initial guess for $\omega \in \mathbb{C}$. With these solutions, we can then construct the Wronskian, and in general, the Wronskian will not vanish when evaluated at some arbitrary intermediate point (between the horizon and spatial infinity). What we want to do is find the value of ω for which the Wronskian does vanish, because for that special value, the two solutions will be linearly dependent, and thus they will both satisfy both boundary conditions (the one at the horizon and the one at spatial infinity). We then search for this special value of ω until we find one which leads to a vanishing Wronskian, and this special value is the quasinormal frequency of the black hole for a given ℓ.

In practice, this calculation requires some fiddling with numerics that most graduate students should be able to do. For example, one cannot start the numerics exactly at spatial infinity (that's not, usually, a point on a computer grid!) or at the horizon (the equations are usually ill behaved there). So one must construct boundary conditions that are sufficiently accurate a finite (small) distance from the horizon and spatial infinity, which usually requires an asymptotic expansion of the solution about those two regular singular points. Of course, when computing each numerical solution, one must ensure that the solutions are not sensitive to the specific point at which one sets the boundary conditions. Moreover, one must ensure that one is using an algorithm to solve the ordinary differential equation that is sufficiently accurate, such that the numerical error is under control. And when evaluating the Wronskian, one must make sure to pick a point at which neither of the two solutions dominates over the other. So an interesting numerical problem this is indeed, but it's nothing insurmountable, and in fact, it is a common exercise for relativity students nowadays.

As the Major explained, there are subtle details involved in the solution of the master equations, and several techniques have been developed. One example is "shooting" methods that are akin to building a railroad track from both ends and hoping that they meet in the middle. More technically, you integrate from the horizon to some matching radius, and from spatial infinity to the same matching radius, and adjust ω so that the solutions meet and are differentiable at that radius. The stability of this method is typically difficult to control numerically, which is why researchers usually opt for the Wronskian method described by the Major. Another popular approach introduced by Edward Leaver in 1985 uses continued fractions. In this approach, one proposes an ansatz for the master functions in terms of the product of a function that incorporates both boundary conditions times a series in $f = 1 - 2M/r$. Substituting this ansatz into the master equations, one finds a recursion relation for the coefficients in the series. For the series to be convergent, however, one must demand that the coefficients satisfy a continued fraction condition, whose numerical algebraic solution yields the quasinormal frequencies. No matter the method, black holes possess a set of unique and complex quasinormal frequencies, and once these are calculated, one can tabulate them or construct fitting functions (more on this later).

But what about perturbations of a spinning Kerr black hole background? The Kerr spacetime is axisymmetric rather than spherically symmetric. This means, in particular, that the components of the metric (e.g., as written in standard Boyer–Lindquist coordinates) contain powers of $r^2 + a^2 \cos^2\theta$ (where $a = |\vec{S}|/(Mc)$ is the Kerr spin parameter, with $\vec{S}$ and M the spin angular momentum and mass of the black hole, respectively) and other similar functions that mix the radial and polar angle dependence. In fact, no coordinate system is known that allows the metric functions to not be rational polynomials that mix the radial and polar directions (or the equivalent combination in a nonspherical coordinate system). Yes, if the dimensionless angular momentum $\chi \equiv c|\vec{S}|/(GM^2)$ of the merger remnant is much less than unity then it almost works. That is, you can expand the field equations in powers of χ, and then at each order, the equations do separate in terms of angular tensors that can be constructed from spherical harmonics. However, to date, no analog of tensor spherical harmonics is known that will allow the Einstein equations to separate exactly in a general axisymmetric background of arbitrary spin magnitude.

Despite this problem, work in numerous creative papers in the 1960s and 1970s demonstrated that a different approach is fruitful. In particular, Saul Teukolsky showed that, although the perturbed Einstein equations are a coupled mess when written in terms of the metric perturbation, they decouple when written in terms of variables called the Newman–Penrose scalars!

Major Payne: Indeed! The five complex Newman–Penrose scalars $\{\psi_0, \psi_1, \psi_2, \psi_3, \psi_4\}$ are contractions of the Weyl tensor (a certain combination of the Riemann tensor, Ricci tensor, and Ricci scalar) with collections of four 4-vectors called tetrads. This is my kind of work ...

Remarkably, the 10 coupled, partial, nonlinear differential equations contained in the perturbed Einstein equations can be combined in a specific way to yield a complex equation (the Teukolsky equation) for the perturbed Newman–Penrose scalars. This equation has impressive generality: depending only on a constant integer s called the "spin weight," the equation can represent a scalar perturbation ($s = 0$), an electromagnetic vector perturbation ($s = \pm 1$), or a gravitational tensor perturbation ($s = \pm 2$). For the gravitational perturbations that are our focus, ψ_0 turns out to represent the waves that are absorbed by the black hole event horizon, and ψ_4 encodes the gravitational waves that escape to large distances; incidentally, this is the quantity that numerical relativists calculate to extract gravitational waves!

The decoupling of the Einstein equations into a single Teukolsky equation is very similar to the second miracle we discussed when considering perturbations of a nonrotating black hole. Recall that in that case, the Einstein equations decoupled for a certain scalar combination of components of the metric perturbation, which we called the Regge–Wheeler and the Zerilli–Moncrief master functions. One can then think of the Newman–Penrose scalars as analogous to those master functions, and the Teukolsky equation as the spinning generalization of the Regge–Wheeler and the Zerilli–Moncrief equations. The main difference, of course, is that the Newman–Penrose scalars are computed from the Weyl curvature tensor, while the Regge–Wheeler and the Zerilli–Moncrief master functions are constructed from the metric perturbation itself and a few derivatives. It is for this reason that while the gravitational-wave polarizations $h_{+,\times}$ are given directly in terms of these master functions (see Equation (3.30)), in the spinning case the Newman–Penrose scalar Ψ_4 is related to the second time derivative of the polarizations.

Another difference is that in the nonrotating case we were able to decompose the metric perturbation in tensor spherical harmonics, so the master functions were only functions of time and radius, but in the spinning case, the metric perturbation cannot be decomposed in the same way, so the Newman–Penrose scalars remain functions of all four coordinates. Can we still separate the Teukolsky equation into an angular and a radial sector? Just as in the nonspinning case, the answer is yes! It turns out that one can decompose the Newman–Penrose scalars into a product of functions of time and radius and functions of angles only, the latter of which are called spin-weighted spheroidal harmonics.[5] Leaver has found that the latter can be represented in closed form as a series with a controlling factor, which depends nontrivially on the cosine angle times $e^{im\phi}$, and a sum over cosines of the polar angle. What is then left, after one Fourier-decomposes the time dependence, is an ordinary differential equation for the radial functions, which looks a lot like the Regge–Wheeler and Zerilli–Moncrief master equations. The boundary conditions on the Newman–Penrose scalars are also similar to the boundary conditions we imposed on the master functions.

[5] Recall that the "spin" here does not refer to the spin angular momentum of the black hole. Rather, it specifies the type of perturbation one is considering. Gravitational perturbations are said to be of spin-weight 2 (because gravitons are spin 2 particles), while an electromagnetic perturbation is of spin-weight 1 (because photons are spin 1 particles) and a scalar perturbation of spin-weight 0. The name "spin-weighted spheroidal harmonics" comes from the fact that they reduce to spheroidal harmonics for scalar perturbations and to spin-weighted spherical harmonics when the spin angular momentum of the black hole is zero.

Captain Obvious: As an aside, you may wonder why it is that the 10 coupled partial differential equations that are contained in the perturbed Einstein equations have reduced to a single equation for ψ_0 or ψ_4. Of the 10 Einstein equations, 4 of them are not evolution equations at all, but rather, they are constraint equations, just like $\nabla \cdot B = 0$ is a constraint equation in electromagnetism. This leaves six perturbed Einstein equations, but four of them describe what we call "gauge" degrees of freedom, i.e., components of the metric perturbation that can be eliminated through a coordinate transformation. This then leaves only two perturbed Einstein equations for the two gravitational-wave perturbations. But ψ_0 and ψ_4 are complex, so they encode four functions, not two! It turns out that there is a way to map ψ_0 to ψ_4 and vice versa through a set of equations called the Teukolsky-Starobinsky identities, so in reality one need only solve for one of them. In detail, it turns out that to avoid numerical instabilities, it is best to work with Fourier transforms if one wishes to extract the gravitational-wave strains $h_{+,\times}$ from ψ_4.

Of course, solving the Teukolsky equation for the Newman–Penrose scalars is significantly more difficult than solving the Regge–Wheeler and Zerilli–Moncrief equations. One complication is that, unlike in the nonrotating case, the Teukolsky equation leads to mode mixing. What does this mean? It means that the Teukolsky equation cannot be solved ℓ mode by ℓ mode, but instead, a given ℓ mode is affected by all other ℓ modes! The root of this complication is, once more, the spin-weighted spheroidal harmonics. Indeed, William Press and Saul Teukolsky have shown that the ℓth spin-weighted spheroidal harmonics is equal to the ℓth spin-weighted spherical harmonics plus a mixing term that depends on other ℓ modes through Clebsch–Gordan coefficients. Moreover, calculating the separation constant for spin-weighted spheroidal harmonics can only be done numerically, as opposed to in the nonrotating case, in which case the separation constant is just $\ell(\ell + 1) - 6$.

All of this implies that solving the Teukolsky equation and finding the quasinormal frequencies associated with rotating black holes can only be done numerically. Fitting functions, of course, have been computed based on numerical results. Defining $\omega_{\rm QNM} = \omega_{\ell mn} - i/\tau_{\ell mn}$, with $\tau_{\ell mn}$ the decay time, and then introducing the quality factor $Q_{\ell mn} = \omega_{\ell mn}\tau_{\ell mn}/2$, one can fit numerical results to

$$M\omega_{\ell mn} = f_1 + f_2(1 - a/M)^{f_3}, \qquad Q_{\ell mn} = q_1 + q_2(1 - a/M)^{q_3}, \tag{3.36}$$

where the constants (f_1, f_2, f_3) and (q_1, q_2, q_3) are fitting coefficients. For example, for the fundamental $(n = 0)$, $\ell = 2$, and $m = 0$ modes, we have that $f_1 = 0.4437$, $f_2 = -0.0739$, $f_3 = 0.3350$, $q_1 = 4$, $q_2 = -1.955$, and $q_3 = 0.142$. Fitting coefficients for many other modes have been calculated and tabulated in the literature, so that in practice, we don't actually have to solve the Teukolsky equation in gravitational-wave astrophysics. It is still good for the soul to remember where these fitting functions come from so that we understand when they are applicable and when they are not. For example, the quasinormal frequencies depend both on the particular background spacetime considered (e.g., a spinning Kerr black hole versus a charged spinning Kerr–Newman black hole, versus black holes in modified gravity), as well as on the particular gravity theory in play, because the field equations of the theory determine the Teukolsky equation (or its analog in modified gravity). These two concepts are not necessarily the same, because there are modified theories of gravity for which the Schwarzschild metric, or even the Kerr metric, are the correct description for the black hole background, yet the equations that control the dynamics of the metric perturbation (the analog to the Teukolsky equation in the modified theory) are not the same. What this means is that ringdown gravitational waves and their characteristic frequencies can be potentially powerful probes of the spacetime, the theory of gravity, and the nature of the central object, allowing for, among other things, tests of the "no-hair" theorems of general relativity.

3.2.1.4 Hybrid Models

But what if you need a model that is valid in the inspiral and the merger and the ringdown? Enter hybrid models. The idea here is to construct a Frankenstein monster from elements of the methods that work in the inspiral, those that work in the merger, and those that work in the ringdown.

The first thing you need to do to construct an inspiral–merger–ringdown (IMR) waveform is to find a way to extend the validity of your approximations as much as possible so that there is hopefully an overlap region between adjacent regimes, e.g., between inspiral and merger methods. One way to do this in the inspiral is to employ semianalytic methods. The idea here is to still find differential equations that describe the dynamics of the system perturbatively in the PN approximation, but then to solve these equations numerically. The simplest example is to take the equations of motion and solve those numerically for the trajectories of the bodies, instead of perturbatively with analytic methods.

But there is more! If we are going to give up the idea of finding an analytic solution, you can in principle make the differential equations you are going to solve numerically as complicated as you want, if in the process you also make them more accurate. Here is where the method of resummation comes in. As explained earlier when we discussed the anharmonic oscillator (see Equation (3.7)), this is the idea of taking an expression written as a sum of terms and rewriting it in a compact form, such that when the latter is reexpanded, one recovers the sum of terms. It turns out that this can be done with the PN Hamiltonian, if one reduces the two-body system to that of a small particle (with mass equal to the binary's reduced mass) orbiting around a single body (with mass equal to the total mass of the binary), and incorporates certain terms in the energy- and angular-momentum-loss rates. The equations of motion can then be computed from the Hamilton equations, enhanced with a (dissipative) radiation-reaction force. The resulting equations of motion are horrendous, but a computer does not care, and so a numerical solution is easily obtained. Once the trajectories are known, one can use these in the far-zone multipolar expansion of the metric perturbation (see, e.g., Equation (3.18)), and you are done.

The Effective-One-Body (EOB) formalism uses this approach to model the inspiral, and then, when the bodies reach the light ring of the effective problem, the waveforms are transitioned to quasinormal ringdown waves. The result is a numerical IMR waveform that agrees by construction with PN theory in the early inspiral and with black hole perturbation theory in the ringdown, interpolating through the region in between. When compared to numerical relativity simulations, one finds that these waveforms are quite good, but can be even further improved by adding fitting coefficients (representing unknown higher PN order terms) to the Hamiltonian and the radiation-reaction force. These coefficients can be fitted to a subset of numerical relativity simulations, resulting in improved EOB waveforms One knows they have been improved not only because (by construction) they fit well with a subset of numerical relativity simulations, but because they also agree with other numerical relativity simulations not used in the EOB model fits.

Another approach is to use guidance from the EOB framework to construct a purely analytic IMR model. The procedure here is similar to the EOB one, in that one employs a PN approximation in the inspiral and quasinormal gravitational waves in the ringdown. The difference is that the PN gravitational waves are prescribed entirely analytically, which of course will produce a model that is not as accurate as the EOB one. But what one does next is to introduce fitting coefficients to the analytic PN waveforms, and then fit these to a set of numerical relativity simulations. These fitting coefficients then improve the accuracy of the gravitational-wave model, allowing one to glue it to the ringdown waveforms effectively. The resulting model is called "Phenom," which here stands for phenomenological, although in reality it is rooted in PN theory and black hole perturbation theory, just like the EOB model is. Both families of models (the EOB family and the Phenom family) are used regularly to analyze gravitational-wave data, leading to parameter estimates that are largely consistent with each other.

3.2.2 Extreme Mass-ratio Binaries

Extreme mass-ratio binaries are a different beast. These binaries are defined as those containing a small compact object that spirals into a much larger compact object, with a mass ratio typically of $m_2/m_1 \lesssim 10^{-4}$. As in the case of comparable masses, one can separate the coalescence into the same three stages as before: an inspiral, a plunge and merger, and a ringdown. In this case, however, the plunge-and-merger phase is more like a pure plunge phase (because the small object is essentially absorbed by the big black hole and there isn't much of a "merger"), and together with the ringdown phase, they contribute very little to the coalescence signal. The EMRI signal is then dominated by the inspiral phase.

Why is that? We already saw that gravitational waves scale with the mass ratio of the system, which then immediately leads to two conclusions: (i) gravitational waves from EMRIs will be naturally weak, (ii) the amount of energy and angular momentum that is transported by these waves away from the binary system is also naturally small. In both cases, the suppression relative to their comparable mass-ratio cousins is proportional to the mass ratio or its square, which in turn implies that to see these systems you'd better hope they exist sufficiently close to Earth (at least within 1 Gpc). Fortunately, the large object has to be a supermassive black hole, and this buys you back some of the amplitude, at the cost of pushing the gravitational-wave frequency below 1 Hz. Another consequence of this suppression is that the radiation-reaction timescale is also larger by a factor of the mass ratio relative to comparable-mass binaries, which then means that EMRIs are not burst sources, but rather are "on" for many years in band. This means the inspiral phase is extremely long, and if one is able to extract all of it, there is more power in this phase of the signal than in the merger and ringdown phases.

Ok, so easy then! "All" we have to do is model the inspiral phase, and we are done. But this is way more easily said than done, because EMRIs are much, much more complex than comparable-mass inspirals. For the latter, we expect the orbital eccentricity to be small because gravitational waves are exceptional at circularizing the orbit. Moreover, if we are dealing with neutron stars, their spin magnitudes are expected to be small, and thus precession can be ignored, at least as a first approximation. This of course is not necessarily true when considering comparable-mass binary black holes, but even then, only a few precession cycles are expected to be in band, so the approximations one uses to model precession can be rougher. For EMRIs, on the other hand, eccentricity is expected to be present, because the waves are so weak that circularization is not expected to be as efficient, as we will discuss in Chapter 5 in more detail. Moreover, the spin of the big object will play a big role in the orbit, causing spin–orbit precession regardless of whether the other object has a spin or not. And finally, the orbit needs to be modeled accurately for years at a time, and this means any errors due to mismodeling that could grow secularly must be minimized as much as possible. And to make matters worse, EMRIs are much more susceptible to the presence of third bodies, which could cause von Zeipel–Lidov–Kozai resonances (see Chapter 5), and to the astrophysical environment, such as an accretion disk on the large black hole or a dark matter overdensity.

Our first gut feeling is probably that none of this is a big issue because, after all, we have developed the PN approximation to high order, so surely we can use it to model EMRIs. Unfortunately, this is not true. All PN approximations end up producing a series in velocity, such as

$$A(t) = \sum_{n=0} a_n(\vec{\theta}) v(t)^n \tag{3.37}$$

for any quantity $A(t)$, where a_n are coefficients that depend on the parameters of the system $\vec{\theta}$, which generically include their mass ratio (at higher orders one also encounters terms that scale as $v^n \log v$, but these are not relevant to our discussion here). When the masses are comparable, the coefficients a_n do not grow very rapidly with n, but when the mass ratio is small, they do. This suggests that the PN series is possibly divergent, as we discussed earlier, at least in the extreme mass-ratio limit.

A better approximation is to expand the Einstein equations in the mass ratio and then try to solve them without ever expanding in small velocities or weak fields. This is the realm of black hole perturbation theory, and more precisely, of the self-force problem. Let us then consider expanding the metric as we did before $g_{\mu\nu} = \bar{g}_{\mu\nu} + h_{\mu\nu}$, where $\bar{g}_{\mu\nu}$ is the background spacetime and $h_{\mu\nu}$ is a spacetime perturbation generated by the small object. This perturbation will contain both a dissipative part (odd under time reversal), which encodes gravitational waves that escape the system, as well as a conservative part (even under time reversal), which modifies the spacetime curvature in which the small object moves. And yes, you heard it right: the small object curves the spacetime geometry, and this curvature (small as it may be) modifies its own trajectory. You can then think of the object as generating a force on itself, a "self-force," that modifies its trajectory, just as we discussed regarding the inspiral within the context of PN theory.

Let us ignore this self-force for now and just consider the evolution of the small object around the large one, as corrected only by the dissipative part of the spacetime perturbation. The orbital evolution of the small object is then basically a sequence of geodesics on the background, corrected by a dissipative radiation-reaction force that takes energy, angular momentum, and Carter constant away from the system.

Captain Obvious: Whoa, whoa, whoa, hold your horses! The authors' haven't told us yet what the Carter constant is! You are probably familiar with the fact that the energy and a certain component of the orbital angular momentum are conserved in geodesics around a spinning black hole. This is simply because the spinning black hole spacetime, the Kerr spacetime, is stationary (so time-translation symmetric) and axisymmetric (so symmetric under rotations about its spin axis).

It turns out that Brandon Carter back in the 1970s proved that the Kerr spacetime is even more symmetric than you would have expected! Indeed, one can show that the Kerr spacetime possesses a "Killing tensor," (named so after the mathematician Wilhelm Killing and completely unrelated to death). It turns out that this Killing tensor generates a third constant of the motion for geodesics around Kerr, which today goes by the name of the Carter constant. In the limit of zero spin (when you reduce the Kerr spacetime to the Schwarzschild spacetime), the Carter constant is just the square of the (azimuthal component of the) orbital angular momentum.

For geodesic motion, these three constants of the motion, the energy, the z components of the orbital angular momentum (the component aligned with the spin of the black hole background), and the Carter constant, can be mapped to the semimajor axis, eccentricity, and inclination angle of the orbit. One can then imagine gridding the three-dimensional space spanned by these three constants, and at each point in this grid, solving for the associated geodesics, as shown in Figure 3.3.

For each of these geodesics, one can then solve the Teukolsky equation to find the Newman–Penrose scalar ψ_0 and ψ_4; the solution is here a bit more complicated than what we did in Section 3.2.1.3, because the Teukolsky equation now has a matter source, and it must be solved with Green's function methods. Because ψ_0 and ψ_4 tell how gravitational waves are absorbed by the background's horizon and escape to infinity, for each geodesic we can then construct the rate of change of the energy, angular momentum, and Carter constant at any point of this three-dimensional space. These rates of change tell us how to move from one grid point to the next by providing a vector flow, as shown in Figure 3.3. With this at hand, one can then solve the geodesic equation one last time, but now promoting the quantities that

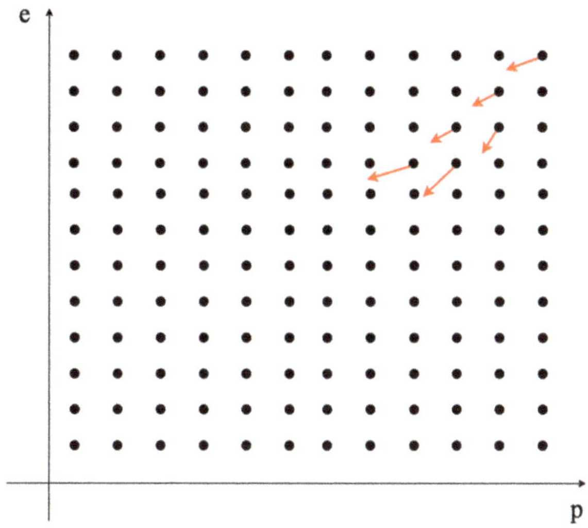

Figure 3.3. Grid of a two-dimensional cross section of the (e, p, ι) space, with the dots representing where geodesics are evaluated, and the arrows indicating how these constants evolve due to dissipative radiation reaction.

were previously constants of the motion to be functions of time, as given by the vector flow. The resulting inspiral trajectory can then be used to compute its associated gravitational waves through a post-Minkowskian multipolar expansion, or better yet, by solving the Teukolsky equation one last time.

The resulting gravitational waves are sometimes called "adiabatic" waveforms because they assume the orbital elements are changing slowly due to gravitational-wave emission. As you can imagine, the process is quite computationally expensive, but the resulting waveforms are not limited by a small velocity expansion. They are, however, limited by an expansion in mass ratio, which was technically only kept to leading order, as conservative self-force effects contribute to second order in the mass ratio in the waveforms. To obtain waveforms accurate to next order in the mass ratio, one must first solve for the first-order conservative self-force, as well as the second-order dissipative correction to the spacetime perturbation. Great progress has been made in recent years to solve both of these problems, and the hope is that accurate waveforms will be available by the time LISA (the Laser Interferometer Space Antenna) flies in the mid-2030s.

3.3 Exercises

1. Let's get a sense of the scale of compact objects.
    (a) What mass must a black hole have in order to be the same size as a neutron star (whose radius is about 12 km)?
    (b) What about for the black hole to be the same size as a white dwarf (whose radius is about 10^3–10^4 km)?
2. Let's now investigate what happens if you get too close to a black hole.

(a) Calculate the distance away from a compact object at which a rod made out of structural steel and of length 10 cm oriented in the direction of the compact object would break. Do this calculation for a black hole of mass $10M_\odot$, a black hole of mass, and a neutron star of mass $1.4M_\odot$ and radius 12 km. (Hint: the ultimate tensile strength of structural steel is about 500 MPa.)

(b) Repeat the same calculation but for a rod made out of carbon nanotubes (which is the material with the highest ultimate tensile strength, 3 GPa, known to humankind), and for a human bone (ultimate tensile strength of 130 MPa), assuming in both cases a length of radius 10 cm.

Do these calculations using Newtonian expressions for the tidal force; if you know general relativity well and want to flex your analytical muscles, feel free to perform the calculation in full detail and determine how much the answer differs from the Newtonian approximation.

3. As a simple example of perturbation analysis, we can once again consider a pendulum. Let θ be the angle from the vertical, so that $\theta = 0$ is the equilibrium state of the system, and consider a small perturbation $|\theta| \ll 1$ from that equilibrium. Thus, the governing equation for θ becomes approximately $\ddot{\theta} + (g/L)\theta = 0$, where g is the gravitational acceleration and L is the length of the pendulum, as we saw earlier in the chapter.
   (a) Put in a trial solution $\theta = \theta_0 e^{i\omega t}$, where ω is the (possibly complex) frequency of the perturbation, and solve for it.
   (b) Consider now an inverted pendulum, where $\theta = 0$ means that the pendulum is directly above the support, and we again consider $\theta = \theta_0 e^{i\omega t}$ with $|\theta| \ll 1$. What is ω in that case, and what does it imply about the system?

4. Let's try to practice and understand resummations.
   (a) Take the expression for the relativistic version of Kepler's third law in Equation (3.11) and perform a (2,2) Padé resummation (i.e., rewrite it as a rational polynomial, with the numerator and denominator polynomials of order 1). Hint: You can find the coefficients of the Padé resummation by requiring that its PN expansion matches that in Equation (3.11).
   (b) Now, evaluate the gravitational-wave frequency in Hertz (and recall that for a quasi-circular binary $f_{\rm GW} = 2f$) using Equation (3.11) and its Padé resummed version at $r_{12} = 6M$ (i.e., at the ISCO of a test particle in a Schwarzschild background), at $r_{12} = 20M$, and at $r_{12} = 100M$ for an equal-mass binary. When considering equal-mass binaries, are the two approximations to $f_{\rm GW}$ more similar to each other when you evaluate them at a given r_{12}? Why or why not?
   (c) Repeat the previous calculation but for an EMRI with mass ratio $q = 10^{-5}$. Are the gravitational-wave frequencies computed with the

PN-expanded and the Padé resummed expressions as similar to each other as in the equal-mass case? What do you conclude from this?

(d) Estimate, as an order of magnitude, the impact of a two-PN term in Equation (3.11).

5. Let's now make sure you understood how to derive some expressions in the book.
   (a) Use the expression for the binding energy in Equation (3.12), the relativistic version of Kepler's third law in Equation (3.11), and the luminosity in Equation (3.14) to calculate the rate of change of the orbital angular frequency.
   (b) Now use this to rederive Equation (3.16) for the amount of time a signal is in band between some initial and some final GW frequencies, $f_{\rm GW,i}$ and $f_{\rm GW,f}$.

6. In this problem, we will investigate how long the gravitational waves emitted by certain systems remain in the sensitivity band of LISA and LIGO.
   (a) Use Equation (3.16) to estimate how long a supermassive black hole binary (with equal masses and a total mass of $m = 10^6 M_\odot$) and an EMRI (with mass ratio $q = 10^{-5}$ and total mass $m = 10^6 M_\odot$) remain in the LISA band. Assume for simplicity that the LISA band extends from 10^{-4} Hz to 10^{-1} Hz. Be careful to not allow the inspiral to proceed past the plunge, which you can estimate in this problem as starting at $r_{12} = 6m$ (i.e., at the ISCO of a test particle on a Schwarzschild background). Is it possible to observe the entire evolution of an EMRI with LISA?
   (b) Given that LISA is most sensitive at frequencies of about 10^{-3}–10^{-2} Hz (although the foreground of unresolved double white dwarf binaries makes the effective sensitivity worse below $\sim$(2–3) $\times 10^{-3}$ Hz), and that the mission duration will probably not exceed 4 yr, calculate how much the frequency changes for an EMRI that starts at 10^{-3} Hz.
   (c) Use Equation (3.16) to estimate how long a comparable-mass black hole binary (with equal masses and a total mass of $m = 50 M_\odot$) and a neutron star binary (with equal masses and a total mass of $m = 3 M_\odot$) remains in the LIGO band. Assume that the LIGO band extends from 10 Hz to 10^3 Hz. Be careful to not allow the inspiral to proceed past the plunge (i.e., past the ISCO of a test particle in a Schwarzschild background) or past the point of surface contact (assuming the radius of all neutron stars is 10 km).
   (d) What is the maximum mass of an equal-mass black hole binary such that LIGO can detect the inspiral?
   (e) How much of the signal of binary neutron stars do we miss by ignoring the merger? (in reality, aLIGO will have sensitivity above 10^3 Hz when at design sensitivity, but you can ignore that for this problem).

7. Dr. I. M. Wrong has decided that all of the black hole binaries that LIGO has detected are primordial black holes (PBHs). A primordial black hole is one that was formed in the early universe, due to the collapse of an overdensity. If this is so, then LIGO may also detect the inspiral of extremely low-mass black holes. Help Dr. Wrong out by computing the properties of such a very low-mass BH binary inspiral.
   (a) Consider an equal-mass PBH binary with total mass $m = 10^{-2} M_{\odot}$ and estimate how long the inspiral would last in the LIGO band (assuming it extends only between 10 Hz and 10^3 Hz).
   (b) Is it reasonable to expect that LIGO can observe the entire inspiral if its duty cycle is 1 hr?
   (c) At what frequency does the PBH merger occur? (Hint: you can estimate this as the time when the horizons touch).

8. Let's now play with asymptotic expansions.
   (a) Start from the definition of the error function in Equation (3.21) and show that a Taylor expansion about $x = \infty$ leads to Equation (3.22).
   (b) Calculate how many terms you need in order to get the Taylor expansion to match the exact numerical answer for the error function to 1% evaluated at $x = 2$.
   (c) Now use integration by parts to find the first two terms in the asymptotic expansion of the error function, and compare that to Equation (3.23).
   (d) Evaluate the asymptotic expansion and determine how good the approximation is with one and with two terms in the asymptotic expansion at $x = 2$.

9. A very important concept in PN theory is that of separation of scales. In the inspiral, the radiation-reaction timescale is always much longer than the precession timescale, which is always much longer than the orbital timescale. Are there any binary systems for which this is no longer true? What method would you use to solve for the trajectories and GWs emitted by such systems?

10. Let's now revisit some Fermi estimates of the ringdown. Consider the eikonal approximation to the QNM frequencies.
    (a) Find the radial location at which the Regge–Wheeler and the Zerilli–Moncrief potentials dominate and then plot these potentials for a few ℓ values to verify your answer.
    (b) Evaluate the orbital frequency of circular geodesics at the previous value of radius that you found.
    (c) How does this approximation for the QNM frequencies do for various values of ℓ relative to the exact QNM frequencies discussed in the text?

Useful Books

Baumgarte, T. W., & Shapiro, S. L. 2010, Numerical Relativity: Solving Einstein's Equations on the Computer (Cambridge : Cambridge Univ. Press)

Bender, C. M., & Orszag, S. A. 1999, Advanced Mathematical Methods for Scientists and Engineers: Asymptotic Methods and Perturbation Theory (Berlin: Springer)

Blanchet, L., Spallici, A., & Whiting, B. 2011, Mass and Motion in General Relativity (Berlin: Springer)

Chandrasekhar, S. 1998, The Mathematical Theory of Black Holes (Oxford: Clarendon)

Choquet-Bruhat, Y. 2009, General Relativity and the Einstein Equations (Oxford: Oxford Univ. Press)

Choquet-Bruhat, Y. 2015, Introduction to Relativity, Black Holes and Cosmology (Oxford: Oxford Univ. Press)

Poisson, E. 2007, A Relativist's Toolkit (The Mathematics of Black-Hole Mechanics) (Cambridge: Cambridge Univ. Press)

Poisson, E., & Will, C. M. 2014, Gravity: Newtonian, Post-Newtonian, Relativistic. (Cambridge: Cambridge Univ. Press)

Rezzolla, L., & Zanotti, O. 2013, Relativistic Hydrodynamics (Oxford: Oxford Univ. Press)

Chapter 4

Gravitational-wave Detection and Analysis

It is sometimes said that in astronomy, "dust" is a four-letter word. The equivalent curse word in gravitational-wave detection is "noise." Nonetheless, curse though we may, we have to understand and, if possible, reduce the noise in detectors to improve our prospects of detecting gravitational-wave phenomena.

In general, the sources of noise are many and complex. For example, if we focus on ground-based detectors, we have to contend with noise from ground motion (which could be from earthquakes, passing trucks, falling trees,[1] or students going into and out of the nearest campus), the frequency of the electrical power used by the instruments, the finite number of photons hitting the detector at a given time, thermal noise, quantum noise, lightning, Schumann resonances, and the background of unresolved sources, among other things. Space-based detectors have other problems, e.g., the electrical charging of test masses. In short, noise is unavoidable, so we'd better learn how to live with it and how to learn from it.

The noise will in general not have "nice" properties. For example, a common assumption in simplified theoretical treatments is that the noise is distributed as a Gaussian. That is, the noise has some mean and standard deviation, and the actual noise in a given realization is just drawn from the equivalent Gaussian. Unfortunately, real noise doesn't always work like that. This is the reason that LIGO and Virgo use a signal-to-noise ratio of 8 as their single-detector threshold; this is eight standard deviations, which would correspond to a probability of $\sim 6 \times 10^{-16}$ for a perfect Gaussian. That would be overkill except that there are various glitches and whatnot that dramatically increase the probability of an excursion that great.

[1] This answers a classic philosophical question: if a tree falls in the forest and no one is around to hear it, does it make a sound? Yes, because LIGO and Virgo will see it in their data. Note that LIGO and Virgo are actually terrible seismometers because instrumentalists have taken pains to reduce such vibrations.

doi:10.1088/2514-3433/ac2140ch4

It would also be wonderful to be able to assume that the noise is stationary, i.e., that the statistical nature of the noise is independent of time. This might not be a terrible approximation over short timescales, but because things do change with time, it is necessary to measure the noise near any given gravitational-wave event. For long-duration observations (as are relevant to continuous-wave sources or will be more important for binaries when observed by future detectors with better low-frequency sensitivity and thus longer in-band time, see, e.g., Equation (3.17)), the nonstationarity of the noise must be incorporated into analyses.

Some part of the noise can be understood and mitigated by environmental sensors around the detectors. For example, lightning strikes, which could plausibly cause correlated noise between detectors, can be sensed electromagnetically and thus "events" coinciding with such strikes can be ruled out. More generally, the data collected by ground-based detectors are rated for quality such that if during some interval there was a large amount of local noise (e.g., a close enough earthquake), then data taken during that interval will be judged unfavorable for reliable detections.

Another path to noise mitigation is the requirement that signals in different detectors be consistent with what one would expect from genuine gravitational waves. For example, in events such as binary coalescences, well-specified portions of the gravitational waveform (such as the point of peak amplitude) need to be received by different detectors at times that are consistent with a single direction. The two LIGO detectors are close to 10 light-milliseconds apart, so if the peak times are reported to be separated by significantly more than 10 ms, the signal is not real, assuming gravitational waves travel at the speed of light, as predicted in Einstein's theory. The amplitudes and phases of the signals seen in different detectors should also be consistent with a genuine astrophysical origin.[2]

But more generally the process of detection involves two major steps: (1) noise characterization, in which we determine the noise relevant to observations that might contain signals, and (2) signal characterization, in which we calculate the signals we expect to see so that we can design optimal strategies for detection. Of course, we also need approaches to detect unexpected signals. We now discuss each of these in turn.

4.1 Noise Characterization

At the most basic level, we can imagine characterizing the noise, i.e., the behavior of a detector when there is no gravitational-wave signal present, as a series of N measurements n_i at times t_i. We can think of n_i as the displacement of the mirrors from some equilibrium state, and we can represent it as a vector $\mathbf{n}$.

[2] The requirements of time coincidence and consistency with an astrophysical origin are in place in part because the early history of gravitational wave astronomy had several spurious announcements of "signals" that would not have passed these tests.

Captain Obvious: It seems like in this chapter, and probably in this chapter only, the authors will use boldface typeset to represent N-dimensional vectors, where N counts the number of measurements. This is unlike in the rest of the book, where the authors have used boldface to mean a three-dimensional, spatial vector. This notation is now fairly standard in statistics, though I remember that this confused me a lot when I was first studying this subject.

One way to think about this boldface notation is to think about how you would code this up on a computer or how you would record data if you were an experimentalist. In either case, every time you record a measurement say of the noise, you could place this measurement in a list, i.e., you could populate a one-dimensional "array." Let's say that after a few minutes you've recorded 100 noise measurements. Then, your array will be a list with 100 entries, which you can also call a kind of vector. Thus, you would write $\mathbf{n} = \{n_1, n_2, \ldots, n_{100}\}$. This is what the authors mean in this section when they use boldface notation.

Our measurements allow us to compute a sample mean $\hat{\mu} \equiv \frac{1}{N}\sum_{i=1}^{N} n_i$. We would also like to compute a sample covariance matrix, but if we use a single set of measurements, $m = 1$, then the sample covariance

$$\hat{C}_{ij} = \frac{1}{m-1}(n_i - \hat{\mu})(n_j - \hat{\mu}) \tag{4.1}$$

between measurements i and j is undefined. If we assume that the noise doesn't change too much in consecutive short segments, then we can use $m > 1$ and obtain the covariance matrix.

Major Payne: The covariance matrix is a highly useful quantity that characterizes the relations between different measurements, each of multiple variables. Suppose that we have m measurements of n variables each; that is, each measurement can be represented by a vector $(x_1, x_2, \ldots, x_n)$ (where each x_i is a different variable), and there are m such measurements. We first compute the sample mean $\mu_i = \frac{1}{m}\sum_{k=1}^{m} x_i(k)$ for each variable i, where $x_i(k)$ is the value of x_i in measurement k. Then, for every pair of variables i, j we calculate $C_{ij} \equiv \frac{1}{m-1}\sum_{k=1}^{m}(x_i(k) - \mu_i)(x_j(k) - \mu_j)$. The (i, j) element of the covariance matrix is C_{ij}, and the $m - 1$ instead of m in the denominator is the usual correction to avoid statistical bias.

This matrix encapsulates significant information about the distribution and the correlations between the variables being measured. For example, the diagonal elements are the variances of each of the variables individually. The off-diagonal elements give information about the correlations between quantities; if, for example, the (i, j) element (with $i \neq j$) is positive, it implies that when variable i exceeds its mean, variable j is likely to exceed its mean as well. The reverse holds when an element is negative. The eigenvectors and eigenvalues of the matrix give the directions and lengths of the principal axes of the distribution of points, which can (if desired) point the way toward an efficient reparameterization in terms of linear combinations of the original variables. Because the matrix is symmetric and real, and is therefore positive semidefinite, computation of the eigenvectors and eigenvalues is straightforward.

If we make additional assumptions about the noise, e.g., that it is Gaussian and stationary, then we can make further progress. For example, noise is called Gaussian if the joint probability distribution between noise measurements is given by

$$p(\mathbf{n}) = \frac{1}{\det(2\pi C)^{1/2}} \exp\left[-\frac{1}{2}\sum_{ij}(n_i - \mu_i)(n_j - \mu_j)C_{ij}^{-1}\right], \tag{4.2}$$

where μ_i are the means and C is the determinant of the covariance matrix, which has components C_{ij}. If in addition the noise is stationary, then C_{ij} depends only on the lag between observations $\tau \equiv |t_i - t_j|$. When the noise is stationary, it is advantageous to Fourier transform the data and go into the frequency domain. In that domain, where we now think about frequencies f_i and f_j instead of times t_i and t_j, stationary noise implies a diagonal covariance matrix in frequency space: $C_{ij} = \delta_{ij}S_n(f_i)$, where δ_{ij} is the Kronecker delta and $S_n(f_i)$ is the power spectral density at frequency f_i. That is, in the Fourier domain stationary noise is uncorrelated between frequencies. The amplitude spectral density is often quoted because gravitational-wave detectors measure the amplitude directly: this is the square root of the power spectral density, $S_n^{1/2}(f)$, and has units of Hz $^{-1/2}$.

Major Payne: We can use some theorems from Fourier analysis to show that if a time series is stationary, then the covariance matrix is diagonal in the Fourier domain. First, we note that the correlation function in the time series between times t and t' is $C(t, t') = E[x(t)x(t')]$, where E indicates that we are taking the expectation value and $x(t)$ and $x(t')$ are the values of the times series at, respectively, times t and t'. If the time series is stationary then $C(t, t')$ is finite and depends only on the lag $\tau \equiv t' - t$. It is the case (see textbooks on Fourier analysis) that the power spectrum $S(\omega)$ is the Fourier transform of the autocorrelation function $R(\tau) = E[x(t)x(t + \tau)]$.

Now let's compute the correlation function $C(\omega, \omega') = E[\tilde{x}^*(\omega)\tilde{x}(\omega')]$ in the frequency domain, where $\tilde{x}(\omega)$ is the Fourier transform of $x(t)$ and the star stands for complex conjugation. Doing so, we have

$$\begin{aligned} C(\omega, \omega') &= E\left[\int x(t)e^{i\omega t}dt \int x(t')e^{-i\omega' t'}dt'\right], \\ &= \int \int C(t, t')e^{i(\omega t - \omega' t')}dtdt', \\ &= \int R(\tau)e^{-i\omega\tau}d\tau \int e^{i(\omega-\omega')t}dt. \end{aligned} \tag{4.3}$$

In the first line, we used the definition of the Fourier transform, while in the second line, we used the definition of the correlation function $C(t, t')$. In the last line, we used $t' = t + \tau$ and the fact that the autocorrelation is defined via $R(\tau) = E[x(t)x(t + \tau)] = C(t, t + \tau)$. The first factor is the Fourier transform of the autocorrelation function, which as we indicated earlier equals $S(\omega)$. The second factor is one of many forms that equal the Dirac delta function (times 2πin this case). We therefore have finally

$$C(\omega, \omega') = 2\pi S(\omega)\delta(\omega - \omega'). \tag{4.4}$$

Therefore, for a stationary time series, the correlation function in frequency space is diagonal. This is one of the main reasons that Fourier methods are used so extensively in the analysis of gravitational-wave data; to the degree that time series can be considered stationary (which is plausible for time series not much longer than a few minutes), the analysis is especially simple in the frequency domain.

Captain Obvious: I like Major Payne's derivation, but if you're willing to take an intuitive leap there is an even simpler way to reach the conclusion that the correlation function in frequency space is diagonal. The intuitive leap is to realize that a stationary signal has a frequency content that is independent of time. Thus, it can be represented as the sum of unchanging sinusoids of different frequencies, which in turn means that the frequencies are independent of each other, i.e., they are uncorrelated.

But as we mentioned earlier, real noise is in general neither Gaussian nor stationary. There may be some particular patches of data that approach this ideal, but because the announcement of a gravitational-wave signal is a big deal, we need to be as sure as possible. Indeed, as most data are not this simple, it is common to have "events" in one detector that look like signals (in that the shapes are sensible), but that could never be signals because they are not coincident with corresponding events in another detector. Of course, it is still possible that two events will appear in two detectors at roughly the same time, and such a possibility is called a false alarm. The confident detection of gravitational waves is all about ensuring that what you've detected is not a false alarm. Is there a robust way to estimate the false-alarm rate or FAR, as it is commonly called?

Yes! One commonly used approach for ground-based detectors uses the key observation that the signals received by different detectors, if they are of a real astrophysical origin, cannot be displaced in time by more than the light-travel time between the detectors. Thus, if we take the actual detector data from, say, detector 1, and shift it by more than a light-travel time to detector 2, then this shifted data will typically not be coincident with a similar event in the data of detector 2. Time slides then allow us to effectively get "noise samples" because a genuine signal would not be consistent with the two data streams shifted in this way.

Because the maximum shifts are small, time slides can produce a very large number of noise realizations. For example, using shifting, a few hours of data around the first detection, GW 150914, was sufficient to simulate hundreds of thousands of years of noise data. A choice one needs to make in such analyses is whether to include the data stretch with the putative signal in the shifting analysis; the conservative choice is to do so, because that maximizes the FAR, but a more representative choice is to not include the signal. In any case, this procedure implicitly incorporates the possibility that there are various glitches in the data that might look like signals.

Time slides are, of course, not the only approach to estimating the FAR. One weakness of time slides is that a given segment of "background" data is likely to contain actual astrophysical signals, even if they are typically faint. Thus, researchers are also developing techniques for Bayesian comparison of the noise-only hypothesis, with the hypothesis that a given stretch of data has at least one signal. These models, in turn, depend crucially on models of noise that can include both "normal" Gaussian noise and various types of blips or glitches. In this Bayesian approach, you then let the data "decide" whether a certain stretch looks more like the Gaussian noise model, whether it looks more like the glitch model, or whether it looks more like a gravitational-wave model.

How do we use this noise data to estimate the FAR? As an illustrative example, suppose that we count the number N of coincident events in the noise samples with a signal-to-noise ratio (or another detection statistic) above some threshold ρ_m. With this at hand, and the key assumption that we can use Poisson statistics,[3] we can calculate the false-alarm probability via

$$\mathcal{F}(\rho_m) = 1 - \exp\left[-T\frac{1 + N(\rho_m)}{T_b}\right], \tag{4.5}$$

where T is the observation time and T_b is the duration of the noise data. You'll have an opportunity to derive this expression in an exercise at the end of this chapter. For example, for the first gravitational-wave observation ever made (GW150914), $T_b \sim 608{,}000$ yr, and the observation time was $T \sim 16$ days, but there were no events in the noise data with a signal-to-noise ratio even remotely close to the detection event (which had $\rho \sim 23.6$). Therefore, one can only place an upper limit on the false-alarm probability by setting $N = 1$ to get $\mathcal{F}(\rho_m) \lesssim 1.5 \times 10^{-7}$. The FAR can then be estimated as the false-alarm probability divided by the observation time, FAR $\approx \mathcal{F}(\rho_m)/T$, which for GW 150914 is then FAR $\lesssim 3 \times 10^{-6}$ per year.

The probability of a given event, or the probability per year, should be enough to characterize how much the event stands out from the noise. However, we have been conditioned to think in terms of Gaussians, which means that we often quote the number of "sigma" of an event, i.e., the number of standard deviations from the mean that would correspond, in a Gaussian, to the probability we found. For a given $\mathcal{F}$, this standard deviation can be computed via

$$\sigma = -\sqrt{2}\,\mathrm{erf}^{-1}[1 - 2(1 - \mathcal{F})], \tag{4.6}$$

which for GW 150814 gives $\sigma \gtrsim 5.1$. These are not exactly the values that the LIGO Scientific Collaboration reported in their discovery paper because they used more sophisticated machinery to compute the FAR and the σ, but this is close enough for our purposes. The FAR is now reported every time the LIGO/Virgo collaboration detects a new event, and it can (and should) be used as a measure of how confident one is that the event is of astrophysical origin and not a noise artifact or glitch.

[3] Why Poisson instead of Gaussian all of a sudden? Because Poisson statistics are particularly well suited to a small discrete number of events, which is the case here as the number N is an integer.

The assumption of stationarity, which drives the use of Fourier transforms, is thought to be reasonable for stretches of data that last for durations up to minutes (comparable to the duration of double neutron star mergers above 20 Hz). However, for significantly longer-lasting signals such as continuous-wave emission or for future detectors with sensitivity at significantly lower frequencies, stationarity may fail. In such cases, the correlation matrix is no longer diagonal and many of the benefits of Fourier analysis are lost. It has therefore been suggested that rather than representing gravitational-wave noise as a sum of sinusoids (which is effectively the assumption underlying Fourier analysis), it would be more appropriate to use sums of other functions. A family of such functions is wavelets, which are short pulses of shapes that are appropriate to the problem at hand (for example, Gaussian pulses or pulses that have sharp rises and exponential decays). A different approach, which is also being used by the LIGO/Virgo collaboration, is to perform analyses directly in the time domain.

4.2 Signal Characterization

Detection and characterization of a signal are much easier when you have an idea of what the signal should be. Consider, for example, a situation in which you are in a group of people who are speaking a language that you don't understand. The conversation will seem to be essentially background noise. But if someone in that group speaks a sentence in a language you do understand, then it leaps out to you. This is essentially because your brain has a template for language that it can match to the sounds you hear; if the sounds match one of your templates, you detect them with high signal to noise!

In the same way, if we have a parameterized description of a signal that we trust, we can attempt to match it to the data. Thus, we now discuss signal detection when we have a template available. We follow that with a discussion of how you might detect a signal when we do not have a template; this is useful for sources with uncertain waveforms as well as for completely unexpected sources.

4.2.1 Detection and Parameter Estimation Using Templates

Suppose that we have parameters, such as the masses and spins of a binary, that we decide to arrange in a list[4] $\vec{\theta}$, and that with particular values for those parameters we can predict the strain vector $\mathbf{h}$. What we mean here is that the response $h(t)$ (including both the + and $\times$ polarizations and the beam pattern functions of the instrument) is a function of time (for a given set of source parameters), so upon

[4] Note that we use different notation for this parameter list $\vec{\theta}$ than we use say for the list of values of the response function $\mathbf{h}$. Although you can think of both of these quantities as "vectors," the dimensions of these vectors is not the same, usually with $\vec{\theta}$ much smaller than $\mathbf{h}$. Moreover, the ith element in $\mathbf{h}$ means here the ith measurement of the response, while the ith element of $\vec{\theta}$ means the ith parameter and thus has nothing to do with the ith measurement.

discretization, it can be written as a vector $\mathbf{h}(\vec{\theta}) = h(t_i; \vec{\theta})$, with each element a different discrete value. We can similarly write the vector for the data stream as $\mathbf{d} = d(t_i)$. If the residuals $\mathbf{r} \equiv \mathbf{d} - \mathbf{h}$ are consistent with the instrumental noise, then $\mathbf{h}$ describes the data well. This process of comparing a template to the data is what is called matched template filtering or matched filtering for short.

To use Bayesian statistics (if you haven't read Appendix A yet, do it now!) we need to define a likelihood function, but the likelihood is the noise model! For stationary Gaussian noise, the likelihood function is proportional to

$$p(\mathbf{d}|\vec{\theta}) \propto e^{-\frac{1}{2}(\mathbf{r}|\mathbf{r})}, \tag{4.7}$$

where

$$(\mathbf{a}|\mathbf{b}) \equiv 2\int_0^\infty \frac{\tilde{a}(f)\tilde{b}^*(f) + \tilde{a}^*\tilde{b}(f)}{S_n(f)}df \tag{4.8}$$

is the noise-weighted inner product. Recall that a tilde denotes a Fourier transform and an asterisk indicates a complex conjugate. Note that this automatically gives less weight to frequencies at which the noise, as captured by the noise spectral density $S_n(f)$, is large. Thus, features with large noise in a small frequency range (e.g., near the frequency of electrical power) play a negligible role in signal analysis.

Major Payne: In our era of machine learning, you might wonder whether some clever computer could come up with a better way to detect a signal of known shape in the presence of stationary Gaussian noise. But no! Norbert Wiener proved (yes, proved, as in a formal mathematical proof) in 1949 that in those circumstances the inner product in Equation (4.8) is the optimal linear filter, in the sense that it produces the largest possible signal-to-noise ratio. Machine learning may lead to new and interesting ways to explore the parameter space, or it may aid when we don't trust (or when we don't have) templates. But in the end, if we trust the templates, then a Wiener filter is the optimal linear way to go.

How do we quantify whether a template is a good match to the signal? Clearly, we wish to find a template that maximizes the likelihood, or put another way, that minimizes $(\mathbf{r}|\mathbf{r})$, which is like the χ^2. Imagine that there is a signal $\mathbf{s}$ contained in some data $\mathbf{d} = \mathbf{s} + \mathbf{n}$ with $\mathbf{n}$ the noise in the data, and imagine that there is a template $\hat{\mathbf{h}}$ with parameters $\widehat{\vec{\theta}}$ that fits the signal exactly. Then, it is useful to define the signal-to-noise ratio ρ via

$$\rho^2 = (\mathbf{s}|\hat{\mathbf{h}}), \tag{4.9}$$

which is sometimes normalized by dividing by $\sqrt{(\mathbf{s}|\mathbf{s})(\hat{\mathbf{h}}|\hat{\mathbf{h}})}$. This quantity tells us roughly what the mean power in the signal is relative to the noise spectral density. If the noise were Gaussian and stationary, the probability that a particular template will produce a spurious signal-to-noise ratio of ρ or higher would be $p = \mathrm{erfc}(\rho/\sqrt{2})$. For example, if $\rho = 3$ then under the stationary Gaussian assumption the probability is $p \approx 99.7\%$ that there is a gravitational wave in the data with the characteristics of the template. In reality, glitches and other noise artifacts increase this probability dramatically, which is why the FAR must be calculated more carefully by time slides or other methods, as described earlier.

To complete the setup of our Bayesian analysis, we need to specify priors for our models (in the case of model comparison, which includes the specific task of judging between the presence or the absence of a gravitational-wave signal in a given data set) and for the parameters in our models (in the case of parameter estimation for a given model). As always, we would like the data to speak for itself as much as possible, which means that once we have decided on a model to explore we would like to select uninformative priors. But the choice isn't as obvious as it may seem. For example, are all black hole masses equally probable, or is it better to choose a distribution that is uniform in log masses? Inclination is another example. We certainly expect that if we were somehow to pick every merging binary out to a fixed radius (e.g., to 1 Gpc) then the binary inclinations would be isotropic. However, binaries that are face-on to us have double the amplitude of binaries that are edge on to us (see, e.g., Equation (3.20)), so when it comes to detected binaries we expect to have an excess of relatively face-on systems. The luminosity distance is yet another example. If the coalescence rates were uniform per comoving volume in the history of the universe, then this would be a good assumption, but there are reasons to think that mergers were more common around redshifts $z \sim 1$ than they are now. Fortunately, none of these choices make a major difference to either detection or parameter estimation.

The data analysis problem is simple then, right? All one has to do is evaluate the likelihood many, many times varying over the parameters of the template, until one maps the likelihood surface over the entire parameter space. If the calculation of the waveform for a given set of parameter values $\vec{\theta}$ were fast, then one would be able to use various techniques to evaluate the likelihood in any portion of the parameter space. However, calculations of waveforms are computationally intensive, even when various tricks are used. For example, it saves an enormous amount of effort to factor binary waveform parameters into intrinsic parameters (those dealing directly with the binary, such as the masses and spins of the two objects) and extrinsic parameters (such as the direction and distance to the binary and the parameters that characterize the detector). Despite these time-saving measures, it may not be practical to compute waveforms on the fly, and as a result, other approaches have been developed.

For binaries, the most common method is to use template banks. That is, a number of waveforms are precomputed using a set of intrinsic parameter combinations with a parameter resolution so that any real waveform within the domain has some minimum overlap (say 97%) with the templates. How do we quantify the overlap? We simply compute the (normalized) inner product between any two such templates: $\mathcal{O}_{1,2} \equiv (\bar{\mathbf{h}}_1|\bar{\mathbf{h}}_2)$, where we have defined the normalized templates $\bar{\mathbf{h}}_A = \mathbf{h}_A/(\mathbf{h}_A|\mathbf{h}_A)^{1/2}$, maximizing over extrinsic parameters. Because binaries with smaller chirp masses drift more slowly through the frequency range of ground-based detectors, this means that there needs to be a larger density of templates at lower chirp masses.

For the detection of a continuous-wave source, similar considerations apply but the approach and the computational difficulty depend strongly on the amount of external information that we apply. If we know the direction and frequency received at Earth of the source as a function of time (this information could, for instance, be obtained for a pulsar by electromagnetic observations), then we simply integrate the signal at a multiple of the frequency. For example, if the source is a mountain on a neutron star then the gravitational-wave frequency should be twice the rotation frequency. If instead the source is an oscillation mode on the star, such as an r-mode, then the multiple could be different.

If we have less external knowledge of the frequency as a function of time, then we need to perform more searching. This can range from imperfectly observed sources (for example, the rotation frequencies of actively accreting neutron stars are often not known precisely, and if the neutron star is in a binary then uncertainties in the binary motion must also be taken into account) to fully blind searches. Given that continuous-wave sources are weak, gravitational-wave observations need to be extremely extended (probably weeks to years) to hope for detection, which means in turn that unless one has a great deal of information about the source the searches will be either insensitive or highly computationally intensive.

For parameter estimation, we can ask ourselves the two key questions from Appendix A: how would we perform our task with unlimited resources, and how can we approximate the right answer within a reasonable time? The "right" approach is conceptually straightforward: after a significant signal has been identified, we "merely" search all of parameter space thoroughly, computing the prior times the likelihood at each parameter combination, and then normalize to get the final posterior. We then obtain the posterior for any subset of the parameters using marginalization (see Appendix A.4).

We are now talking about parameter estimation for an event that has already been deemed significant, so the computational effort is reduced greatly because only the data segment in question needs to be considered and because the initial fast template-matching gives us a head start on the exploration of parameter space. Even with these advantages, however, it is not feasible to perform a brute-force analysis; there are simply too many parameters. Thus cleverer approaches are needed. We won't discuss them in detail, but example approaches include nested sampling and Markov Chain Monte Carlo (or MCMC) techniques.

4.2.1.1 The Fisher Approximation

Suppose that the noise is Gaussian and stationary, that the signal to noise is large, that the likelihood surface is unimodal (which means that there is only one peak in the likelihood), and that we somehow know where this peak is located in parameter space. Because it is easier to find a maximum of a function (in this case the posterior) than it is to fully characterize the function, we can think about exploring the region near the maximum of the likelihood and using that to estimate the results of our parameter estimation. We can then effectively Taylor expand around the peak (at which all first derivatives are zero) to second order in all parameters, to form the Fisher information matrix:

$$I_{ij} = -E\left[\frac{\partial^2}{\partial\theta_i\partial\theta_j}\ln P(\mathbf{d};\,\theta)\right]. \tag{4.10}$$

Here $E[\cdot]$ is the expectation value, the model parameters θ_i and θ_j are in the $\vec{\theta}$ vector, P represents the posterior, and $\mathbf{d}$ is the data set. In the limit of strong signals, the inverse of I gives the covariance matrix, which in turn tells us the uncertainties in parameters, and their pairwise correlations, around the best fit.

When can we feel that the Fisher estimate is adequate? The path to the Fisher information matrix through a Taylor expansion gives us the hint: when the log posterior surface near the maximum is adequately expressed using a Taylor expansion to second derivatives (rather than also needing higher derivatives), within the parameter ranges indicated by the variances obtained from the covariance matrix, then the Fisher matrix method gives a reasonable answer to how well one would do in estimating parameters. Clearly, this method is great to get a quick-and-dirty estimate or projection of parameter estimation, but it cannot be used on real data. This is not just because with real data we don't know whether the likelihood is unimodal (usually it's not!), but more importantly, because we don't know a priori where the maximum of the likelihood function is, so we don't know where to expand about! Real parameter estimation on real data therefore requires an exploration of the likelihood surface. For that, and also for a more reliable projection of how well parameters can be estimated, one needs a full Bayesian analysis of the likelihood.

Putting those concerns aside for a moment, the Fisher matrix does provide a quick and dirty way to estimate how well one can estimate parameters given a gravitational-wave template. Imagine that we have a signal that is identical to one of our waveforms, $\mathbf{s} = \hat{h} = \mathbf{h}(\hat{\vec{\theta}})$, such that the data $\mathbf{d} = \hat{h} + \mathbf{n}$. Then, if we investigate the likelihood close to its peak, around $\Delta\vec{\theta} \equiv \hat{\vec{\theta}} - \vec{\theta} \ll \hat{\vec{\theta}}$, we have

$$\begin{aligned} \log[p(\mathbf{d}|\theta)] &\propto -\frac{1}{2}(\mathbf{r}|\mathbf{r}) = -\frac{1}{2}\left(\mathbf{s} - \mathbf{h}\left(\hat{\vec{\theta}} + \Delta\vec{\theta}\right)\middle|\mathbf{s} - \mathbf{h}\left(\hat{\vec{\theta}} + \Delta\vec{\theta}\right)\right), \\ &= -\frac{1}{2}\left(\mathbf{s} - \hat{h} - \mathbf{h}_{,i}\,\Delta\theta^i \mid \mathbf{s} - \hat{h} - \mathbf{h}_{,j}\,\Delta\theta^j\right), \\ &= -\frac{1}{2}\left[(n|n) - 2(n \mid \mathbf{h}_{,i})\Delta\theta^i + (\mathbf{h}_{,i} \mid \mathbf{h}_{,j})\Delta\theta^i\Delta\theta^j\right], \end{aligned} \tag{4.11}$$

where in the second line we are using the Einstein summation convention, where we have Taylor expanded about $\Delta\theta^i = 0$ and where we have defined the short-hand $\mathbf{h}_{,i} = \partial\mathbf{h}/\partial\theta^i\,|_{\hat{\theta}}$. The maximum likelihood occurs when $\partial_k \log[p(\mathbf{d}|\theta)] = 0$, which occurs when

$$\Delta\theta^j = (n \mid \mathbf{h}_{,i})[\Gamma^{-1}]^{ij}, \tag{4.12}$$

where we have defined $\Gamma_{ij} \equiv (\mathbf{h}_{,i} \mid \mathbf{h}_{,j})$. Notice that the indices in the Fisher matrix range over the parameters of the waveform, and thus, they have nothing to do with spacetime coordinates. The variance-covariance matrix, $C^{ij} \equiv E[\Delta\theta^i \Delta\theta^j]$, is then given by

$$C^{kl} = E\Big[(n \mid \mathbf{h}_{,i})[\Gamma^{-1}]^{ik}(n \mid \mathbf{h}_{,j})[\Gamma^{-1}]^{jl}\Big] = E\Big[\Gamma_{ij}[\Gamma^{-1}]^{ik}[\Gamma^{-1}]^{jl}\Big] = [\Gamma^{-1}]^{kl}, \tag{4.13}$$

where we have used the identity $E[(n \mid \mathbf{h}_{,i})(n \mid \mathbf{h}_{,j})] = (\mathbf{h}_{,i} \mid \mathbf{h}_{,j})$. The diagonal elements of C^{ij} tell us the square variance in the parameters, while the off-diagonal components tell us the correlation between parameters. Usually, one uses the normalized waveform $\bar{\mathbf{h}}$ we introduced earlier to compute the variance-covariance matrix, so that the off-diagonal elements have a maximum range of +1 (indicating a perfect correlation) to −1 (indicating a perfect anti-correlation). Notice that the calculation of C^{ij} and Γ_{ij} does not require any data, and thus, this is simply an estimate of how well one could measure parameters in a sufficiently loud signal, given a template model that matches it perfectly.

Major Payne: In fact, there is a theorem by Harald Cramér and C. R. Rao that says that the best one can hope to ever do in estimating a given parameter, say θ^A, in a signal that is buried in noise is exactly $\Delta\theta^A = (C^{AA})^{1/2}$, where no summation is implied here. This is known in the literature as the Cramér–Rao bound. Therefore, if your parameter estimation pipeline predicts the measurement of a parameter with a mean square error smaller than the Cramér–Rao bound, you should suspect that something went really wrong!

Let's stop here for a second to make a quick estimate based on a Fisher analysis of how well we can measure parameters in gravitational-wave astrophysics. For this estimate, let us assume the simplest signal: a quasi-circular early inspiral of compact objects with no spins to leading order in post-Newtonian theory in general relativity. In particular, we will ignore the merger part completely. A (very rough) waveform template for such a signal is

$$\tilde{h}(f) = \mathcal{A} f^{-7/6} e^{i\Psi(f)}, \tag{4.14}$$

where $\mathcal{A}$ is a constant Fourier amplitude (which depends on the luminosity distance, the mass, the beam pattern functions, and the inclination angle, but here we keep it as an overall constant), and the Fourier phase is

$$\Psi(f) = 2\pi f t_c - \phi_c + \frac{3}{128}(\pi \mathcal{M} f)^{-5/3}, \tag{4.15}$$

with the chirp mass $\mathcal{M}$, the "time of coalescence" t_c, and the "phase of coalescence" ϕ_c. Mathematically, t_c and ϕ_c correspond just to a constant frequency and phase shift, which one is always allowed to choose to maximize the match of a template with a signal. Our parameter vector will be $\vec{\theta} = (\ln \mathcal{A}, t_c, \phi_c, \ln \mathcal{M})$, which then naturally decomposes into a set of three extrinsic parameters ($\ln \mathcal{A}$, t_c, ϕ_c) and only one intrinsic parameter ($\ln \mathcal{M}$).

With this waveform template in hand, the Fisher prescription is simple: compute all of the components of the Fisher matrix, invert it and then look at the square root of the diagonal elements. But that's a lot of work! Can we estimate the accuracy to which we can measure any one of these parameters with our Fermi tricks? The answer turns out to be yes!

For this Fermi estimate, the first thing we need to notice is that gravitational-wave interferometers are much more sensitive to the phase evolution of the wave than to its amplitude. This should be familiar from Chapter 1, but you can also see it from the Fisher analysis or even the likelihood function discussed in this chapter. In data analysis, we calculate the "inner product" of the waveform with a signal contained in the data, and this inner product is an integral in Fourier space. If the phase of the waveform and the signal do not match exactly, then the integrand of the inner product will be proportional to the sine or cosine of the phase difference. The integration over a wide frequency band then makes the result very small, because heuristically the sine or cosine of the phase difference will average out. For this reason, to a very good degree of approximation, if a parameter enters in the phase of the waveform, then we can just look at the phase to estimate how well it can be measured.

The second ingredient we need is to realize that how well a parameter can be estimated must depend on how loud the signal is! Indeed, parameter estimation typically scales as the reciprocal of the signal-to-noise ratio ρ. With this in mind, let us then require that a variation of one of the parameters in the phase be equal to ρ^{-1}. Focusing on the chirp mass, for concreteness, we then have

$$\frac{1}{\rho} \sim \delta_{\mathcal{M}}\Psi(f) = \frac{\partial \Psi}{\partial \mathcal{M}}\delta\mathcal{M} = -\frac{5}{128}(\pi\mathcal{M}f)^{-5/3}\frac{\delta\mathcal{M}}{\mathcal{M}}. \tag{4.16}$$

We then see that $(\delta\mathcal{M}/\mathcal{M})_{\rm Fermi} \sim (128/5)(\pi\mathcal{M}f)^{5/3}\rho^{-1} \sim 7 \times 10^{-5}$, when we evaluate this Fermi estimate at $\rho = 10$, $f = 100$ Hz (a typical frequency near the maximum sensitivity of advanced LIGO), and for a chirp mass of $\mathcal{M} \sim 1.2 M_{\odot}$ (consistent with an equal mass neutron star binary with $m_1 = 1.4 M_{\odot} = m_2$, because as we recall $\mathcal{M} = \eta^{3/5} m$, where $\eta = m_1 m_2/m^2$ is the symmetric mass ratio and $m = m_1 + m_2$ is the total mass).

Is this even close to the right answer? To find out, we need to do a Fisher estimate. The first thing we'll do is compute the signal-to-noise ratio, because this will be useful in the Fisher estimate. Using our definition in Equation (4.9), but assuming the signal is equal to the template, we have

$$\rho^2 = (\hat{h} \mid \hat{h}) = 4\mathcal{A}^2 \int \frac{f^{-7/3}}{S_n(f)} df \equiv 4\mathcal{A}^2 f_0^{-4/3} I(7), \tag{4.17}$$

because the phases cancel exactly, and where we have defined the so-called noise-weighted moments

$$I(q) \equiv \int \frac{x^{-q/3}}{S_n(x)} dx, \tag{4.18}$$

with the dimensionless frequency $x \equiv f/f_0$ for some characteristic f_0, so that these moments are dimensionless. With this at hand, the Fisher matrix is

$$\Gamma_{ij} = \rho^2 \begin{pmatrix} 1 & 0 & 0 & 0 \\ 0 & 0 & -2\pi J(4) & (5/128)u_0^{-5}J(12) \\ 0 & -2\pi J(4) & 4\pi^2 J(1) & -(5\pi/64)u_0^{-5}J(9) \\ 0 & (5/128)u_0^{-5}J(12) & -(5\pi/64)u_0^{-5}J(9) & (25/16384)u_0^{-10}J(17) \end{pmatrix}, \tag{4.19}$$

where we have further defined the normalized moments $J(q) \equiv I(q)/I(7)$ and the characteristic velocity $u_0 = (\pi \mathcal{M} f_0)^{1/3}$. These normalized noise-weighted moments of ours depend on the noise spectral density, so they will be different when computed with an advanced LIGO noise curve versus a third-generation detector noise curve or a LISA noise curve. However, they do generically have the property of increasing with increasing q; for example $J(q) = \mathcal{O}(1)$ for $q < 11$, and then increases rapidly with q for the advanced LIGO noise curve.

We can extract several conclusions from the Fisher matrix. The first thing we notice is that every element of the Fisher matrix is proportional to ρ^2, and therefore, the inverse of the Fisher matrix (the covariance matrix) will be proportional to $1/\rho^2$ and the square root of the variance to $1/\rho$, as previously anticipated. The second thing is that the Fourier amplitude is not correlated at all with $(t_c, \phi_c, \ln \mathcal{M})$, which makes sense because these parameters only appear in the Fourier phase. And the third thing to notice is that some of these components are multiplied by u_0 to a negative power, and the more negative this power, the larger the term is. This is because u_0 is a characteristic velocity, so it is smaller than unity during the early inspiral; indeed, evaluating u_0 for the same system we picked for our Fermi problem, we find $u_0 \sim 0.12$, and so $u_0^{-5} \sim 3 \times 10^4$ and $u_0^{-10} \sim 10^9$.

The above considerations imply that to a very good approximation, the $(\ln \mathcal{M}, \ln \mathcal{M})$ component of the Fisher matrix is much larger than any other component. This means that we can approximate the entire Fisher matrix just through this one component, and if so, the inverse is trivial! We then find that

$$C_{ij} \sim \delta_i^{\mathcal{M}} \delta_j^{\mathcal{M}} \frac{16{,}384}{25} \frac{u_0^{10}}{\rho^2 J(17)}, \tag{4.20}$$

and so the error in the chirp mass can be obtained from $(\delta \ln \mathcal{M})_{\rm approx} = (\delta \mathcal{M}/\mathcal{M})_{\rm approx} \gtrsim \sqrt{C_{\ln \mathcal{M} \ln \mathcal{M}}} \sim (16{,}384/25)^{1/2} u_0^5/(\rho\sqrt{J(17)}) \approx 1.5 \times 10^{-5}$. This estimate is to be compared to the exact answer (found by inverting the full Fisher matrix), which gives us $(\delta \mathcal{M}/\mathcal{M})_{\rm exact} \gtrsim 2.1 \times 10^{-5}$, and thus we see the approximation is highly accurate. What's more, we see that our Fermi estimate above $((\delta \mathcal{M}/\mathcal{M})_{\rm Fermi} = 7 \times 10^{-5})$ is also really good! Ultimately, of course, to do proper parameter estimation, we must carry out a Bayesian analysis and construct posterior probability distributions. But before embarking on such a daunting task, it is usually a good idea to do estimates such as the ones presented above.

4.2.1.2 Expected Angular Dependence of Stochastic Sources

Another type of template has to do with the expected angular correlation from a stochastic background of gravitational waves. Rather than looking for events that are localized in time (as is the case for late inspiral and mergers or binaries at high frequencies) or in frequency (as is the case for continuous sources), we need to search data for a persistent signal that spans a wide range of frequencies. As always, a key feature that distinguishes noise from signal is that a true gravitational-wave signal will be correlated between different detectors, whereas noise will not be. But it is always good to look for additional ways to establish that a signal is real because this increases the sensitivity of detections.

Consider, for example, stochastic backgrounds that can be treated as isotropic over the whole sky of the observer (characteristic of cosmological populations but not Galactic populations, such as double white dwarfs). Using this assumption, it has been shown for pulsar timing arrays (which look for subtle timing deviations from a set of very precisely timed pulsars) that the correlation between the time-dependent amplitudes of the signals to two pulsars that are separated by an angle ζ is

$$\chi(\zeta) = \frac{1}{2} + \left(\frac{1 - \cos\zeta}{2}\right)\left[-\frac{1}{4} + \frac{3}{2}\ln\left(\frac{1 - \cos\zeta}{2}\right)\right], \qquad (4.21)$$

which we plot in Figure 4.1. One initially nonintuitive feature of this correlation diagram is that in the limit of zero angular separation the correlation is 1/2 rather than the 1 we might have expected. The reason is that we assume that even pulsars in the same direction have distances whose difference is many times the wavelength of the gravitational waves (which is at most several light-years), and thus on average we expect a correlation of only 50%. Thus if there are many separate angular baselines between pairs of pulsars, adherence to this angular dependence is a check on whether the correlations are due to an isotropic population.

Major Payne: It is all very well to assume that the sources are isotropic, but for some important cases, it is far from obvious that this is true! Consider a population of supermassive black hole binaries, which collectively produce a background in the $\sim 10^{-6}$–10^{-9} Hz range that is expected to be detected using pulsar timing arrays. Because $dE/df \propto \mathcal{M}^{5/3}$ at a given frequency f, where we recall that $\mathcal{M}$ is the chirp mass of the binary, there is a strong weighting toward more massive binaries. Such binaries are less common than the less massive binaries, which means that statistical fluctuations play a larger role than they would for more numerous populations. Thus, although I grudgingly admit that isotropy is a good place to start the calculations, we have to be careful.

This expected angular dependence serves as a template by which we can distinguish pulsar timing noise from a genuine signal and is a key tool in the arsenal of pulsar timing arrays. In effect, we use the angular correlation function in Equation (4.21) like the signal template h that we discussed earlier.

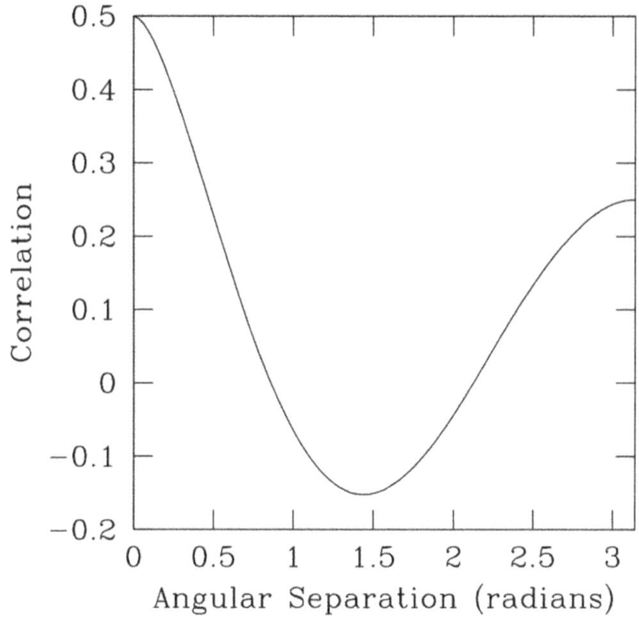

Figure 4.1. Expected correlation between the gravitational-wave induced timing residuals of two pulsars, as a function of their angular separation. This is the Hellings–Downs curve, which assumes that the gravitational wave background is isotropic on the sky. One might have guessed that the expected correlation should tend to 1 as the angular separation tends to zero, but the assumption is that the pulsars are also at different distances from us, and that those distances differ by many times the wavelength of the gravitational radiation.

4.2.2 Detection of Events without Reliable Templates

For binaries and continuous-wave sources, the waveform can be parameterized straightforwardly; the challenge is to perform the analysis efficiently. In contrast, by definition, gravitational-wave bursts are (usually short) events for which we have at best an imperfect model for the waveform. It is therefore not possible to use matched filtering. Instead, more approximate and therefore less sensitive approaches are necessary. In the methods described below, the correlation of signals between detectors (with a consistent delay based on time of arrival) is key to distinguishing signal from noise.

For some source categories, it is believed that we have at least a partial understanding of the expected waveform. For example, core-collapse supernovae involve a remarkable diversity of physics, but simulations have improved steadily to the point where one has a general idea of the expected gravitational radiation. There are also electromagnetic events that could provide clues about possible gravitational counterparts; for instance, highly magnetic neutron stars (called "magnetars") occasionally undergo giant flares that can release $\sim$few $\times$ 10^{47} ergs, as we mentioned in Chapter 2. Quasi-periodic oscillations have been seen in the X-rays from these events, so it would be logical to look for gravitational radiation near those frequencies or small-integer multiples of them.

However, for the majority of sources, there is little information about expected waveforms that analysis cannot be precise. Indeed, if there are genuinely unknown types of sources, we want to be able to detect them without prejudice. With that in mind, researchers have developed a variety of approaches to searches for bursts. These approaches essentially fall into two categories: (i) use of generic templates, and (ii) searches for excess signal.

In the first approach, one uses the fact that gravitational-wave bursts have short durations, so they can sometimes be modeled using a Gaussian in time, or a Gaussian times a sine. More generally, one can decompose any time series in sums of wavelets, which as described earlier are finite-duration pulses of various shapes. In this approach, one uses a model (e.g., a sum of wavelets) with a small number of parameters to fit the data. One must always be careful about false positives, and to that end it is necessary to perform comparisons with the data, most of which will not have genuine gravitational-wave events, to determine whether instrumental artifacts or glitches can mimic the apparent signal.

In the second approach, one searches for excess signal by analyzing "whitened" data, i.e., data that are reweighted by the inverse of the expected noise as a function of frequency. One can then search for excess power in the frequency domain or excess signal in moving time averages over some specified intervals. Another method has been to look instead for a linear increase in counts as a function of time, which models the rise of a signal toward a peak.

It is worth keeping in mind that although the lack of specificity of burst searches means that they are less sensitive than searches based on accurate templates, there are already several binary coalescences that have been detected using burst searches. In fact, the first gravitational-wave detection was very well recovered with a Morlet wavelet reconstruction, implemented through Bayesian analysis. This proves that existing algorithms can detect sufficiently loud events.

4.2.3 Stochastic Backgrounds

Stochastic backgrounds (or foregrounds) comprise a large number of sources that are individually too weak to be detected or are indistinguishable from other sources. These could be binaries of various types; for example, a white dwarf binary stochastic foreground is expected to be detected using the space-based detector LISA, and pulsar timing arrays will likely see a background of supermassive black hole binaries. Another possibility is the tens of supernovae per second in the observable universe. Or, there could be other source categories with many individual but weak sources. As with burst sources, if one has multiple detectors then consistency between detectors is key to distinguishing between genuine signals and instrumental noise.

Many types of stochastic background can be modeled assuming that the amplitude is a power law in frequency. For example, for binaries, $h \propto f^{2/3}$ as long as the observation frequency is below the maximum attainable by the source and is such that the binaries will evolve in the age of the universe. This comes from

Equation (1.6), which gives $h \sim a^{-1}$, because for a binary $a \sim f^{-2/3}$. Another example is cusps from cosmic strings or early-universe sources such as those that might emerge from phase transitions. With that assumption, if one has a particular power law in mind, then the search is specific enough to be relatively sensitive. If the power law is free then the lack of specificity decreases the power of the search. In that case, detection relies heavily on correlated variation between detectors.

If the individual sources or events have in-band durations much larger than the interval between them, then we can assume that the background is steady. This is not a terrible approximation around ~20 Hz for double neutron star coalescences throughout the visible universe. If instead the durations are smaller than the interval between them, then the background is not steady; that situation has been dubbed "popcorn noise". This is applicable to double black hole coalescences above tens of Hertz.

The optimal analysis differs between the two cases. For example, a steady background, with many sources per frequency bin, can reasonably be considered Gaussian (due to the central limit theorem). Therefore, cross-correlation analyses (i.e., comparison of variations in the gravitational-wave signals of different detectors) can be efficient. In the popcorn noise limit, cross-correlations are no longer optimal and other techniques must be employed. For example, Bayesian analyses can incorporate a probability that in a given short interval there is a source of a given type, with most such intervals not including that source. Interestingly, and apparently by coincidence, current observations suggest that the amplitude of the stochastic background from all of the unresolved double neutron star events in the universe is comparable to the amplitude of the stochastic background from all of the unresolved double (stellar-mass) black hole events in the universe. There are intriguing signs based on early LIGO/Virgo detections that both backgrounds could be detected within a few years.

4.3 Exercises

1. Derive Equation (4.5) with the help of the following hints:
   (a) Start with the Poisson distribution, from which we know that if you have seen n_b events in a time T_b, the probability distribution for the number m of events that you would expect in that time is

$$P(m|n_b, T_b) = \frac{m^{n_b}}{n_b!}e^{-m}. \tag{4.22}$$

   (b) If the expected number of events in time T_b is m, then the expected number in your observation time T is $m(T/T_b)$.
   (c) Thus, again using the Poisson distribution, the probability that you would see zero events in time T is $\exp[-m(T/T_b)]$, and therefore the probability that you would see one or more events is $1 - \exp[-m(T/T_b)]$.

(d) Hence, to compute the probability, given n_b events observed in time T_b, that you see one or more events in time T, you integrate the product $P(m|n_b, T_b)\exp[-m(T/T_b)]$ over all m from 0 to infinity.
(e) As a final hint, we will assume that $T \ll T_b$, and will use the limit $\lim_{n\to\infty}(1 + 1/n)^{-xn} = e^{-x}$.

2. The simplest continuous source of gravitational waves is one whose frequency does not change significantly during the duration of the observation (although of course more sophisticated and time-consuming searches can accommodate a very slowly changing frequency). In the following three parts, assume that if the source frequency changes by less than $1/T$ during an observation of duration T, then its signal has a constant frequency for the purposes of data analysis.
   (a) Consider a circular binary with two 10 $M_\odot$ black holes. Assuming a three-year observation period (comparable to the minimum expected lifetime of LISA), what is the orbital (not gravitational-wave) frequency needed for the binary to have an effectively constant frequency during the observation?
   (b) Say that you are looking at an isolated millisecond pulsar with a rotation frequency of 500 Hz and a period derivative $\dot{P} \approx 10^{-21}$. Assuming that we know the direction to the pulsar and can therefore correct for Doppler shifts, how long can you observe the pulsar before you need to take into account its spin down?
   (c) Now suppose that we are searching gravitational-wave data for a pulsar like the one in the previous problem, but that we do not know the direction to the pulsar. Given Earth's motion around the Sun, how long could we observe the pulsar before the frequency drifted out of its original bin (i.e., before the frequency shift exceeded $1/T$)? Does the answer depend on the direction to the pulsar, e.g., whether the pulsar is in Earth's orbital plane or whether at the beginning of the observation Earth is moving straight toward, straight away from, or transverse to the pulsar?

3. Suppose that you measure quantities x_1 and x_2 that are in fact uncorrelated; in any one of your m measurements, where m is so large that $m - 1 \approx m$, x_1 is drawn randomly and uniformly from the range $[-1, 1]$, and x_2 is separately drawn randomly and uniformly from the range $[-1, 1]$. Any two measurements are uncorrelated; for example, if $x_1 = 0.3$ in a given measurement, then the next measurement of x_1 is still drawn randomly and uniformly from $[-1, 1]$. Compute the expected value of each of the four elements of the sample covariance matrix.

4. Dr. I. M. Wrong doesn't understand why everyone is so worried about white dwarf noise. After all, with so many white dwarf–white dwarf binaries in a given bin, the total flux in gravitational waves will be very stable; in particular, it is obvious that from frequency bin to frequency bin, the flux

will vary so little that even a weak additional source will show up easily. This follows trivially from taking the square root of the flux to get a measure of the amplitude.

Evaluate Dr. Wrong's claim by doing the following model problem. Let there be N sources in a given frequency bin. Suppose that they are all equally strong, but have random phases between 0 and 2π.

(a) What do you find when you add the complex amplitudes based on those random phases?
(b) Now, take the squared magnitude of the total amplitude as a measure of the typical flux. What is this typical flux?
(c) Determine the mean and standard deviation of the flux that results.

You should find that, unlike what happens when you add sources incoherently (i.e., square the amplitudes, then add), the standard deviation of the flux is comparable to the flux, so Dr. Wrong's idea fails.

Useful Books

Auger, G., & Plagnol, E. 2017, An Overview of Gravitational Waves: Theory, Sources and Detection (Singapore: World Scientific)

Blair, D. G. 2012, Advanced Gravitational Wave Detectors (Cambridge: Cambridge Univ. Press)

Creighton, J. D. E., & Anderson, W. G. 2011, Gravitational-Wave Physics and Astronomy: An Introduction to Theory, Experiment and Data Analysis (New York: Wiley)

Dym, H., & McKean, H. P. 1972, Fourier Series and Integrals (New York: Academic)

Jaranowski, P., & Krolak, A. 2009, Analysis of Gravitational-Wave Data (Cambridge: Cambridge Univ. Press)

Maggiore, M. 2007, Gravitational Waves: Volume 1: Theory and Experiments (Oxford: Oxford Univ. Press)

Reitze, D., Saulson, P., & Grote, H. 2019, Advanced Interferometric Gravitational-wave Detectors (Singapore: World Scientific)

Schutz, B. F. (Berlin: Springer)

Saulson, P. 2017, Fundamentals of Interferometric Gravitational Wave Detectors (Singapore: World Scientific)

Chapter 5

Gravitational-wave Astrophysics

Gravitational waves promise to tell us remarkable new things about the universe. In the next four chapters, we will discuss what we have learned, and what we expect to learn, about astrophysics, cosmology, nuclear physics, and fundamental physics. In this chapter our focus is on the astrophysics of binary coalescences but at the end, we have some discussion about other possible sources.

5.1 Binaries

To poorly paraphrase Tolstoy in Anna Karenina: from the standpoint of gravitational radiation, all widely separated binaries are alike; each close binary coalesces in its own way. That is to say, when the two members of a binary are far enough from each other, their nature doesn't matter; in this phase, their slow inspiral can be represented in a way that obeys universal scaling laws related to their mass and mass ratio. However, when the binary members are close to each other, as measured, for example, by their separation divided by the larger of the two stellar radii, or their separation in units of the gravitational mass, then details do matter.

From the standpoint of astrophysics, even widely separated binaries can have different characteristics. For example, if there is a lot of other mass around (e.g., in the form of gas or stars), then the binary can inspiral at a different rate than predicted using gravitational radiation alone, or it can even be widened. As a result, it is possible at least in principle to learn about the origin, evolution, and environment of binary coalescences using observations of gravitational waves.

In this section, we therefore need to make a choice about how to divide our studies. Should it be by type of object (black hole, neutron star, white dwarf, main-sequence star, other)? Should it be by mass (massive black holes versus all others)? Should it be by mass ratio (comparable mass, as has been seen in most events reported to date, versus much more extreme mass ratios)? Here we choose an organization by mass, but other possibilities are equally valid.

doi:10.1088/2514-3433/ac2140ch5

5.1.1 Stellar-mass Binaries

If we focus initially on black holes with masses of a few to a few tens of solar masses, there are essentially three paths that have been considered to binarity and mergers: (i) primordial black holes, (ii) evolution from isolated massive star binaries, and (iii) dynamics in dense stellar systems (see Figure 5.1 for elaborations on the latter two). Or, to quote Shakespeare in *Twelfth Night*: "Some are born great, some achieve greatness, and some have greatness thrust upon 'em." In principle, each channel, sometimes with subchannels, has unique signatures in the form of the expected rates, masses, mass ratios, spins, and eccentricities. In practice, however, as each new black hole coalescence has been announced, advocates of each channel have found ways to accommodate the observations. This is but one way in which observation drives theory.

We won't discuss primordial black holes in this book except to say that there are ideas for how to produce them in the early universe and that there are constraints from various observations on their abundance, in most of the mass range from the mass of a mountain to millions of solar masses. In principle, they could account for many observed double black hole coalescences, but the uncertainties in the physics of their production are so large that genuine predictions are challenging. We will therefore focus on the two main channels that have been considered for double compact object coalescence: isolated binaries and dynamics in a dense stellar system.

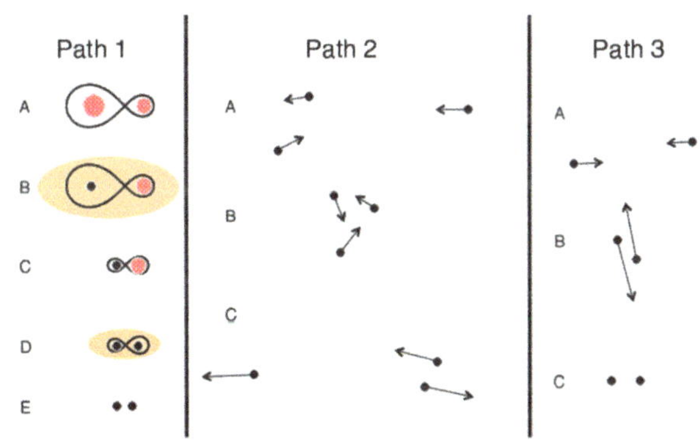

Figure 5.1. Three example paths to the formation of a double black hole binary. Path 1: Two massive stars form in a binary; here the figure "8" represents the gravitational equipotential that crosses at the inner Lagrange point. At time A, the stars are both within that equipotential and therefore do not transfer mass; here the star on the left has a higher mass than the star on the right. With time increasing downward, we go from the first giant stage (B), where there is typically stable Roche lobe overflow, to after the first star has evolved to a black hole but the second star is still on the main sequence (C) to the second common-envelope stage (D) to the final double black hole system (E). Path 2: a dynamical encounter that initially (time A) contains a wide black hole binary (left) and a single black hole (right). Time B shows a short-lived stage in which all three stars interact together, and time C shows the outcome, where the binary (right) is tighter than it was, and all three stars have acquired kinetic energy. Path 3: a lower-probability event in which two initially unbound black holes (time A) pass close to each other (time B) and lose enough energy to gravitational interaction in the initial close encounter that they become bound (time C). Miller (2016). With permission of Springer.

5.1.1.1 *Isolated Binaries*

Isolated binaries may seem to be the most natural path. If we want to end up with a coalescing pair of black holes or neutron stars, what could be simpler? We just start off with two stars in a binary that are massive enough to evolve to neutron stars or black holes when they run out of fuel to burn. Then, gravitational radiation takes over, and eventually, they merge. Is there a problem with this idea?

There is indeed! We need to have the objects merge within the age of the universe (i.e., roughly 10 billion years), and gravitational-wave emission is a terribly inefficient way to make objects inspiral. When we plug masses, semimajor axes, and eccentricities into Peters' formula (Equations (2.8) and (2.9)), we find that for a binary with reduced mass μ and total mass M the inspiral time is

$$T_{\text{inspiral}} \approx 4 \times 10^{17}\ \text{yr} \left(\frac{M_\odot^3}{\mu M^2}\right)\left(\frac{a_0}{1\ \text{au}}\right)^4 (1 - e_0^2)^{7/2} \tag{5.1}$$

for an initial semimajor axis a_0 and eccentricity e_0. The prefactor depends somewhat on the eccentricity, but this estimate is accurate to ~30% for any e_0 as long as the separation is large compared with the gravitational radius because the prefactor depends somewhat on the eccentricity.

This equation implies that for a small e_0, $T_{\text{inspiral}} < 10^{10}$ years requires $a_0 \lesssim 0.02$ au for two neutron stars, and $a_0 \lesssim 0.1$ au for two 10 $M_\odot$ black holes. This is challenging because normal stellar evolution leads to a phase in which the star becomes a giant, with a stellar radius of more than 1 au, which would envelop the binary. As a result, this channel isn't as simple as two well-separated stars in a binary undergoing undisturbed evolution that results in a coalescing compact object binary. Luckily, we have observed a number of double neutron star binaries in our Galaxy that are close enough to merge within tens of millions to a few billions of years. And by now, LIGO and Virgo have observed the merger of neutron stars! This tells us that there must be some pathway to form these binaries in the first place, even if during the stellar evolution the binary is enveloped.

Captain Obvious: But you might wonder: what about binary white dwarfs? There are examples of double white dwarf systems with binary periods of minutes, which implies a separation on the order of a tenth the radius of the Sun. Thus, something has to operate to shrink the orbit to such small separations in this case, too. It is believed that, in the white dwarf case, a major mechanism is magnetic braking.

To understand magnetic braking, we begin with the knowledge that main-sequence and giant stars have winds, which are driven by their radiation interacting with the atoms (and for cooler stars, even with the molecules) that exist in stellar atmospheres. For lower-mass stars on the main sequence, the resulting mass flow rate isn't much; for example, the Sun's rate is only around 10^{-14} $M_\odot$ yr^{-1}. However, stars also have magnetic fields, and the particles in the wind can be stuck to the field lines out to many times the star's radius. The angular momentum per unit mass (often called the "specific" angular momentum, where in this and other contexts "specific" just means "per unit mass") scales as Ωr^2, where Ω is the angular velocity and r is the distance from the center of the star. As a result, if the particles in the wind are tied to the field lines they can remove angular momentum very efficiently. This is thought to be the major reason why the spin rate of single stars decreases by a factor that can be 10 or more over their lifetimes.

In a binary, the loss of angular momentum to magnetic braking will shrink the orbit. This could drive the system into a common-envelope phase, which could yield a close double white dwarf system. Or, in more massive stellar systems, this can drive the system into Roche lobe overflow, which is nonconservative (meaning that a significant fraction of the mass lost from one star is ejected from the system rather than accreting onto the companion). Under the right circumstances, this can lead to orbital decay and the formation of a double compact object binary that will merge within billions of years or less, without any common-envelope phase.

The "classical" pathway to a compact object coalescence from an isolated binary involves the so-called common-envelope mechanism, but more recently, people have proposed another path called chemically homogeneous evolution that avoids the giant phase altogether. Let's discuss each of these in a bit more detail.

Common envelope—In this picture, first one star and later the other in the binary do expand into a giant, and during that phase, they envelop their companion. Both stars are therefore in a single, common, envelope. The companion (whether or not it has evolved into a compact object) then spirals inward because of drag through the envelope. The idea is that the two original main-sequence stars can start relatively far apart (but not significantly farther than the size of the giants) and be dragged in to a close enough separation that as compact objects they can spiral together in a few billion years or less. One challenge in this picture is that you want the orbit to shrink, but not too much; if the two objects spiral together before they are both compact objects, then you get a single object, not a coalescing binary. Thus, something has to halt the inspiral during the common-envelope phase. This could happen, e.g., because enough energy is injected by the inspiral or by accretion onto the compact object to eject the envelope entirely, but given that it is difficult to find direct evidence of systems in the common-envelope phase, we must rely mostly on an extremely uncertain theory. It matters, too: different treatments of common envelopes have given rate estimates for double black hole coalescences that differ by more than two orders of magnitude!

Captain Obvious: Shouldn't the supernovae that produce the two compact objects disrupt the binary? The answer is that it can, but it does not have to. We can determine the criterion for this just based on energetic considerations. There are two basic kick mechanisms: one based on mass loss and the other based on any speed of the remnant relative to the original center of mass of the presupernova star.

First, the mass loss. Supernova ejecta come out at around 10^4 km s^{-1}, which is so much faster than any orbital speeds that we can effectively assume that the mass vanishes instantly. Suppose that prior to the supernova the orbit was circular. This means that for an initial total mass between the two stars of M_0 and a semimajor axis a_0, the orbital velocity was $v_{\rm circ,0} = (GM_0/a_0)^{1/2}$. If the mass loss due to the supernova reduces the total mass to $M_f < M_0$, and if there is no kick relative to the original center of mass, then the separation between the stars and their relative speed remains the same as before the supernova but the mass is less. But because of the change in the mass of the system, the orbit must become eccentric, and thus, even with the same separation and relative velocity as before, the semimajor axis changes. Given that the escape speed for a mass m and separation a is $v_{esc} = (2Gm/a)^{1/2}$, we see that if there is no kick and more than half the original mass is lost ($M_f < M_0/2$), then the system becomes unbound because $v_{circ,f} \geqslant v_{esc}$. This is sometimes called the Blaauw kick (after Adriaan Blaauw). Because the second supernova in a massive binary typically removes more than half the remaining mass, the Blaauw kick could be a major obstacle to binary formation. This may be why most neutron stars are single even though $\gtrsim$80% of stars that become neutron stars are originally in binaries. The effect of the Blaauw kick can be reduced if, e.g., (a) earlier slow wind loss gradually reduced the mass of the progenitor so that the supernova does not remove more than half the mass of the system, or (b) the core collapse produces a black hole directly, without a supernova.

However, the observed motions of many neutron stars and some black holes in our Galaxy indicate that upon formation they do get a kick relative to the center of mass of the presupernova star. The origin of the kick is debated; candidates include asymmetric neutrino emission (neutrinos carry away ~99% of all of the energy in a supernova!) or a "gravitational tugboat" in which asymmetric ejection of matter can simply accelerate the neutron star gravitationally. Observationally, the kick can be tens of km s^{-1} for black holes and hundreds of km s^{-1} for neutron stars. One of the uncertainties in compact object formation via isolated binary evolution is the direction of the kick. Is it randomly directed? Is it typically directed along the orbital axis, or in the direction of the orbit, or away? Different answers could significantly impact the expected rates.

Chemically homogeneous evolution—It was suggested in the 2010s that the giant phase could be avoided altogether. A massive single main-sequence star goes through a giant phase because its core hydrogen runs out (having converted to helium) and a disequilibrium in the nuclear fusion of a hydrogen shell on top of the helium causes the star to puff out. But if the star is stirred substantially (e.g., by rapid rotation that could stem from the tides induced by a nearby massive companion), then fresh hydrogen is continually cycled through the core, and the star does not become a giant. In this picture, two massive stars could form very close to each other and become black holes or neutron stars that are close enough to each other to merge in billions of years or less without having to be dragged closer. No stars are known to be in this phase, so this is again purely theoretical and thus has large uncertainties, but it does provide another possibility.

So what are the masses of the neutron stars and black holes that are formed through this channel? Typically, the final mass of a compact object is much less than the mass of the original star. This is because mass is lost to winds, Roche lobe overflow and common-envelope processes, and supernovae. Wind losses are difficult to quantify for massive stars in part because there aren't many massive stars to observe. For a fixed luminosity, stellar winds are stronger when the opacity in the matter is larger. In the case of stellar atmospheres, the opacity is provided primarily by atomic lines and edges, and there are far more of these for heavier elements. Therefore, it should be that stars with a higher fraction of elements heavier than helium (i.e., higher-"metallicity" stars in astronomical lingo) have stronger winds. If so, then stars with lower metallicity should hang on to more of their mass when they collapse and, when they form black holes, those holes should be more massive.

And of course, like anything else in astrophysics, there is (at least) one more complication worth mentioning. Many "isolated" binaries aren't actually isolated! Observations suggest that $\gtrsim 10\%$ of the stars that become neutron stars and black holes have more than one companion. As an example of why this might matter, please look at the discussion of the von Zeipel–Lidov–Kozai effect in Appendix B. The importance of the effect is that if a stellar system has an inner binary and an outer tertiary star significantly farther away, and if the tertiary plane is strongly inclined relative to the binary plane, then over many orbital periods of the binary and tertiary, the inclination and the eccentricity of the binary can undergo large and cyclical swings. Thus, for example, even if the binary initially has too large a separation to be able to merge in billions of years or less, if the eccentricity can be induced to be high enough, then during the high-eccentricity parts of the cycle, the binary could shrink significantly (through gravitational-wave burst emission at pericenter) and merge. In principle, this is a way for a high-mass system to coalesce without any common-envelope phase.

5.1.1.2 Dynamical Processes

This would be a good time for you to read Appendix B, which covers some of the basics of stellar dynamics, the topic of this section in the context of compact object formation channels. As we discuss in that appendix, dense stellar systems are fertile breeding grounds for compact binaries. If interactions of single stars with binaries are common enough, then compact objects that are heavier than the average star tend to exchange into the binaries and then become tighter with subsequent interactions (again, see Appendix B). Because black holes have more than 10 times the average stellar mass in an old stellar system, such as a globular cluster, they tend to exchange into binaries. Neutron stars are about three times heavier than the average star, so they also tend to exchange into binaries, but the tendency is not as strong as it is for black holes.

Captain Obvious: What are globular clusters? They are old self-contained stellar systems that have tens of thousands to millions of stars and have central densities that are tens of thousands to millions of times greater than the $\sim 0.1\ \mathrm{pc}^{-3}$ number density of stars in the Sun's vicinity. Globulars orbit their host galaxy, some at distances that are more than ten times greater than the typical radius of the visible disk or elliptical part of the Galaxy. Elliptical galaxies tend to have about 10 times as many globulars per stellar mass in the Galaxy as spiral galaxies (like our Milky Way) do. The other type of dense stellar system is nuclear star clusters, which are around the nucleus of a galaxy. These can have masses of millions to tens of millions of solar masses and sometimes have central massive black holes. In such systems, although star–star collisions are still rare (distances are huge in astronomy!), binary–single and binary–binary interactions are vastly more common than they are in lower-density parts of galaxies

Suppose that a binary and a single interact and, after the interaction, the binary has tightened. In the Newtonian point-mass limit, which is an excellent approximation in most cases, the total energy of the binary–single system is conserved.

Thus, if the binary tightens, which therefore makes its energy more negative, then the binary's center of mass, as well as the single star, acquires extra speed: a Newtonian kick. It turns out that for very hard binaries (in which the gravitational binding energy of the binary is much greater than the kinetic energy of the single relative to the binary), the kick speed is proportional to the binary orbital speed. As a result, as the binary tightens, the kicks get larger. Studies indicate that in systems such as globular clusters (which have typical masses $\sim(10^4\text{–}10^6)\ M_\odot$ and typical escape speeds of a few to tens of km s^{-1}), the binary is likely to be kicked out of the cluster before it merges. It therefore might merge after it gets kicked out, but at least in globulars, it is unlikely (although not impossible) that it will be able to merge, stay in the cluster, undergo dynamical interactions, and eventually have additional mergers. Another important effect is mass segregation (see Appendix B.2), in which the heaviest objects in a cluster sink to the center of the cluster. This is believed to lead to the formation of dense black hole subclusters in the centers of globular clusters, but when binaries form, the hardening of those binaries acts as an energy source for the stars and black holes in the cluster and prevents the formation of extremely dense collections of stars. Despite this, the rate of coalescences per star in a globular cluster is likely to be much greater than the rate per star in the "field," which is another name for the bulk of the host galaxy. This is because in the field only a small fraction of massive binaries produce merging compact objects, but a fairly large fraction of black holes in globulars can merge, even if they do so outside the cluster.

Does this mean that interactions in dense stellar systems dominate the channels for the production of double compact object coalescences? Not necessarily! The problem is that only a small fraction of stars are in systems dense enough to promote common dynamical interactions. For context, in our Galaxy only a fraction $\sim 10^{-4}$ of stars are in dense systems, and in other galaxies that fraction might get up to $\sim 10^{-3}$. Thus, to compete with massive stellar binary channels, the dynamical pathways have to be at least a thousand times more efficient. This is difficult, although not impossible. As one example of such a calculation, pre-LIGO/Virgo estimates of the double neutron star coalescence rate focused entirely on the double neutron star systems in the field of our Galaxy and ignored the double neutron star systems known in globular clusters that will merge in a Hubble time. Why is that?

We can understand this omission by performing a simple estimate. There are $\sim$200 globular clusters observed around the Milky Way galaxy, and on average they might contain as many as a few hundred neutron stars. Even if there are then a hundred neutron star–neutron star binaries per cluster, and they all merge within a $\sim 10^{10}$ yr Hubble time, this gives a total rate of only $\sim 200 \times 100/(10^{10}\ \text{yr}) = 2 \times 10^{-6}$ yr^{-1} per Milky Way Equivalent Galaxy (or MWEG, as it is commonly known). The rates estimated from the binaries in the disk of our Galaxy, and from gravitational-wave observations of neutron star–neutron star coalescences, are about 10 times larger than this. Thus, ignoring double neutron star mergers in dense stellar systems, such as globular clusters, introduces only a small error. However, when it comes to other types of compact object mergers (black hole with a neutron star or black hole with black hole) the situation is not as well understood, because we don't see any black hole–neutron star or black hole–black hole systems in our Galaxy.

Yet another possibility, although one with much smaller probability, is that two initially unbound compact objects could pass close enough to each other that the gravitational radiation they emit during their closest passage carries away enough energy to bind the objects together. They would then coalesce quickly. The reason that this is a very low-probability event is that the objects would have to come very close to each other to radiate the required energy, and thus the cross section for the process is tiny (see Equation (B.8) and the associated problem). If this happens, it is much more likely to occur during the many chaotic interactions involved in a binary–single interaction than during a random encounter between two single objects. In any case, high stellar densities are obviously important for this mechanism to have any chance.

Another location where dynamical processes might be important is at first surprising: accretion disks around supermassive black holes in the centers of galaxies. When we think about such disks, our first inclination is to consider them to be made of gas, along with radiation and magnetic fields. But because these accretion disks are at the centers of galaxies, there are two general ways in which stars, and thus possibly black holes, could reside in the disks and thus potentially pair up to form gravitational-wave sources. That is, you could have stellar-mass black hole binaries inside an accretion disk! The first is the capture of stars by gas drag due to repeated passages through the disk (see Section 5.1.3 for a calculation). The second is the formation of stars in the outer accretion disk; if the inspiral time to the supermassive black hole is long enough, and the stars are massive enough, many stars could become black holes by the time they reach the center. A challenge for the production of stellar-mass black hole binaries in this path is that if stars and black holes are dragged efficiently toward the supermassive black hole, they might never interact. There are various ideas related to locations in the disk where the inflow might stall (these are called "migration traps"), which are currently being investigated.

If interactions and black hole mergers can take place in accretion disks, then there is one respect in which they have a big advantage over other dynamical scenarios. Recall that in, for example, globular clusters, Newtonian three-body interactions or the gravitational-wave kick upon the merger of two black holes is very likely to kick the remnant out of the system, which means that it is improbable that it could have additional interactions. In contrast, the interactions that might happen in accretion disks are deep in the gravitational potential of the supermassive black hole. For example, it has been suggested that promising migration traps (where black holes linger and can interact) might be at $\sim 10^3 GM/c^2$ from the supermassive black hole. At that radius, the orbital speed is $v_{\rm orb} \approx c/(1000)^{1/2} \approx 10^4$ km s^{-1}. Newtonian three-body kicks will not come close to these speeds, and kicks from gravitational radiation are typically only hundreds of km s^{-1}. Thus, the remnant will stay basically where it is, ready for another interaction. As with all of the scenarios discussed here, considerable work remains to be done on the accretion disk channel.

5.1.2 Massive Binaries with Comparable Masses

Many to most galaxies have a central black hole, and their masses are known to span at least the range ~few × 10^4 $M_\odot$ to few × 10^{10} $M_\odot$. Galaxy collisions are relatively common in the universe, which means that it is possible to have the two massive black holes in a colliding pair of galaxies coalesce at some point. Moreover, unlike for stellar-mass binaries (where the mass ratio is less than $\sim$100:1), it is possible to have a binary of a massive black hole and a stellar-mass object with a mass ratio of 10^6:1 or greater. Although the scale-free nature of general relativity implies that mergers between big black holes are fundamentally the same as mergers between small black holes, there are astrophysical processes that are important in the realm of the massive that can be ignored for stellar-mass coalescences.

Let us focus first on the formation channels for massive binaries with comparable mass ratios, for which the basic question is: will the central massive black holes of galaxies merge with each other due to the collisions of galaxies in less than a Hubble time? If so, this will provide a strong source of gravitational waves.

The first consideration is that, clearly, a typical galaxy–galaxy collision will not be so precisely head on that the central black holes hit each other directly; after all, even a $2 \times 10^{10}\ M_\odot$ black hole has a horizon radius of $2M \lesssim 10^{16}$ cm (or around 10^{-2} pc), which pales in comparison to the $\sim 10^{23}$ cm (or about 3×10^4 pc) radius of its typical host galaxy. Collisions will be at least somewhat oblique, so on the first pass, the black holes will miss each other by a lot, maybe several kiloparsecs. Indeed, even after the Galaxy collision has relaxed into quasi-equilibrium a few hundred million years after initial contact, the two black holes will still typically be separated by kiloparsecs.

Now suppose that the initial collision phase is over and the system has relaxed into some kind of equilibrium. The two supermassive black holes are still far away from each other, say a couple of kiloparsecs, which means that if nothing else happens, then it will take essentially forever for them to merge due to the emission of gravitational radiation. However, there are a lot of stars around; after all, this is the collision of two galaxies! As we discuss in Appendix B, an object such as a supermassive black hole that is much more massive than stars will experience dynamical friction that will drag it to the center of the system. Indeed, a supermassive black hole is typically surrounded by a dense "bulge" of stars that has hundreds of times the mass of the black hole, which turns out to mean that until the bulges of a pair of supermassive black holes overlap, dynamical friction operates on the whole bulge—black hole system and dramatically accelerates the inspiral of the two bulges (and thus the two black holes) toward each other. As a result, each of the two supermassive black holes will, separately, spiral toward the center of mass of the system.

There are, however, some important subtleties, all of which are driven by the following significant question: can the supermassive black holes actually get close enough to each other to spiral together within the age of the universe? The collection of considerations that arise is often called the "final-parsec problem." A summary of this problem is as follows. It's straightforward for the supermassive black holes to spiral to within a few parsecs of each other. At that point, however, they start running out of stars with which to interact, so their settling slows down. But at this stage, they are still so far from each other that gravitational radiation is ineffective. The star–star dynamics can supply new stars to interact with the supermassive black hole binary, but this process might be too slow to bring the binary close enough that gravitational radiation can take over. Thus, you are in a bit of a stalled scenario, and from the above considerations, it's not clear how to bring the black holes closer to each other: this is the final-parsec problem. A lot of work by many people has led to the conclusion that other effects probably bring the binary together, but if they don't, then massive black hole mergers might not be as common as we would hope.

Our starting point in our exploration of the details of these statements is to note that, beginning in the 1990s, various empirical relations have been discovered that relate the mass M of a supermassive black hole at the center of a galaxy with properties of the Galaxy itself. Given that M is orders of magnitude less than the mass of the Galaxy it is far from obvious that any relation must exist, but dozens of plausible theoretical explanations have now been proposed. The tightest of these relations, and the most useful for our purposes because it links directly to the stellar dynamics of interest, is the "$M - \sigma$" relation between M and the velocity dispersion σ of the central bulge of the Galaxy. There are battles about the precise form of the relation, but for our purposes, we will use the simplified but largely accurate form

$$M_{\rm BH} = 10^8\ M_\odot \left(\frac{\sigma}{200\ {\rm km\ s^{-1}}}\right)^4, \qquad \text{or} \qquad M_{\rm BH,8} = \sigma_{200}^4. \tag{5.2}$$

Here we use what will turn out to be extremely useful shorthand: $M_{\rm BH} = 10^8\ M_{\rm BH,8} M_\odot$ and $\sigma = 200\ \sigma_{200}$ km s^{-1}.

Next we will assume that if the supermassive black hole were not present, then the velocity dispersion σ would be constant throughout the bulge, which might typically be hundreds to thousands of parsecs in radius; we will implicitly assume that the mass distribution is roughly spherically symmetric but that the density can vary with radius. The velocity dispersion is roughly equal to the typical orbital speed, which means that if the mass contained within some radius R is $M(r < R)$, then $\sigma = [GM(r < R)/R]^{1/2}$, and thus, $M(r < R) = \sigma^2 R/G$. This means that $dM/dr = \sigma^2/G$. Because in general $dM/dr = 4\pi r^2 \rho(r)$, where $\rho(r)$ is the mass density, this also means that for a constant velocity dispersion $\rho(r) = \sigma^2/(4\pi G r^2)$.

Incidentally, because constant velocity dispersion means constant kinetic energy with radius, and this is like a constant temperature with radius, the $\rho \propto r^{-2}$ distribution is often called an "isothermal sphere" or a "singular isothermal sphere" because $\rho \to \infty$ as $r \to 0$. Of course, in reality, the $1/r^2$ dependence must break down at small distances (the density doesn't become infinite) and at large distances (because the mass doesn't increase indefinitely with increasing r), but $\rho \propto r^{-2}$ is often accurate over a wide range of radii.

We can now define the radius of influence $r_{\rm infl}$ as the radius inside of which the mass of stars equals the mass of the black hole: $r_{\rm infl} = GM_{\rm BH}/\sigma^2$. Using the $M - \sigma$ relation, this gives

$$r_{\rm infl} = \frac{GM_{\rm BH}}{\sigma^2} = \frac{G\ 10^8 M_\odot}{(200\ {\rm km\ s^{-1}})^2} M_{\rm BH,8}\, \sigma_{200}^{-2} \approx 10\ {\rm pc}\ \sigma_{200}^2 = 10\ {\rm pc}\ M_{\rm BH,\,8}^{1/2}. \tag{5.3}$$

Using our density formula, the mass density at the radius of influence is then $\rho(r_{\rm infl}) \approx 7 \times 10^3\ M_\odot\ {\rm pc}^{-3}\ \sigma_{200}^{-2} = 7 \times 10^3\ M_\odot\ {\rm pc}^{-3}\ M_{\rm BH,\,8}^{-1/2}$.

Now we can ask: if we have two supermassive black holes of, say, equal mass, how long will it take to sink to the center at that mass density and velocity dispersion? From Equation (B.4), we find that the time needed for a supermassive black hole to sink in this field of stars is about $t_{\rm rlx,ener} \approx 2 \times 10^4$ yr $M_{\rm BH,\,8}^{1/4}$. Yes, we're being a bit sloppy (two $10^8\ M_\odot$ black holes have a mass of $2 \times 10^8\ M_\odot$ rather than $10^8\ M_\odot$), but the point is that this is extremely short compared with the age of the universe. So, no problem, right? Once you get your supermassive black holes to the radius of influence, they just zip right in.

No! The problem is that the interaction of the stars with the supermassive binary kicks the stars out. By the time the binary gets down to the radius of influence or below, it has nearly run out of stars. If nothing else happened, then the binary would orbit in a region without stars and would therefore not shrink anymore. Maybe that's okay because gravitational radiation will take it the rest of the way? Let's see. First, how much will the binary shrink due to interactions with stars in the radius of influence? We can judge that by noting that if the binary kicks all of the stars out with zero energy at infinity, then the reduction in the energy of the binary is just the gravitational binding energy of the collection of stars. If there is a mass $M_{\rm BH}$ within a radius $r_{\rm infl}$, then the gravitational binding energy is roughly $GM_{\rm BH}^2/r_{\rm infl}$ with a prefactor that isn't too much different from unity. This is about the gravitational binding energy of the binary, so as the energy scales as $1/r$, this means that you can shrink the binary orbital radius by about a factor of 2. What does this imply about the time to spiral in due to gravitational radiation? From Equation (5.1), if we assume a circular binary then the inspiral time from half of the radius of influence turns out to be

$$T_{\rm insp} \approx 2 \times 10^{17}\ {\rm yr}\ M_{\rm BH,\,8}^{-1}. \tag{5.4}$$

Yikes, that's a long time! Thus, if this were the end of the story, then supermassive black hole binaries wouldn't coalesce. Instead, they would just orbit each other.

What else might be going on? One thing that should occur to us is that although the stars interior to the supermassive black hole binary might have been kicked out, the mutual interactions of the stars farther away will cause them to diffuse into the binary region on a relaxation time. Once more, using Equation (B.4), but this time for a presumed-standard $1\ M_\odot$ star, we find that the time needed is $t_{\rm rlx,ener} \approx 2 \times 10^{12}$ yr $M_{\rm BH,\,8}^{5/4}$. That tells us that the time is still too long for $M_{\rm BH} = 10^8\ M_\odot$ supermassive black holes, but that for $M_{\rm BH} < 10^6\ M_\odot$, the time is short enough that over billions of years, stars could diffuse inward, be kicked out by the binary, and therefore tighten the binary to the stage that gravitational radiation could take over.

But what about the more massive black holes? For this there are a few possibilities that have been raised. One possibility has to do with the possible existence of a special kind of orbit called "centrophilic," because it loves to seek the center. The argument goes as follows. The energy relaxation timescale is rather conservatively long in the sense that it assumes that the stars are more or less distributed spherically. However, we're thinking about the result of a collision of the two galaxies that originally hosted the supermassive black holes. That will lead to a nonspherical distribution of stars, at least until everything settles down. Nonspherical distributions of mass can lead to a wide variety of orbits including orbits that radically change their angular momentum from one orbit to the next. For example, if the distribution is roughly a triaxial ellipsoid, then some orbits are "centrophilic" because they preferentially shoot stars through the center of the mass distribution. Thus, the mass supply rate to the central region where the binary orbits can be much greater than we guessed based on spherical energy relaxation.

Another possibility that has been raised is related to gas. This story goes as follows. When we calculated how much the binary would shrink due to interactions with $r < r_{\rm infl}$ stars, we used energy conservation. That's valid for stars, which only interact gravitationally; they can't radiate their orbital energy. But what if a significant fraction of the mass is in gas rather than in stars? Then, the gas can interact gravitationally with the binary, but then it will shock with itself and radiate energy. Thus, energy can actually be lost from the system, so the gas can come back and interact more than once. This could shrink the binary more effectively.

A third possibility we will mention has to do with the idea that maybe the black hole binary doesn't shrink readily! Galaxies will often merge more than once. If a stalled supermassive black hole binary is at the center of a galaxy and a new supermassive black hole binary is brought in, three-body effects such as von Zeipel–Lidov–Kozai cycles can shrink the binary further or increase its eccentricity substantially, possibly leading to eventual mergers.

Thus, the belief and hope are that supermassive black holes commonly merge in the universe, which would be exciting as a source of gravitational waves and as a way to learn about strong gravity and associated extreme physics. In fact, the detection or nondetection of massive black hole binaries could inform us about whether any of the possibilities listed above resolve the final-parsec problem, or whether this problem remains and massive black holes do not frequently merge.

5.1.3 Massive Binaries with Extreme Mass Ratios

From comparable-mass binaries, we turn to the opposite case: EMRIs (recall that this stands for extreme mass-ratio inspirals), where an object of stellar mass (this could be a normal star, white dwarf, neutron star, or stellar-mass black hole) orbits a supermassive black hole. These systems, as well as supermassive black hole binaries, are prime targets of LISA, which we remember stands for the Laser Interferometer Space Antenna and is planned to fly in 2034. One reason for such interest is that EMRIs can spend thousands or even millions of orbits in the extreme gravity of a massive black hole, in contrast to comparable-mass systems, which spend just a few orbits. In fact, EMRIs can outlive the mission lifetime of space-based instruments! Thus, careful observation of EMRIs can lead to the precise characterization of strong gravity and tests of general relativity. Moreover, because EMRI orbits can be highly inclined to the spin axis of a massive black hole and can have clearly measurable eccentricity in the LISA band, the measured orbits can probe specific aspects of massive black hole spacetimes that are not easily accessed with comparable-mass binaries. Finally, the specific nature of the orbits (how inclined and how eccentric) will give us a new way to probe the dynamics of galactic centers and the formation channels of these systems.

With this motivation in mind, there are three basic paths to EMRIs that have been discussed in the literature. Let's tackle each of those below.

5.1.3.1 Two-body Relaxation

Building off of Appendix B, the most-discussed path to EMRIs involves two-body relaxation. For concreteness, suppose that we are thinking about a $\sim 10\ M_\odot$ black hole that eventually spirals into a massive black hole. The low-mass black hole wanders in the galactic center region, sinking gradually against the lighter stars, and also has its angular momentum perturbed by two-body interactions. If its angular momentum is small enough, then at each pericenter passage it can lose energy to gravitational radiation and gradually spiral into the massive black hole. However, this might not lead to detectable inspirals.

The reason is that for EMRI mass ratios (e.g., 10^{-5} for a 10 $M_\odot$ black hole spiraling into a $10^6\ M_\odot$ black hole), only a small fraction of the orbital energy is lost with each pericenter passage, even for pericenter distances that are just a few gravitational radii of the massive black hole. As a Fermi estimate and using what we've learned so far, this is because the energy carried away by gravitational waves in a highly eccentric pericenter passage is $\Delta E \sim \dot{E}\Delta t \sim \dot{E}\, r_p/v_p \sim \eta^2 M v^8$, where r_p and v_p are the pericenter distance and velocity respectively, η is the symmetric mass ratio, and we have approximated the orbital velocity as $v^2 \sim M/r_p$. To determine the fraction that ΔE is of the Newtonian gravitational binding energy of the orbit, we note that if the orbit were circular with radius r_p, then the binding energy would be of order $\sim \eta M v^2$. However, the orbit is highly eccentric, with $1 - e \ll 1$, and because the binding energy scales as $1/a$, where a is the semimajor axis, and because $r_p = a(1 - e)$, the Newtonian binding energy is $E_b^{\rm Newt} \sim \eta M/a \sim \eta M v^2(1 - e)$, and thus, $\Delta E/E_b^{\rm Newt} \sim \eta v^6/(1 - e)$. If $\eta \sim 1 - e$ (which would be accidental but plausible) and $v \sim 1/2$ (reasonable near the massive black hole), then $\sim 1\%$ of the energy is lost per orbit and several tens of orbits are needed to reduce the apocenter distance by a factor of 2. During that time, the gravitational perturbations from other stars can change the angular momentum of the orbit. Indeed, the smaller black hole could be perturbed directly into the massive black hole rather than completing the thousands of orbits needed for detection. Details matter a lot, but it has been estimated that only a few percent of coalescences of this type might produce detectable signals, with the rest leading to merger but not a strong signal.

Detection of EMRIs with LISA likely requires that the gravitational-wave frequency be above $\sim$few $\times\ 10^{-3}$ Hz. The reason is that below that frequency, an unresolvable foreground of double white dwarf binaries in our Galaxy will mask out weaker signals such as EMRIs. At those frequencies, EMRIs from two-body relaxation are likely to have eccentricities of several tenths. Moreover, because the orbits can come from anywhere, we expect that the orbits will have an arbitrary inclination relative to the rotational plane of the supermassive black hole.

5.1.3.2 Tidal Separation of Binaries

Suppose that our small black hole is in a binary with another stellar-mass object. The center of mass of the binary is perturbed by distant interactions in the same way that the orbit of a single black hole is perturbed. But when the binary gets close enough to a massive black hole, the tidal field of the massive black hole can separate the binary: one of the components of the binary is flung out at high speed, whereas the other becomes tightly bound to the massive black hole. The critical radius at which this occurs, which is typically several to tens of astronomical units, is much larger than the radius needed for gravitational radiation to be important. Thus, in this scenario, the bound small black hole starts its inspiral with a much larger pericenter than in the two-body EMRI path (and, as it happens, a much smaller apocenter also). As a result, circularization is effective, and we expect the black hole to enter the LISA frequency range with a nearly circular orbit. However, because the binary originated at a large distance from the massive black hole, we again expect arbitrary inclinations relative to the black hole spin plane.

Captain Obvious: We can estimate the distance from a supermassive black hole at which a star, or a binary, or any other self-gravitating object or system can be tidally ripped apart by simply comparing the gravitational acceleration of the object with itself to the tidal acceleration across the object due to the supermassive black hole. For example, suppose the object (which, again, could be a binary) has a total mass m and size R and is a distance r from a supermassive black hole of mass M. Then, the gravitational acceleration of the object with itself is $a_{\rm self} \sim Gm/R^2$ and the tidal acceleration due to the supermassive black hole is $a_{\rm tide} \sim GMR/r^3$. Setting the two equal to each other gives $Gm/R^2 \sim GMR/r^3$, or $m/R^3 \sim M/r^3$, or $r \sim R(M/m)^{1/3}$, in the limit that $r \gg R$. Note that m/R^3 is proportional to the average density of the object and M/r^3 is proportional to the density we would get if we smeared the mass of the supermassive black hole over the full volume inside radius r; this is reminiscent of (and related to!) our previous insight in Chapter 2 that the maximum orbital, rotational, or vibrational frequency of a gravitationally self-bound object depends only on its average density. Putting in numbers we find that a clone of the Sun would be disrupted at about 0.5–1 au from a $10^6\ M_\odot$ supermassive black hole (the exact value depends on details), and a binary with a total mass of 20 $M_\odot$ and a semimajor axis of 1 au would be tidally separated at about 50 au from the same supermassive black hole.

The relatively small apocenter of orbits in the tidal separation channel means that the angular momentum perturbation per orbit is smaller than it is in the two-body relaxation channel. Moreover, the relatively large pericenter of orbits from tidal separation means that the angular momentum of the orbits is also larger. Thus, many more orbits are required to change the angular momentum significantly than in the two-body relaxation channel, and so there is a better prospect that the EMRI will persist for the thousands of orbits needed for detection. It is therefore possible that, even if the overall rate of tidal separations is much smaller than the rate of two-body relaxation events, the rate of detectable EMRIs is dominated by the tidal separation scenario.

5.1.3.3 Inspiral through an Accretion Disk

A final major path for an EMRI involves an active galactic nucleus with an accretion disk. As we discussed when we talked about mergers of two stellar-mass black holes, if the gaseous disk is embedded in a quasi-spherical distribution of stars and black holes, then as the stars (and much less likely, the holes, as Captain Obvious will surely mention) cross the disk in their orbits, dynamical friction with the gas can in principle settle some fraction of the stars in the disk. Another way to get black holes in the disk is for them to form there; in standard disk scenarios the outer part of the disk is actually unstable to star formation, and if some fraction of those stars form black holes after millions of years, then the holes will be in the disk from the beginning. Once the holes are moving with the disk, they follow the disk toward the supermassive black hole until they get close enough that gravitational radiation becomes the primary driver of inspiral. Such orbits will follow the gas and will therefore be nearly circular and have nearly zero inclination. One might worry that gas drag would significantly influence the inspiral near the massive black hole. For standard disk structure, the final months of inspiral will not be affected in a detectable way, but if the disk density is much greater than expected in standard assumptions, then migratory effects could be seen.

Captain Obvious: We need to dig deeper into some of the astrophysics here! For example, you might think that supernovae would disrupt the accretion disk, or that it would take forever for disk crossings to cause a star to settle into the disk, or that a star formed in the disk would spiral into the supermassive black hole too fast for it to form a black hole. So let's investigate using typical numbers based on disk theory.

In an accretion disk around a $10^8\ M_\odot$ supermassive black hole, stars might form at $\sim 10^3$ gravitational radii and beyond. From this distance, it will take several millions of years to spiral into the supermassive black hole (based on the plausible assumption that a star will be dragged with the gas in the disk). There is, therefore, enough time for a massive star to evolve into a black hole or neutron star because that evolution also takes a few million years.

At a distance of $\sim 10^3$ gravitational radii, the orbital speed is $\sim c/(10^3)^{1/2} \approx 10^4$ km s^{-1}. This is comparable to the speed of ejecta of a supernova, and much larger than the postsupernova kick speeds, so much of the debris will rain back down on the disk and the neutron star or black hole that is formed will remain tightly bound to the disk. Moreover, the typical thickness of the disk is ~ 0.01 times the orbital radius, which means that, at worst, a small fraction of the disk will be affected.

What about the capture of black holes or stars by the accretion disk? The rule of thumb is that to capture something, it has to interact with of order its own mass. That is, the amount of mass the black holes encounters in its Bondi–Hoyle radius (which, we recall from Appendix B, is roughly Gm/v^2 for a mass m and speed v) needs to be of order the black hole mass for the black hole to slow down significantly and be captured. Again, for our illustrative situation, the surface mass density of the disk is a few $\times\ 10^5$ g cm^{-2}, and at orbital speeds of 10^4 km s^{-1} the area within the Bondi–Hoyle radius for black holes of mass $10\ M_\odot$ is $\sim 10^{19}$ cm^2 and $\sim 10^{10}$ orbits are needed to grind down the typical orbit. At ~ 3 years per orbit for our case, that's too long. A small fraction of holes might orbit nearly in the disk plane, which would speed up the drag. However, a more realistic path is for a massive star (area $\sim 10^{24}$ cm^2) to cross the disk; this can be captured by $\sim 10^5$ passages, which is a few hundred thousand years.

In summary, the three major paths discussed for forming EMRIs are expected to leave very different signatures in the LISA frequency band: (1) two-body relaxation produces significant eccentricity and arbitrary inclination, (2) tidal separation yields low-eccentricity circular orbits and arbitrary inclination, and (3) drag through accretion disks leads to nearly circular orbits and nearly zero inclination. There is, therefore, hope that the origin of EMRIs will be evident on a case-by-case basis. It is further hoped that there will be a nonzero number of EMRIs seen with LISA, and there may well be, but rate estimates are highly uncertain.

5.2 Nonbinary Sources

5.2.1 Continuous Sources

We have plenty of evidence, in our Galaxy as well as in gravitational-wave observations, that merging binaries are abundant. What about other sources?

We will begin with continuous-wave radiation, which as you recall requires that there be some nonaxisymmetry around the rotation axis. What is it, exactly, that could produce the required nonaxisymmetry? We will divide the possibilities into two categories. "Lumps" will be nonaxisymmetries that are fixed relative to the star, whereas "waves" will be nonaxisymmetries that move relative to the star.

5.2.1.1 Lumps

Suppose we use the spherical cow approximation and assume that a neutron star is a perfect self-gravitating incompressible fluid with no magnetic field. Set that star into a uniform rotation that starts slowly and gradually speeds up, maintaining equilibrium at all times (this is sometimes called a "sequence of stationary configurations.") Our intuition would suggest (correctly) that the equilibrium shape should be an oblate spheroid, with the smallest axis pointing in the direction of the angular momentum. But in the late 19th century, Jacobi demonstrated that with rapid enough rotation, the equilibrium shape is a triaxial ellipsoid with the rotation axis being the axis of the maximal moment of inertia. To make the statement more formally: for a self-gravitating fluid of fixed mass and uniform density that rotates as a rigid body, there is an angular momentum above which the shape that minimizes the energy is a triaxial ellipsoid rather than an oblate spheroid (and the bifurcation to ellipsoids is, formally, a second-order phase transition). Such a figure of rotational equilibrium is called a Jacobi ellipsoid.

We can very roughly motivate the transition to an ellipsoid at a high-enough angular momentum in the following way. The total energy can be broken into the negative gravitational potential energy and the positive kinetic energy. Given a specified uniform density and mass, the gravitational potential energy is most negative for a sphere.[1] The rotational kinetic energy for a given angular momentum

[1] We can motivate this statement by supposing that we start with a nonrotating fluid that has some high points and some low points. If we take fluid from a high point and use it to fill in a low point, then the fluid is now closer to the center of mass and thus has a more negative gravitational potential energy. Only when all the fluid is at the same level do we have the minimum energy.

J and moment of inertia I is $E_{\rm kin} = J^2/(2I)$. Thus, for an angular momentum J, the rotational kinetic energy is less positive when the moment of inertia is larger. This pushes the fluid toward a more extended system. When J is larger, the rotational kinetic energy is a larger component of the total energy, so an increase in the moment of inertia becomes more important. At large enough J, this increase points to an ellipsoid rather than to an oblate spheroid.

This obviously suggests an interesting path to gravitational-wave emission: if there are circumstances in which a neutron star can be spun rapidly enough to become ellipsoidal, then gravitational waves could be emitted at a high rate. For example, if for some reason there are core-collapse supernovae in which the cores collapse with high angular momentum, then perhaps there could be a brief stage in which the collapse actively radiates gravitational waves.

But as we mentioned briefly in Chapter 2, the problem is that the rigidly rotating Jacobi ellipsoids are not the only possible configuration for a uniform-density self-gravitating fluid of a given mass and angular momentum. Indeed, we can imagine a spectrum, at one end of which is a rigidly rotating shape and at the other end of which is a shape that does not move but has internal flows that carry the angular momentum. That other end of the spectrum is the Dedekind ellipsoids. Dedekind ellipsoids do not have a changing-mass quadrupole; instead, the only gravitational radiation could come from the current quadrupole, which is much weaker. It was demonstrated in 1974 that the action of gravitational radiation is to drive a fluid from a Jacobi configuration (if that was the initial condition) to a Dedekind configuration and thus to decrease the rate of emission of gravitational waves. Countering this are viscosity and internal magnetic fields, which try to enforce rigid-body rotation. A series of papers in the 1970s examined the evolution of a rapidly rotating fluid body under several of these forces and did not reach a clear conclusion, but it seems unlikely that this will be a major source of gravitational radiation.

Another possibility for lumps is that a star that would otherwise be axisymmetric has a substantial magnetic field that is misaligned with the rotation axis. The magnetic stresses would produce triaxiality, which would then lead to gravitational radiation. Is there evidence that such misalignment happens? Yes! At the simplest level, rotation-powered pulsars have to have somewhat misaligned magnetic fields, as otherwise we wouldn't see pulsations. Thus, to the degree that magnetic fields produce lumps, rotating neutron stars emit gravitational waves, with a luminosity that can be determined using Equation (2.13) if the star can be represented as an ellipsoid.

We can make a simple estimate based on a comparison of stresses, that is, of the magnetic field strength needed to potentially produce a given ellipticity. Magnetic stresses are of the order of B^2 for a magnetic field strength B; for example, the magnetic energy density is $B^2/(8\pi)$. These should be compared with the mass–energy density of the surrounding matter to determine, in a very approximate way, the fractional perturbation induced by the magnetic field. For a neutron star of mass M and radius R, the mass–energy density is $Mc^2/\left(\frac{4}{3}\pi R^3\right)$ and thus if $B^2 = 6Mc^2/R^3$, then the total magnetic energy equals the mass–energy. For $M = 1.4\ M_\odot$ and $R = 12$ km, equality occurs when $B \approx 3 \times 10^{18}$ G. This tells us that for a volume-averaged field

$B \sim$ few $\times\ 10^{12}$ G (similar to the inferred surface magnetic field of the Crab pulsar and similar young pulsars), the magnetic energy in the star is a fraction $[B/(3 \times 10^{18}\ \mathrm{G})]^2 \approx 10^{-12}$ of the total mass–energy. We therefore expect perturbations and ellipticities due to magnetic fields to be of this order or less (indeed, if the magnetic field is axisymmetric around the rotation axis, there will be no ellipticity induced).

At that level, our equations from Chapter 2 tell us that it will be very difficult to detect the continuous gravitational waves from the lump. Moreover, if the exterior magnetic field is significantly stronger than a few $\times\ 10^{12}$ G, we would expect the star to spin down rapidly due to magnetic braking.

Captain Obvious: How rapidly? We can get a scaling by noting (see various textbooks on neutron stars suggested at the end of this chapter) that for an isolated neutron star spinning down due to magnetic dipole radiation, the surface magnetic field strength B is related to the rotation period P and period derivative $\dot{P}$ by the rough relation $B(\mathrm{Gauss}) \approx 10^{12}[P\dot{P}/(10^{-15}\ \mathrm{s})]^{1/2}$. For example, for $P = 0.1$ s (giving a gravitational-wave frequency of 20 Hz, about the minimum that can be seen by current ground-based detectors) and $B = 10^{13}$ G, $\dot{P} \approx 10^{-12}$ and thus the pulsar will spin down significantly in $\sim P/\dot{P} \sim$ few thousand years. That's an eyeblink for an astronomical system, and it would be shorter for a smaller P or higher B.

Thus, there would only be a small window of time in which the star is rotating rapidly enough that its lump could produce observable gravitational waves; one possibility is a newly born neutron star that has a strong magnetic field and initially rapid rotation. Another possibility that has been discussed is a neutron star with a moderate exterior magnetic field (so that the star does not spin down too rapidly) and a strong-enough interior magnetic field that the ellipticity can be pronounced. It is not known whether this magnetic configuration is stable, and as of the writing of this book, the best guess is that it is difficult to have internal magnetic fields much more than one order of magnitude stronger than the external magnetic field.

The last possibility we will discuss relates to "mountains" on accreting neutron stars. Neutron stars in so-called low-mass X-ray binaries (LMXBs, in which the companion star has mass $M < M_\odot$, and could be a main-sequence star, a red giant, or a white dwarf) accrete at a high rate, roughly $(10^{-10}$–$10^{-8})\ M_\odot\ \mathrm{yr}^{-1}$ for the brighter sources. Various phenomena, including regular pulsations during the accretion-powered emission, regular pulsations during thermonuclear X-ray bursts, and quasi-periodic oscillations in the brightness of the accretion-powered emission can be used to infer the spin frequencies of these sources, which span a broad range, from 45 Hz to 620 Hz. There is no particular evidence of clustering in spin frequency, but no sources close to the maximum possible frequency (which is $\nu_{\max} \approx 1/(2\pi)(GM/R^3)^{1/2}$, or $\approx$1650 Hz for $M = 1.4\ M_\odot$ and $R = 12$ km) are seen. This suggests that some braking torque is probably operating to offset the spin-up produced by the accreting matter (the other possibility is that there simply has not been enough time to spin up sources to higher frequencies). Magnetic torques appear to be able to explain all the observations, but people have also explored the possibility that gravitational radiation can play a role. If so, then nonaxisymmetries must be generated somehow. One possibility is pooling of accreted matter in buried magnetic fields, and people have also investigated the possibility that nonaxisymmetric accretion could lead to nonaxisymmetric compositions.

We could try, of course, to approach the question from another angle. Forget about how it happens; how large a deformation could be sustained? At some point, the strength of the material is insufficient to hold up a lump. This, for example, is why Earth doesn't have any mountains much higher than 10 km; above that height, the granite that holds up the mountain crumbles. The calculations of the maximum possible quadrupole for a neutron star are unfortunately very involved, with a number of subtleties, but as of the writing of this book the best guess appears to be that the maximum fraction ϵ of the quadrupole moment that is nonaxisymmetric (sometimes called the ellipticity) is a few $\times 10^{-7}$. That's not very much, but there are upper limits to the ellipticity of some millisecond pulsars that are $\epsilon < 10^{-8}$, so there is some room for discovery.

5.2.1.2 Waves

All of the "lump" mechanisms would imply a gravitational-wave frequency of twice the spin frequency. However, fluids also can oscillate in many ways. For example, water waves are an example of "g-modes" because the restoring force comes from gravity, and for "p-modes" the restoring force comes from pressure gradients in a perturbation. But those modes require an external excitation, so although they can be excited due to an energy release (e.g., a quake on a neutron star or in a more extreme circumstance the merger between two neutron stars), they are not expected to be persistent. However, it has also been proposed that there are wave instabilities that can produce gravitational radiation. The most discussed of these are Rossby waves, or r-modes.

> **Captain Obvious:** These modes are related to weather patterns on Earth. On a rotating body (such as Earth or a neutron star), fluid that moves in latitude experiences a Coriolis restoration force. The result is circular patterns of fluid movement, centered on the rotational equator. The lowest-order (and thus probably highest-amplitude) mode has a frequency as seen at infinity that is two-thirds of the rotation frequency (in the simplest case; various effects can produce a somewhat different frequency).
>
> Because the wave has a changing-mass quadrupole, it emits gravitational radiation. The loss of energy to gravitational radiation causes the wave to move more slowly. But because the wave frequency is less than the rotational frequency, this means that the wave was moving backward as seen in the rotating frame; as a result, the loss of energy causes the wave to move more rapidly as seen in the rotating frame. As a result, the fluid of the star is unstable to r-modes. In a perfect fluid with no magnetic field, this instability operates for all angular velocities!

Because the pattern frequency of r-modes moves (in the simplest approximation) at two-thirds of the stellar rotational frequency, the gravitational waves would appear at four-thirds of the rotational frequency. Therefore, if for some source we know the spin frequency (e.g., by measurement of coherent pulsations) and can measure periodic gravitational waves, we can determine whether it is lumps or Rossby waves that are present.

As you can derive in answer to one of the exercises, if lumps or waves completely balance the accretion torque, then we should easily be able to detect the highest-flux low-mass X-ray binaries in the sky. That's an optimistic conclusion! However, we have to be careful. If neutron stars were really perfect fluids with no complications, and if these modes could reach high amplitudes, then we'd never see isolated neutron stars with frequencies of hundreds of Hertz. That is, the r-modes would have spun down all pulsars! But this is not what we observe, and therefore, something else must be going on. For example, in addition to the lowest-order modes, fluids have many other modes of oscillation. Perhaps there is nonlinear saturation of the modes at low amplitude because of coupling between a large number of modes. Other ideas have included viscous effects that damp the modes (the viscosity is interestingly temperature dependent and might in some circumstances lead to limit-cycle behavior), effects related to the interface between the liquid core and solid crust, and magnetic couplings. It is fair to say that at this point there is no consensus about the strength of Rossby waves or the role they could play in neutron star spindown and gravitational radiation.

5.2.2 Burst Sources

As we discussed in Chapter 2, "burst" refers to events of limited duration that do not have to have any special periodicity. Data analysis for these will be very challenging indeed, but because they are by definition associated with violent events, we could potentially learn a great deal from the detection of gravitational radiation.

Captain Obvious: An example of a violent event is the explosion of a star, but how can we estimate the fraction of the mass–energy in a core-collapse supernova that comes out in gravitational waves? One path we might try is the following. We know that $h_{ij} \sim \ddot{I}_{ij}/r \sim (1/r)(\mathfrak{M}\mathfrak{L}^2)/\mathfrak{T}^2$ for a characteristic mass $\mathfrak{M}$ and length scale $\mathfrak{L}$ over a time $\mathfrak{T}$. When the core of a massive star collapses, it is close to the Chandrasekhar mass, or roughly $\mathfrak{M} \sim 1\ M_\odot$. Core bounce occurs when the core reaches roughly nuclear density and repulsion between nucleons sets in; this happens at $\mathfrak{L} \approx 15$ km, which is $\mathfrak{L} \sim 10\ M_\odot$ in geometrized units. The timescale is on the order of the freefall time, which might be roughly $\mathfrak{T} = 10^{-4}$ s for that mass and radius, or $\mathfrak{T} \sim 30\ M_\odot$ in geometrized units. At a distance of $r = 10$ kpc from us (so that we could see most of our Galaxy), which is roughly $r = 2 \times 10^{17}\ M_\odot$ in geometrized units, we then have $h \sim 10^{-18}$. To compute the amount of mass–energy emitted in the collapse, we need to integrate the flux, so you'd get something like $E_{GW} \sim \dot{E}\mathfrak{T} \sim 4\pi r^2 \dot{h}^2 \mathfrak{T} \sim 4\pi r^2[(1/r)(\mathfrak{M}\mathfrak{L}^2)/\mathfrak{T}^3]^2\mathfrak{T} \sim 4\pi(\mathfrak{M}^2\mathfrak{L}^4)/\mathfrak{T}^5$. Using our previous numbers, we get $E_{\rm GW} \sim 10^{-2}\ M_\odot c^2$ using these assumptions.

Major Payne: Aha! Your sloppiness has caught up with you. Detailed simulations suggest that the actual mass–energy emitted in gravitational waves from core-collapse supernovae is more like $10^{-10}\ M_\odot c^2$ to $10^{-8}\ M_\odot c^2$, depending on details. So why does your morally incorrect hand-wavy approach fail so badly? The biggest issue is that, as we showed in Chapter 1, spherically symmetric collapse emits zero gravitational radiation. If the core of a massive star, just prior to core collapse, is not rotating, then the initial collapse will be spherically symmetric to a high degree of precision. Only later, after asymmetries have developed due to convection, instabilities, and nonspherical accretion is there significant gravitational radiation.

If the precollapse core was rotating, then collapse will produce a time variation of the mass currents, and this will generate gravitational waves in less than a millisecond after core bounce. However, if the precollapse core was rotationally coupled to the star via convection, then its initial rotation period was hours (if the star's size was close to that of a main-sequence star) to years (if the star was a giant). Then, angular momentum conservation during the collapse doesn't lead to rotation even close to Keplerian, and the signal is extremely weak. So there! I do grudgingly admit that your terrible calculation has the benefit that it provides us with some context; without such calculations, we don't have a framework to judge the energies obtained from the best simulations. But in this case, you have to be a lot more careful.

The most commonly discussed burst sources are core-collapse supernovae. Simulations of these are extraordinarily challenging due to the range of scales and the details of physics that are important: convection, general relativity, radiation transfer, neutrino transport, and a good treatment of magnetic fields seem essential, as Major Payne clearly described. At the current best guess that $\sim 10^{-10}\ M_\odot c^2$ to $10^{-8}\ M_\odot c^2$ is emitted in gravitational waves, most of it at hundreds of Hertz, supernovae outside our Galaxy will be undetectable, and even supernovae in our Galaxy are not guaranteed detections until we get to the next generation of ground-based gravitational-wave detectors. If there are some situations in which core collapse is accompanied by rapid rotation and the development of an ellipsoid, then much more energy could be radiated as gravitational waves, but even current numerical simulations are insufficient to tell us whether this happens.

Other categories of burst events?—To paraphrase the Bard: "What's in a name? That which we call a merger by any other word would smell as sweet." Leaving aside the problem that smells don't travel well in space, the point is that there are some mergers whose character could be represented as bursts. For example, a sufficiently high-mass binary will have so few cycles in the sensitivity range of a ground-based detector that it is effectively a burst. Even for lower-mass binaries, if they have high eccentricities, then for much of the coalescence, the signal is essentially bursts (near pericenter passage) separated by long intervals without any detectable signal. As we discussed in Chapter 4, unmodeled bursts may be best extracted with wavelet methods, and their detection could reveal important astrophysical properties about the sources that generate them.

5.2.3 Stochastic Backgrounds

Stochastic backgrounds can be pictured as the superposition of a large number of sources that are individually unresolvable. For example, pulsar timing arrays are expected to detect a stochastic background of supermassive black hole binaries long before they merge, and as we have mentioned before, the Galactic population of double white dwarf binaries is expected to provide a foreground that will effectively be a source of noise for LISA up to ~ (2–3) $\times 10^{-3}$ Hz. The most spectacular stochastic source would arguably be cosmological, e.g., from the inflationary era of the early universe.

In this section we will begin with some basic principles associated with stochastic collections of binary sources. We will then apply them to supermassive black hole binaries, double white dwarf binaries, and binaries of two stellar-mass black holes or two neutron stars. We will postpone our discussion of early-universe sources of stochastic gravitational waves to Chapter 6.

5.2.3.1 Binaries as Stochastic Gravitational-wave Sources

Suppose that we have a large set of circular binaries (thousands, millions, billions ⋯ gadzillions!) that evolve purely by gravitational radiation and are in a statistical steady state, meaning that the distribution of signals is stationary even as the individual binaries evolve. Then the individual binaries shrink at the rate given by the Peters formulae; from Equation (2.8), $da/dt \propto a^{-3}$ for zero eccentricity. Because for a circular binary the gravitational-wave frequency is $f_{\rm GW} = (1/\pi)(GM/a^3)^{1/2} \propto a^{-3/2}$, this means that $df_{\rm GW}/dt \propto f_{\rm GW}^{11/3}$. With the mass and other constants put back in, we find

$$\begin{aligned}\frac{df_{\rm GW}}{dt} &= \frac{96\pi}{5}\left(\frac{\pi G\mathcal{M}}{c^3}\right)^{5/3} f_{\rm GW}^{11/3} \\ &\approx 5.8 \times 10^{-23}\ \text{Hz s}^{-1}\left(\frac{\mathcal{M}}{10^8\ M_\odot}\right)^{5/3}\left(\frac{f_{\rm GW}}{10^{-8}\ \text{Hz}}\right)^{11/3} \\ &\approx 1.8 \times 10^{-18}\ \text{Hz s}^{-1}\left(\frac{\mathcal{M}}{0.5\ M_\odot}\right)^{5/3}\left(\frac{f_{\rm GW}}{10^{-3}\ \text{Hz}}\right)^{11/3}.\end{aligned} \tag{5.5}$$

Here as before $\mathcal{M} = \eta^{3/5} m$ is the chirp mass, where we recall that $m = m_1 + m_2$ is the total mass and $\eta = m_1 m_2/m^2$ is the symmetric mass ratio, and the latter two scalings are appropriate for supermassive black holes and double white dwarf binaries, respectively.

If the duration of an observation is T, then the frequency resolution is $df = 1/T$; thus, reasonable frequency resolutions are $\sim 10^{-8}$ Hz (for a few-year mission such as LISA) to $\sim 10^{-9}$ Hz (because observations of some pulsars stretch over decades). Because the time spent by a binary between frequencies $f_{\rm GW}$ and $f_{\rm GW} + df$ is $dt = (dt/df)df$, and because in steady state the number of sources in that frequency range is proportional to dt, the number of sources between $f_{\rm GW}$ and $f_{\rm GW} + df$ is proportional to $f_{\rm GW}^{-11/3}$. Indeed, our assumption that the distribution is stationary means that if the overall coalescence rate is $\dot{N}$, then the expected number of binaries in a given frequency bin is $N = \dot{N}dt$ (of course, a given bin might have more or less than the average). If there are more than two sources in a given frequency bin (two because there are two gravitational-wave polarizations), then it becomes challenging to detect all the sources individually unless they all have very strong signals.

Captain Obvious: To see that it is theoretically possible to detect several sources separated by much less than the nominal $df = 1/T$ frequency resolution, we can recall the argument in Section 1.3 that it is possible for gravitational-wave detectors to measure mirror displacements that are much smaller than the wavelength of the light used for the detections. In the case of detection of sources with frequencies closer than the nominal resolution, we note that for a very strong source, its frequency can be measured to higher precision than $df = 1/T$; df is more properly viewed as the width of the frequency in a Fourier spectrum of the source, but the centroid of the peak in Fourier space can be determined more precisely. Similarly, if the sources are strong enough, two sources can be distinguished from one even if their frequencies are close. In practice, however, the LISA sources that might fall into the category of (a) very strong, (b) very close in frequency, and (c) not changing in frequency over several years are not easy to envision: supermassive black hole binaries will chirp, and nonchirping double white dwarf binaries would have to be improbably close to have very large signal-to-noise ratios. So the "maximum two sources per frequency bin" rule is still pretty good in practice.

5.2.3.2 Supermassive Black Hole Binaries

Let's now consider supermassive black hole binaries. As a very rough guess, let's imagine that there are roughly 10^{10} "major" galaxies in the universe with central black holes of masses $\sim 10^7\ M_\odot$ or above, and that on average, in the $\sim 10^{10}$ year history of the universe, each major galaxy has had one merger with a similarly major galaxy and that the central black holes merged as well. Those assumptions tell us that the average merger rate is about $\dot{N} = 1\ \text{yr}^{-1}$. Suppose that we observe for $T = 30\ \text{yr} \approx 10^9\ \text{s}$, so that our frequency resolution is $\Delta f \approx 10^{-9}$ Hz. Then, using Equation (5.5), we find that $dt/df_{\rm GW} \approx 3 \times 10^{16}\ \text{yr Hz}^{-1}(\mathcal{M}/10^7\ M_\odot)^{-5/3}(f_{\rm GW}/10^{-8}\ \text{Hz})^{-11/3}$. This suggests that the number of binaries in a $df_{\rm GW} = 10^{-9}$ Hz interval around a frequency $f_{\rm GW}$ is

$$
\begin{aligned}
N &= \dot{N}dt = \dot{N}df_{\rm GW}\left(\frac{dt}{df_{\rm GW}}\right),\\
&= (1\ \text{yr}^{-1})\ (10^{-9}\ \text{Hz})\ (3 \times 10^{16}\ \text{yr Hz}^{-1})\left(\frac{\mathcal{M}}{10^7\ M_\odot}\right)^{-5/3}\left(\frac{f_{\rm GW}}{10^{-8}\ \text{Hz}}\right)^{-11/3},\\
&= 3 \times 10^7\left(\frac{\mathcal{M}}{10^7\ M_\odot}\right)^{-5/3}\left(\frac{f_{\rm GW}}{10^{-8}\ \text{Hz}}\right)^{-11/3}.
\end{aligned}
\tag{5.6}
$$

This is clearly $N \sim 3 \times 10^7$ sources for $f_{\rm GW} = 10^{-8}$ Hz and $N \sim 1$ source for $f_{\rm GW} = 10^{-6}$ Hz. Therefore, roughly, at frequencies below $\sim 10^{-6}$ Hz, this is an unresolved stochastic background and above that frequency, we would expect to be able to see individual sources (but they will be few and difficult to find).

Of course, to make these estimates we have used approximations that might not be valid. For example, it turns out that the signal is actually dominated by the orbits of black holes considerably more massive than $\sim 10^7\ M_\odot$; $10^9\ M_\odot$ is more like it. There are far fewer heavy black holes than light black holes, so the number of binaries that matter is also much smaller and there are better prospects to see individuals. One thing to note is that because the frequency dependence is so strong, the frequency at which we transition from unresolved to resolved is pretty well determined: for example, if our rate $\dot{N}$ changes by a factor of 10, the frequency boundary changes by a factor of $10^{3/11} < 2$. But in spite of all of that, the above reasoning is sufficiently sound to give us a rule of thumb and help us understand why there can be an unresolved background of supermassive black hole binaries.

5.2.3.3 Double White Dwarf Binaries

As we have mentioned before, the foreground noise produced by double white dwarf binaries in our Galaxy is expected to limit the effective sensitivity of LISA up to a few millihertz. We can motivate that in the following way. It is estimated that there are a few tens of millions of double white dwarf systems in our Galaxy that will merge in the $\sim 10^{10}$ year age of the universe, which suggests a merger rate of one per few hundred years. It is also estimated that the rate of white dwarf supernovae is one per few hundred years, so if most of these involve two white dwarfs merging rather than a white dwarf accreting matter from a nondegenerate companion, then the two rates are consistent. Let's take as our fiducial rate $0.003\ \text{yr}^{-1}$. We will also assume that the typical chirp mass is about $0.5\ M_\odot$ (if we take two white dwarfs each of mass $0.6\ M_\odot$, which is average, the chirp mass is $0.52\ M_\odot$). Finally, given that LISA's lifetime is expected to be a few years, we will assume that the duration of the observation is ≈ 3 yr, or $\approx 10^8$ s and thus that the frequency resolution is about $df_{\rm GW} = 10^{-8}$ Hz.

Using Equation (5.5), then we have an expected number of sources in our Galaxy per 10^{-8} Hz frequency bin of

$$
\begin{aligned}
N &= \dot{N}dt = \dot{N}df_{\rm GW}\left(\frac{dt}{df_{\rm GW}}\right),\\
&= (3\times 10^{-3}\ \text{yr}^{-1})\ (10^{-8}\ \text{Hz})\ (1.8\times 10^{10}\ \text{yr Hz}^{-1})\left(\frac{\mathcal{M}}{0.5\ M_\odot}\right)^{-5/3}\left(\frac{f_{\rm GW}}{10^{-3}\ \text{Hz}}\right)^{-11/3} \\
&\approx 0.5\left(\frac{\mathcal{M}}{0.5\ M_\odot}\right)^{-5/3}\left(\frac{f_{\rm GW}}{10^{-3}\ \text{Hz}}\right)^{-11/3}.
\end{aligned}
\tag{5.7}
$$

This suggests that $f_{\rm GW} \sim 10^{-3}$ Hz is the approximate dividing line between unresolved binaries at low frequencies (where there are on average multiple double white dwarf binaries per $df = 10^{-8}$ Hz frequency bin) and resolved binaries at high frequencies. The best current estimates put the dividing line at $\approx 2\times 10^{-3}$ Hz, so our estimate is fairly close.

Captain Obvious: It might occur to you that in this estimate we've only considered double white dwarf binaries in our own Galaxy. There are hundreds of billions of galaxies in the universe, so shouldn't we multiply our number by that factor? Wouldn't that mean that we'd have to go to really high frequencies to resolve all of those white dwarf binaries? Yes, but there's another consideration that means we don't have to. That is the consideration of the amplitude of the signal. Although there are an enormous number of extragalactic white dwarf binaries, they are distant enough that their amplitude is lower than the LISA noise and therefore we don't need to concern ourselves. The amplitude question here, like the question of the expected amplitude from the population of double supermassive black hole binaries, requires models of the number and distribution of the binaries.

5.2.3.4 Binaries with Stellar-mass Black Holes and/or Neutron Stars

At frequencies greater than about 10 Hz relevant to second-generation ground-based detectors, individual frequency bins will not contain multiple binaries. However, there will be binaries distant enough that they cannot be detected by current instruments because their signals will be too weak. Thus, whereas previously we discussed genuinely unresolvable stochastic signals, here we are thinking about signals that can't be detected individually because of signal-to-noise issues. As a result, future more sensitive detectors will see individual signals, and thus, the "stochastic background" from this contribution will be reduced.

As we mentioned in Chapter 4, there is an interesting difference between the character of the stochastic background from individually undetectable double black hole binaries and from individually undetectable double neutron star binaries. The locally estimated rate of double black hole coalescences is roughly 20 $\mathrm{Gpc}^{-3}\ \mathrm{yr}^{-1}$. If we assume that the rate per galaxy has remained constant during the history of the universe, and if we integrate out to a redshift of $z = 5$, then the relevant volume (the comoving volume, which is a volume that factors out the expansion of the universe) is about 2000 Gpc^3 and thus the total rate is about $4 \times 10^4\ \mathrm{yr}^{-1}$ or $\sim 10^{-3}\ \mathrm{s}^{-1}$. If we instead assume that the rate per galaxy throughout the universe averages 10 times what it is locally (the motivation is that massive star formation was $\sim$10× more common at $z = 1$ than it is now), then this goes up to $\sim 4 \times 10^5\ \mathrm{yr}^{-1}$ or $\sim 10^{-2}\ \mathrm{s}^{-1}$. Because the locally estimated rate of double neutron star coalescences is roughly 500 $\mathrm{Gpc}^{-3}\ \mathrm{yr}^{-1}$ or 25× larger than the locally estimated double black hole rate, the equivalent range for neutron stars is $\sim$0.02–0.2 s^{-1}.

Casting Equation (5.1) in terms of frequencies, one can estimate that the inspiral time is

$$T_{\mathrm{inspiral}} \approx 200\ \mathrm{s} \left(\frac{\mathcal{M}}{M_\odot}\right)^{-5/3} \left(\frac{f_{\mathrm{GW}}}{20\ \mathrm{Hz}}\right)^{-8/3}. \tag{5.8}$$

Multiplying this time by the rates for double black holes and double neutron stars, we get an estimate of how many such signals are, on average, simultaneously present above a frequency f_{GW}. For black holes it is

$$N \approx (10^{-3} - 10^{-2})\left(\frac{\mathcal{M}}{20\ M_\odot}\right)^{-5/3} \left(\frac{f_{\mathrm{GW}}}{20\ \mathrm{Hz}}\right)^{-8/3} \tag{5.9}$$

and for neutron stars it is

$$N \approx (4 - 40)\left(\frac{\mathcal{M}}{M_\odot}\right)^{-5/3} \left(\frac{f_{\mathrm{GW}}}{20\ \mathrm{Hz}}\right)^{-8/3}. \tag{5.10}$$

Here we have normalized the chirp mass to values that are roughly appropriate for black holes ($\sim$20 $M_\odot$) and neutron stars ($\sim$1 $M_\odot$). This justifies the statement that we made in Chapter 4 that in current ground-based detectors, we will typically have several to many double neutron star signals in the band at once, whereas we will typically have one or zero double black hole signals in the band at the same time. Above 100 Hz we will normally have only one double compact binary at a given time.

Note, incidentally, that although above $\sim$20 Hz we expect multiple double neutron star signals at any given time, this does not mean that they are unresolved! The reason is twofold. First, to be unresolved they would have to be at the same frequency at the same time, so in reality, we'd need to ask how many neutron stars are in a single frequency bin at a given moment, and the answer is far fewer than one. Second, because double neutron star binaries don't all have identical chirp masses, their signals sweep through the detection band at different rates and can therefore be distinguished from each other.

5.3 Exercises

1. We mentioned random walks in the text. These have broad applicability in physics and astrophysics, and in this problem, you will derive an important property of such motion. Suppose that a particle moving in three dimensions does so with steps of a fixed length d, but with each step in a completely random direction that is uncorrelated with any previous steps. Derive the expected distance of the particle from its starting point after N steps. Hint: for consecutive vector steps **A** and **B**, the expected net square distance after both steps is $\langle A^2\rangle + 2\langle \mathbf{A}\cdot\mathbf{B}\rangle + \langle B^2\rangle$.

2. Consider a population of binaries, each of which has reduced mass μ and total mass m. Suppose they are all circular, and that the population is in steady state, meaning that the number in a given frequency bin is simply proportional to the amount of time they spend in that bin. Also assume that the only angular momentum loss process is gravitational radiation, rather than mass transfer or other effects. For each of the following parts, derive the answers in general and then apply the numbers to white dwarf–white dwarf binaries, where we assume that both masses are 0.6 $M_\odot$.
   (a) Using the Peters equations for circular orbits of point masses, derive the frequency $f_{\rm min}$ such that the characteristic inspiral time $T_{\rm insp} \sim 1/[d\ln f/dt]$ is equal to the Hubble time $T_H \sim 10^{10}$ yr. What is the frequency specifically for a white dwarf–white dwarf binary?
   (b) Below $f_{\rm min}$, the distribution dN/df of sources with frequency will depend on their birth population. Above it, gravitational radiation controls the distribution. Derive the dependence of dN/df on f for $f > f_{\rm min}$ (the normalization is not important).

3. Dr. I. M. Wrong has heard about the possibility of seeing continuous-wave emission from lumpy neutron stars using high-frequency ground-based

detectors. Indeed, it is obvious that lumpy white dwarfs will be good sources for low-frequency space-based detectors.

To evaluate this suggestion, do the following. Recall that $h \sim (1/r)(\partial^2 I/\partial t^2)$. For a rotating source only, the nonaxisymmetric part contributes, so let's say that $I \sim \epsilon MR^2$, where $\epsilon \ll 1$. Also, let $\partial^2/\partial t^2 \sim \Omega^2$, where Ω is the rotation frequency. Supposing that the ellipticity ϵ is the same for neutron stars and white dwarfs, how does the maximum possible h at a given r for white dwarfs compare with the maximum possible h for neutron stars? Hint: how fast can they rotate? Combine this with our calculation that gravitational waves can just barely (if at all!) be detected from rotating neutron stars, and the fact that high-frequency detectors have a better h sensitivity than low-frequency detectors, to evaluate Dr. Wrong's idea.

4. Suppose that a neutron star with zero magnetic field accretes matter from a companion at a rate $\dot{M}$. The rate at which angular momentum is accreted is $\sqrt{12}\,\dot{M}GM/c$. In the following, assume that the mass M and moment of inertia I of the star do not change significantly in the period of interest. In addition, we are working under the assumption that a lump, rather than a wave, produces gravitational radiation.
   (a) Suppose that torques from gravitational radiation balance the accretion torques at a frequency $\Omega_{\rm equil}$. As a function of $\dot{M}$, M, I, and $\Omega_{\rm equil}$, calculate the required ellipticity ϵ. Approximately what is the value of ϵ if $\dot{M} = 10^{17}$ g s^{-1}, $M = 3 \times 10^{33}$ g, $I = 10^{45}$ g cm^2, and $\Omega_{\rm equil} = 2\pi \times 300$ Hz?
   (b) Suppose that an accreting nonmagnetic neutron star has an observed electromagnetic flux F, which we assume is produced by accretion onto the surface: the luminosity is $L = 0.2\dot{M}c^2$, where the 0.2 is the efficiency of energy release. Show that, if the torque from the accreting matter is balanced exactly by gravitational radiation, that the flux in gravitational radiation has a simple relation to the electromagnetic flux, and find that relation. Assume isotropic emission of both.
   (c) The highest-flux accreting neutron star in the sky is Sco X-1, with a flux of about 3×10^{-8} erg cm^{-2} s^{-1}. If other torques are negligible and its gravitational radiation emerges at 1000 Hz, roughly what strain amplitude would we measure? At that frequency, the target sensitivity in standard configuration for LIGO 1 is $\sim 10^{-22}$ Hz $^{-1/2}$ and for advanced LIGO is $\sim 10^{-23}$ Hz $^{-1/2}$. Approximately how long would it take each of them to measure the signal?

5. Suppose that in a fraction ϵ of core-collapse supernovae, remarkable things happen (very fast rotation, or extremely strong magnetic fields, or something else) that lead to the event releasing 0.1 $M_\odot c^2$ in gravitational radiation at 200 MWEGs per cubic megaparsec and that each galaxy, on average, has a core-collapse supernova per 100 years. Calculate the value of ϵ needed so that we would expect to see one such event per year at a signal-to-noise ratio of 10 in advanced LIGO, assuming that the needed amplitude is $h = 10^{-21}$.

6. Suppose that a supernova anywhere within 10 kpc of us (that is, somewhat past the distance to the Galactic center) will produce a gravitational-wave signal that will be detectable by advanced LIGO. Planned third-generation detectors such as the Einstein Telescope and the Cosmic Explorer could be as much as 20× more sensitive than advanced LIGO. Assuming that the rate of supernovae is proportional to the number of stars, estimate to within a factor of 3 the increase in rate one would expect with these third-generation detectors. Hint: the answer to this question depends more on astronomy than on detectors.

7. Consider all of the binaries in the visible universe (not just compact object binaries!), and assuming that at high-enough frequencies their orbital evolution is driven by gravitational radiation, calculate the frequency above which there is on average fewer than one binary instantaneously resident in a 10^{-8} Hz frequency bin (i.e., roughly a three-year observation period). Hint: not all binaries in the universe can get to a given frequency.

Useful Books

Colpi, M., Casella, P., Gorini, V., Moschella, U., & Possenti, A. 2009, Physics of Relativistic Objects in Compact Binaries: From Birth to Coalescence (Berlin: Springer)

Dewitt, B., & Dewitt-Morette, C. M. 1973, Les Houches Lectures : 1972 (Boca Raton, FL: Routledge)

Hansen, C. J., Kawaler, S. D., & Trimble, V. 2004, Stellar Interiors—Physical Principles, Structure, and Evolution (Berlin: Springer)

Kouveliotou, C., Wijers, R. A. M. J., & Woosley, S. 2012, Gamma-ray Bursts (Cambridge: Cambridge Univ. Press)

Lipunov, V. M. 2011, Astrophysics of Neutron Stars (Berlin: Springer)

Mészáros, P. 1992, High-Energy Radiation from Magnetized Neutron Stars (Chicago, IL: Univ. Chicago Press)

Miller, M. C. 2016, GReGr, 48, 95

Prialnik, D. 2009, An Introduction to the Theory of Stellar Structure and Evolution (Cambridge: Cambridge Univ. Press)

Rezzolla, L., & Zanotti, O. 2013, Relativistic Hydrodynamics (Oxford: Oxford Univ. Press)

Rezzolla, L., Pizzochero, P., Jones, D. I., Rea, N., & Vidana, I. 2018, The Physics and Astrophysics of Neutron Stars (Berlin: Springer)

Ryden, B., & Peterson, B. M. 2020, Foundations of Astrophysics (Cambridge: Cambridge Univ. Press)

Shapiro, S. L., & Teukolsky, S. A. 1983, Black Holes, White Dwarfs and Neutron Stars: The Physics of Compact Objects (New York: Wiley)

Thorne, K. S., & Blandford, R. D. 2017, Modern Classical Physics: Optics, Fluids, Plasmas, Elasticity, Relativity, and Statistical Physics (Princeton, NJ: Princeton Univ. Press)

Unno, W., Osaki, Y., Ando, H., & Shibahashi, H. 1979, Nonradial Oscillations of Stars (Tokyo: Tokyo Univ. Press)

Chapter 6

Gravitational-wave Cosmology

When a new window is opened to the universe, one thrill is the contemplation of currently undetected or even unknown sources. Cosmology is flush with such possible gravitational-wave sources: Cosmic strings! Early-universe phase transitions! Primordial black holes! Direct signals from the inflationary era! In addition to these spectacular possibilities, any of which might result in a Nobel Prize if unambiguously detected, there are prospects for using gravitational waves to measure the universe in ways that might help resolve disputes about the expansion of the universe and could give us important hints about the nature of dark energy. In this chapter, after giving the briefest possible introduction to the cosmology relevant for our purposes, we will talk both about potential cosmological sources of gravitational waves and how we can use coalescences that involve neutron stars and/or black holes to obtain cosmological distances that will provide critical input to theories.

6.1 The Bare Bones of Cosmology

Cosmology is a vast and important subject that is discussed in great detail in any number of excellent textbooks; you should read those books, and you can find some suggestions at the end of this chapter. Here we give only the barest outline of the aspects of cosmology that are most critical to our subject. We will simply state results rather than trying to derive or motivate them.

- **Isotropy and homogeneity**. Locally, and now, our surroundings are obviously not isotropic (looking toward the Sun is very different from looking elsewhere) and are also obviously not homogeneous (your average density is $\sim 10^{30}$ times the average density in the universe). However, on large enough scales the universe approaches isotropy and homogeneity, which dramatically simplifies theoretical treatments.
- **The universe is expanding**. On large scales, the average motion of things in the universe (stars, galaxies, aliens) follows that expansion, although locally the

doi:10.1088/2514-3433/ac2140ch6

motion can deviate (e.g., we orbit the Sun, and the solar system orbits our Galaxy, rather than everything drifting away).

- **The scale factor and the redshift**. The expansion at a given time is usually parameterized with a "scale factor," commonly denoted a (yes, this is different from the Kerr angular momentum parameter; we only have so many letters). To picture the scale factor, suppose that we have two objects that are moving entirely with the expansion of the universe, without any "peculiar motion" relative to the overall flow. Then the scale factor a, at any given time, is proportional to their separation. It is standard to say that $a = 1$ now, which means that $a < 1$ in the past and a will be >1 in the future. The wavelength of photons, gravitational waves, and other relativistic entities in the universe is proportional to a; thus, as the universe expands, photons, and gravitational waves stretch by the same factor. This is often indicated by the redshift z, where $z = 0$ now and $z > 0$ in the past. Using the $a = 1$ now convention, $a = 1/(1 + z)$.
- **The evolution of anisotropies.** Inhomogeneity and anisotropy have increased with time because structure formation (mostly due to gravitational effects) has occurred and is occurring. One consequence is that when $a \ll 1$, the universe was very smooth and uniform. For example, at the era at which the cosmic microwave background (CMB) was emitted ($z \approx 10^3$ and thus $a \approx 10^{-3}$), the typical fractional deviation from the average density and temperature was $\sim 10^{-5}$.
- **The matter–energy content of the universe.** There are numerous matter–energy components to the universe, including photons, gravitational waves, neutrinos, electrons, baryons, dark matter, and dark energy. The energy density of these components evolves differently from each other as the scale factor changes. Relativistic components (such as photons and gravitational waves, and such as baryons when the universe was extremely hot) evolve as a^{-4} with the scale factor a. Nonrelativistic components (such as baryons now) evolve as a^{-3}. The energy density of dark energy, if it behaves as a cosmological constant, is constant: a^0. We can even define an effective density related to the spacetime curvature, which scales like a^{-2}. This means that the dominant component changes as the universe expands; now it's dark energy, but at redshifts $z \sim 1 - 4000$, it was nonrelativistic matter (mainly dark matter) and at yet higher redshifts it was radiation.
- **Hubble things.** At any given time, the rate of expansion can be expressed using the Hubble parameter, which is defined via $H = \dot{a}/a$. The value now, which is given the special symbol H_0, is called the Hubble "constant." This might suggest that its value does not change with time (because that's what "constant" means …), but that's misleading; instead, it means that at a given time, H is the same everywhere in space.
- **The critical energy density.** At a given time, H can be used to define a critical energy density $\rho_c \equiv 3H^2/(8\pi G)$, where G is Newton's constant. If the total energy density from all components other than the spacetime curvature equals ρ_c, then the universe is spatially Euclidean, in the sense (for example)

that the interior angles of even large triangles, measured at a given time, add to 180°. Current measurements indicate that the total energy density is consistent with being equal to ρ_c, and the smaller a was, the closer the equality was. We can therefore treat the early universe as if it had $\rho = \rho_c$ exactly.

- **The density parameter**. For a given component (e.g., dark energy, or baryons, or gravitational waves), it is convenient to indicate the ratio of the energy density in that component to the critical energy density, and this ratio is indicated using Ω. For example, if the energy density in dark matter is $\rho_{\rm DM}$, then $\Omega_{\rm DM} \equiv \rho_{\rm DM}/\rho_c$. Similarly, for gravitational waves, $\Omega_{\rm GW} \equiv \rho_{\rm GW}/\rho_c$ and so on. If we also define an equivalent energy density ρ_k for spacetime curvature, then the sum of the Ω values for all components is unity: $\Omega_{\rm DE} + \Omega_{\rm DM} + \Omega_{\rm bary} + \ldots + \Omega_{\rm GW} + \Omega_{\rm k} = 1$.

With this background, you now know cosmology! Well, not really. If you want (and you do want) to learn more, you should dive into textbooks that cover this topic in detail (again, see the suggestions at the end of this chapter). But for us, this will be enough to move forward with our discussion of gravitational waves in cosmology.

6.2 Cosmological Sources of Gravitational Waves

6.2.1 Generic Stochastic Backgrounds from the Early Universe

We will begin by thinking of generic possibilities, where rather than specifying a particular generation mechanism, we just imagine that at some phase in the universe gravitons are emitted, and we ask about their properties now. With that in mind, suppose that a graviton of frequency $f_{\rm GW*}$ is emitted when the scale factor of the universe is a_*. Then, today we observe the frequency to be $f_{\rm GW} = f_{\rm GW*}(a_*/a_0)$, where a_0 is the current scale factor (again, it is normal to set $a_0 = 1$). In the radiation-dominated era, the temperature simply scales as a^{-1}, so the current frequency of the graviton at our current temperature $T_0 \approx 3$ K is just T_0/T_* times the frequency at emission:

$$f_{\rm GW} \approx 10^{-13} f_{\rm GW*}\left(\frac{1\ {\rm GeV}}{k\ T_*}\right), \tag{6.1}$$

where T_* is the temperature at a_* and k is the Stefan–Boltzmann constant; the numerical factor is because $T \approx 10^{13}$ K when $kT \approx 1$ GeV. Note that a detailed calculation would have to take into account the full set of relativistic particles at scale factor a_* (which share in the energy, and thus, help determine the temperature).

What value of $f_{\rm GW*}$ should we take? Based on general causality considerations we know that the lowest frequency possible is the Hubble parameter H_* at that epoch. The frequency could be higher, though, so we'll take $f_{\rm GW*} = H_*/\epsilon$, with $\epsilon < 1$. In the radiation-dominated epoch of the universe (which as we noted before is valid for $z > 10^4$, which will be true for almost all processes we consider), the expansion rate of the universe scales as the square of the temperature, i.e., $H_* \sim T_*^2$. Thus, because $f_{\rm GW*} \propto H_* \propto T_*^2$ and $f_{\rm GW} \propto f_{\rm GW*}/T_*$, then $f_{\rm GW} \sim T_*$. With the constants put in,

$$f_{\rm GW} \approx 2 \times 10^{-8}\ {\rm Hz} \left(\frac{1}{\epsilon}\right)\left(\frac{kT_*}{1\ {\rm GeV}}\right). \tag{6.2}$$

If ϵ is not too much less than unity, this tells us the energy scale probed at a given present-day gravitational-wave frequency. For example, LISA (at $\sim 10^{-4}$ Hz) would probe the tens of TeV scale and ground-based detectors (at $\sim 10^2$ Hz) would probe the EeV scale. As we discuss in more detail below, particularly interesting temperatures might be $T_* \sim 10^2$ GeV corresponding to $f_{\rm GW} \sim 10^{-5}$ Hz (because this is the temperature expected for the electroweak phase transition, such that at lower temperatures, the electromagnetic force separates from the weak nuclear force), and $T \sim 10^2$ MeV corresponding to $f_{\rm GW} \sim 10^{-8}$ Hz (because this is the temperature expected for the quark–hadron phase transition, such that at lower temperatures quarks are bound in hadrons rather than being free).

6.2.2 Observational Limits on the Gravitational-wave Background

There exist some fairly strict limits on the energy density in gravitational waves from the early universe. The first we will consider comes from big bang nucleosynthesis (BBN). This is the very successful model that relates the overall density of baryons in the universe to the abundances of light elements. To understand the idea, we can think about different effects that occur as the universe expands and therefore cools and becomes more rarefied. At a high-enough temperature and density, the universe is a "soup" of quarks, gluons, photons, electrons, positrons, neutrinos, and possibly other fundamental particles depending on how hot it is. When the temperature and density drop enough (below about $kT_* \sim 150$ MeV), the quarks and gluons are bound into individual baryons (three-quark bags, of which the most familiar and by far the most numerous are protons and neutrons). However, at that temperature, the protons and neutrons cannot yet come together to form more complex nuclei because even if they bond temporarily, the energy of photons is high enough that the nucleus is photodisintegrated.

When the universe has cooled enough that photodisintegration is less common (starting when the temperature is $kT_* \sim 1$ MeV), light nuclei can form. In addition to ordinary hydrogen nuclei (which are just individual protons), deuterium, helium-3, and helium-4, and trace amounts of lithium and beryllium form at this stage. This is therefore the period of nucleosynthesis, or BBN because it is expected to have occurred shortly after the big bang, at redshifts of roughly 10^7–10^8. The Russian–American physicist George Gamow initially hoped that all the elements could be synthesized via this mechanism, but the bottleneck is that there are no stable nuclei with five or eight total protons plus neutrons, which means that combinations of hydrogen (the most common nucleus) and helium-4 (by far the most stable nucleus in this mass range) cannot lead quickly to heavier elements. After, characteristically, tens of minutes, any remaining free neutrons have decayed and BBN ceases. Extensive studies beginning in the 1970s demonstrated that the relative primordial abundances of the light elements depend almost entirely on the energy density of the universe and the number density of baryons.

What this means is that it is possible to put constraints on the energy density in gravitational waves at the time of BBN. If the energy density were too high, then the rate of expansion of the universe during this epoch would be inconsistent with what is needed to match observations of light-element nucleosynthesis. We can then project that limit to today, and we find that the constraint on the current energy density of gravitational waves is

$$\int_{f=0}^{f=\infty} d\ln f\ h_0^2\ \Omega_{\rm GW}(f)\ \lesssim 10^{-6}, \tag{6.3}$$

where the fraction of the critical density in a logarithmic frequency range $d\ln f$ is $\Omega_{\rm GW}(f)d\ln f$, and $h_0 \equiv H_0/100$ km s^{-1} Mpc^{-1}. Note that this is an integral constraint that applies to the overall gravitational-wave energy density. Normally, we would expect that gravitational waves from the early universe would have a broad frequency distribution, but in principle, we could imagine some strange situation in which $\Omega_{\rm GW}(f) \gg 10^{-6}$ in some narrow frequency interval. Depending on the situation then, this constraint can translate into an averaged bound on the gravitational-wave energy density throughout multiple frequency decades or a bound on the density in a narrow temperature range.

The second limit we will consider here on the energy density in gravitational waves from the early universe comes from the CMB. As we discuss in more detail below, gravitational waves would produce temperature fluctuations in the CMB and also contributions to the polarization of the CMB. The limit depends on the gravitational-wave frequency and is a minimum of $\Omega_{\rm GW}(f)h_0^2 \sim$ few $\times\ 10^{-16}$ at a frequency of $\approx 10^{-17}$ Hz (which is roughly the frequency corresponding to the peak amplitude of temperature fluctuations in the CMB, which maps to a current length of $\sim$100 Mpc). At higher frequencies, gravitational waves (which, recall, are metric perturbations) redshift away rapidly as the universe expands.

Yet another constraint comes from pulsar timing. As we have discussed in Chapter 4, a stochastic background of gravitational waves would be manifest as spatially correlated timing residuals between the pulsars included in a given network. The current bound from pulsar timing is $\Omega_{\rm GW} \lesssim 10^{-10}$ in the frequency range $\sim 10^{-8}$–10^{-7} Hz accessible with these observations. It is entirely possible that by the time you read this book, pulsar timing arrays have detected gravitational waves, but this discovery is not to be confused with what we are talking about here. The most likely pulsar timing discovery will be of a background of gravitational waves produced by supermassive black hole binaries and not of a cosmological source.

6.2.3 Possible Source 1: Quantum Fluctuations from the Era of Inflation

The current consensus is that the seeds of structure (i.e., the inhomogeneities that, through gravitational interactions, became clusters, galaxies, stars, etc.) were quantum fluctuations in the early universe. These fluctuations are believed to have been stretched out by a period of "inflation," during which the scale factor increased exponentially by at least a factor of 10^{26}. Among other consequences, this flattened out the spatial geometry of the universe (so that Ω_k was practically zero after inflation) and also meant that a large portion of the universe that had been in causal contact (which in practical terms meant that it was able to interact and come into equilibrium) was spread out and no longer in causal contact. This idea addresses the "flatness problem" (why is the spatial geometry of the universe so close to Euclidean?) and the "horizon problem" (without such a period of inflation, different parts of the CMB wouldn't have had time to interact and thus there is no reason that it would be so uniform in temperature). It also explains why we don't have magnetic monopoles, because even if they existed prior to inflation they would have been spread out by inflation and the temperature after inflation would have been too low to regenerate them.[1]

A prediction of inflation, although unfortunately one that is extremely accommodating in the sense that upper limits don't have much impact, is that gravitational waves would have been generated in the inflationary epoch. The hoped-for signature of these gravitational waves is in the polarization field of the CMB. Briefly, scattering of photons off of electrons generically leads to polarization after scattering. If the scattering happens before the universe becomes essentially transparent to photons (which happens when the universe cools enough that electrons pair with nuclei rather than moving around freely), then lots of scattering leads to negligible net polarization. However, if in the period when the universe is becoming transparent to photons there is a quadrupolar temperature asymmetry, so that photons from some directions are more common than from other directions, scattering can lead to a net polarization. In practice, this means that sensitive detectors can see what is effectively a vector field of polarization on the sky, where each direction has a net linear polarization of the photons.

You may remember that if you have a vector field (think of a field of arrows) on a two-dimensional surface such as a plane or (relevant for us) a spherical surface, you can decompose it into a contribution that has a divergence but no curl (this is called the E mode) and a contribution that has a curl but no divergence (this is called the B mode). It turns out that although simple density perturbations in the early universe (called scalar perturbations) can only produce E-mode polarization, primordial gravitational waves can produce both E-mode and B-mode polarization.

Captain Obvious: Why is that? The density perturbation is a function of time and spatial coordinates, which we call a "scalar" like a temperature field, so there is only one direction that it cares about: the direction in which it changes the most. And before Major Payne chimes in, yes, this is the gradient of the scalar. The polarization of the CMB effectively "latches" onto this special direction. We call these E modes because of how they transform under a parity transformation (essentially like the electric field transforms in electromagnetism).

Gravitational waves are also perturbations (in this case, of the fabric of spacetime), but rather than scalars they are "metric waves"! Indeed, as we've discussed before, gravitational waves are metric perturbations that come in two (transverse-traceless) polarizations

[1] Although, as Martin Rees has pointed out, because we don't know whether magnetic monopoles exist at all, a mechanism to get rid of them might not be the most convincing driver of a hypothesis.

in general relativity (e.g., + and ×). Therefore, the polarization of the CMB doesn't have to latch onto the gradient of the gravitational-wave modes, as their direction of propagation also matters. Gravitational waves can therefore "scramble" the CMB polarization producing not just E modes but also B modes (again, modes that are even under a parity transformation, like magnetic fields in electromagnetism).

Thus, if other contaminating factors can be eliminated, a detection of B-mode polarization would be a signature of gravitational waves. In turn, this would give us a unique tool to understand better the physics of inflation.

The caveat is that other contaminating factors can be significant. In 2014, the BICEP-2 team announced the detection of B-mode polarization from the CMB, which could have been produced by inflation-era gravitational waves. However, it was subsequently discovered that the signal was almost certainly produced by scattering off of Galactic dust (which as we noted before is a four-letter word in astronomy). Another contaminating influence that can generate B modes in the more recent universe is gravitational lensing. Notwithstanding these complications, there are numerous current experiments that are focused on the characterization of polarization from the CMB, in part to search for B modes. There is no consensus on the expected amplitude of the B mode, which means that nondetections will not conclusively rule out the idea of inflation.

6.2.4 Possible Source 2: Phase Transitions in the Early Universe

As we already mentioned, there are phase transitions in the early universe that might produce gravitational waves. But unfortunately the best current guess is that these gravitational waves will be extremely weak. To set the stage and to understand this dispiriting conclusion, we need to talk a bit about phase transitions.

A familiar example is the freezing of liquid water. This takes the water from one state to another and does so in such a way that there is a discontinuous change in a thermodynamic variable: the density of ice is less than that of liquid water just above freezing. This discontinuity is the hallmark of a first-order phase transition.

But not all phase transitions are first order. There are many circumstances in which the transition to a new phase occurs gradually, without any discontinuous changes; these are often called "crossovers."

The reason this matters is that, for the production of significant amounts of gravitational waves in an early-universe phase transition, the transition has to be abrupt, i.e., it needs to be first order. Otherwise, the change to the new state as the universe expands is smooth and gradual. The bad news is that, based on calculations using the standard model of particle physics, both the electroweak transition and the quark–hadron transition are crossovers and are therefore not expected to produce much in the way of gravitational waves.

The glass-half-full perspective on this is that if, contrary to expectations, there is compelling evidence of gravitational waves from these epochs, that will be a major discovery that indicates new physics. Let's adopt this perspective and consider how, qualitatively, gravitational waves would be generated if the transitions are first order.

A key point is that a phase transition does not happen everywhere at once. Instead, based on random chance, the transition starts at some seed points. It then expands outward from those seed points, in a way that (given the extremely homogeneous state of the early universe) is very close to spherical, just like a bubble. However, as we know, spherical symmetry means no gravitational waves. The production of gravitational waves then comes down to three possible mechanisms:

1. **Bubble collisions**. If two expanding bubbles collide with each other, then spherical symmetry is broken at the interface. This is a mechanism that would even work if there were no fluid between the bubbles, i.e., in a vacuum.
2. **Generation of turbulence**. Because there is, in fact, fluid between the bubbles (which is by assumption in the previous state), the expansion of the bubbles can generate turbulence. Turbulence does not have spherical symmetry and therefore can generate gravitational waves.
3. **Collision of acoustic waves**. An alternate definition of a first-order phase transition is that it is one that releases latent heat; for example, freezing of liquid water releases energy when the liquid becomes solid, and that energy has to be removed from the system to complete the freezing. As the bubble expands, this latent heat drives the bubble expansion to be faster, which means that significant sound waves can be produced, and those can collide with the sound waves from other bubbles. As before, this breaks the spherical symmetry and can lead to the generation of gravitational waves.

No gravitational waves have been detected that would correspond to the electroweak or the quark–hadron phase transition. This nondetection is expected if, as currently believed, the transitions are crossovers. However, so that we don't get too depressed about the prospects for this epoch of the universe, we can return to our earlier comment that periods of phase transitions may be favorable for the production of primordial black holes. This is because phase transitions lead to states that have lower energy density than would the original state given the external conditions. Thus, if there is a density perturbation of sufficient amplitude and scale during a period of a phase transition, the contraction of the perturbation will be less opposed by increasing pressure gradients during a phase transition than it would during a generic portion of the history of the universe.

Captain Obvious: In the early universe, when matter and energy were distributed nearly homogeneously and when (as indicated earlier) the mass–energy density of the universe was close to the critical energy density, we can determine the most likely scale for the collapse of a region with greater than the average density. We start by noting that the larger the mass is, the easier it is for some fractional overdensity to collapse; this is essentially because it's then easier to overcome temperature, turbulence, etc., which would tend to expand the overdensity, and thus, smooth it out. Therefore, we can look at the largest scale possible, which is the horizon scale. A time t after the big bang, the horizon scale is $r_H = ct$. The mass contained in a sphere of radius r_H is then $m_H = (4\pi/3)r_H^3\rho_c$, where we are assuming that the average density is the critical density $\rho_c = 3H^2/(8\pi G)$ for the Hubble parameter $H = \dot{a}/a$. That is, the mass in a horizon volume is $m_H = 1/(2G)(\dot{a}/a)^2(ct)^3$. Thus, the horizon radius in gravitational units is $r_H/(Gm_H/c^2) = 2[(\dot{a}/a)^2t^2]^{-1}$. It can be shown that in the radiation-dominated portion of the universe (which lasts from the big bang to a redshift of about 5000), $a \propto t^{1/2}$ and thus $\dot{a}/a = 1/(2t)$, so that gives us $r_H = 8Gm_H/c^2$.

Given that for a nonrotating black hole the event horizon radius is $r = 2Gm/c^2$, that means that r_H is in the vicinity of the radius of a black hole with mass about m_H. This rough correspondence has led a surprising number of science popularizers to gush that this basically means that we are living in a black hole. Nonsense! A black hole has a lot of stuff in one place and much less stuff around it. The early universe is nearly homogeneous, so it's not like a black hole at all; matter is not drawn to a particular point. Indeed, it is reminiscent of the classic philosophical quandary of Buridan's donkey, which is situated precisely between two equally enticing piles of hay and dies of hunger as a result. But if there is extra mass in some region, it expands more slowly than the universe as a whole and therefore increases in relative density until it finally collapses. If the overdensity is large enough, it might collapse into a black hole. Therefore, as a first guess, we can think that if collapses to primordial black holes do occur, they tend to occur at a mass that is roughly equal to the horizon mass at the epoch of formation.

6.2.5 Possible Source 3: Cosmic Strings

Yet another exciting but speculative cosmological source of gravitational radiation is cosmic strings. If these objects exist, they are large, effectively one-dimensional (the length could be astronomical; the thickness would be less than that of a proton) strings that would have enormous mass per length μ (more on this below). One suggested origin comes from a phase transition shortly after inflation; the suggested analogy is to the cracks that form when liquid water freezes into ice, which makes this a "topological defect." It has also been proposed that fundamental strings (a.k.a. superstrings) could have been stretched during inflation to the size of a galaxy.

Cosmic strings are expected to form cusps (very sharp bends along their length) or kinks (self-intersections or intersections with other cosmic strings). Because cosmic strings are expected to have extremely high tension along their length, cusps or kinks would move close to the speed of light. As a result, they would produce gravitational radiation that would be (1) highly beamed and (2) spread out over a large range of frequencies. The consequence is that gravitational waves from cosmic strings are expected to produce, effectively, a stochastic background that has an amplitude that is a power law over a wide range of frequencies. Current limits from pulsar timing arrays are $G\mu/c^2 \lesssim 10^{-11}$, which implies $\mu \lesssim 10^{17}$ g cm^{-1}. This is still a fantastically high linear mass density; it means that a segment of such a string that spanned the solar system (with an assumed size of $\sim 10^5$ au) would have a mass of $\sim 100\ M_\odot$!

If any of these scenarios bears fruit, and in fact, there is a cosmological background of gravitational waves detected with planned instruments, this will obviously be fantastic news. However, what if it isn't seen? That won't be a surprise, but there has been discussion about missions to go after weaker backgrounds. It is often thought that the 0.1–1 Hz range is likely to be least "polluted" by foreground vermin (i.e., the rest of the universe!). Either way, whether we see a background or "merely" detect a large number of other sources, gravitational-wave astronomy has wonderful prospects to enlarge our view of the cosmos.

6.3 Measuring the Universe Using Gravitational Waves

Distance measurements have led to a surprising number of revolutions in our understanding of the universe. From the realization that the universe is much larger than our solar system, to the understanding that our Galaxy is but one of many, to the revelation that the universe expands and, later, that its expansion is accelerating, distance measurements have often been fundamental. At the most precise level, puzzles will always exist. For example, as of the writing of this book there is an intriguing difference between the measured value of the Hubble constant using different methods, which span the range from ≈ 67 km s^{-1} Mpc^{-1} to ≈ 74 km s^{-1} Mpc^{-1}, with individually quoted uncertainties much smaller than 7 km s^{-1} Mpc^{-1}. It is a reasonable bet that these differences will ultimately be resolved by better understanding of the systematic errors inherent in each method. However, there is also a chance that the different results stem from new physics. As a result, electromagnetic observations of many types are improving their samples and their precisions, so within a few years, we might get clarity about the differences.

Into this mix, we can now throw gravitational-wave measurements. In this section, we will talk about two ways that distance measurements using gravitational radiation can help us measure the universe. First, they can provide new measurements of the Hubble constant. Second, when gravitational-wave sources at cosmological distances (redshifts not much less than 1) can be measured regularly they will help us track down the mystery of dark energy.

6.3.1 Measuring the Hubble Constant Using Gravitational Waves

Recall that the Hubble constant is the Hubble parameter $H = \dot{a}/a$ measured at the current epoch. The Hubble–Lemaître law says that, locally, the redshift is linearly proportional to the distance D, namely $z \sim H_0 D$. Thus, to get an estimate for the Hubble constant, we need one or a set of observations for which we have both the redshift and the distance. With electromagnetic observations, it is straightforward to measure the redshift if the host galaxy can be identified. That's because it is comparatively easy to identify atomic lines in electromagnetic spectra, after which a comparison of the observed wavelength with the known rest-frame wavelength (known, because of atomic physics and measurements in laboratories) gives the redshift. Redshifts are commonly obtained in this manner to fractional precisions of 10^{-3} or better, although obviously the challenge is greater for fainter sources.

Major Payne: It again falls to me to explain things properly. As you will see below, the authors are, as usual, being sloppy when they simply talk about "distance" rather than being more specific. Distance is a unique concept on small scales, but at the scales of cosmology, there are different types of distance depending on the application. One of the most commonly used for gravitational-wave applications is the "luminosity distance", which is often written as D_L. This is defined as the distance in a Euclidean universe between us and an isotropic source of bolometric luminosity L such that we would measure a bolometric flux F. That is, $D_L = [L/(4\pi F)]^{1/2}$. If the source is close enough that redshift can be neglected then this is the standard "distance" with no ambiguity. But because light redshifts, at cosmological distances the flux decreases faster than it would in a Euclidean universe; thus, D_L increases without limit as the redshift increases. As an example of a different distance measure, we can consider the angular diameter distance, often written as D_A. This is the distance in a Euclidean universe between us and an object of known transverse size such that the latter appears to have a given angular diameter. At small distances, D_A is very close to D_L, but at large enough redshift D_A actually decreases with increasing redshift. You can motivate this by thinking about an extreme example: a meter stick that emitted light when the universe was two meters in diameter would appear to take up half the sky! Thus, cosmologists usually specify the type of distance they mean.

In contrast, the distance to electromagnetic sources is more challenging to measure and involves the idea of a "distance ladder". We can start with direct measurements of distance on the scale of the solar system; historically this involved measurements of distances on the Earth followed by the use of, e.g., parallax from different spots on Earth during transits of Venus across the Sun's disk, but now we rely on radar measurements of the distances to planets. The next step out involves parallax; because of Earth's motion around the Sun, the direction to closer stars appears to shift more, annually, than the direction to farther stars, and thus we can get a geometric measurement of the distances to stars. But this only allows us to measure distances to partway out in our Galaxy, i.e., to kiloparsecs at most; on cosmological scales of megaparsecs to gigaparsecs, we need something different.

There are a number of techniques that have been used, but the workhorse in cosmology is "standard candles." These are sources whose intrinsic luminosity can be estimated from their properties. For example, more massive stars have lower average densities than less massive stars, which means (as we discussed earlier) that their oscillation frequency is lower. More massive stars are also more luminous than less massive stars. Some massive stars (Cepheid variables are the best-known examples) oscillate, which means that the flux we receive from them also oscillates. If the oscillation period is longer then their intrinsic luminosity is greater, in a way that has been calibrated by observing Cepheid variables with distances established by parallax (i.e., the closest ones). Ironically, therefore, variable stars can make good standard candles! This is essentially because Cepheids (and their variable friends) are (1) very luminous, which means that we can see them far away (to several tens of megaparsecs), and (2) distinguishable from other stars because they are considerate enough to change regularly in flux.

That's the distance ladder: from directly measured distances to parallax to standard candles. A measurement of the flux F from a standard candle of inferred luminosity L gives the distance D (more precisely, the luminosity distance, as Major

Payne reminded us) because $F = L/(4\pi D_L^2)$ if we know L. However, just as standing on a ladder becomes more precarious on higher rungs, the distance ladder becomes more susceptible to error when the rung is farther removed from direct measurement. For example, new and more precise measurements of parallaxes shift the entire scale (this happened in the 1990s with the parallax mission Hipparcos). On top of that, if there are better measurements and understanding of, e.g., Cepheid variables, then distances based on them have an additional shift. And when we want to measure distances of hundreds of megaparsecs to gigaparsecs, which are beyond the reach of Cepheids, we need to use brighter sources (such as a particular type of supernova) whose standard "candleness" needs to be established using distances based on Cepheids and other sources. In summary, distances in electromagnetic astronomy are hard.

For gravitational-wave sources such as binaries, in contrast, the luminosity distance (modulo the uncertain orientation of the binary) comes directly from their amplitude. It has therefore been pointed out that in this sense binaries are self-calibrated sources of gravitational radiation, that is, they are standard sirens! Indeed, as we have mentioned before, in 2017, the LIGO/Virgo collaboration detected what is most probably a binary neutron star inspiral and merger, and shortly after, the Fermi telescope (and then many other telescopes) detected electromagnetic emission. This allowed for the electromagnetic localization of the Galaxy from which the gravitational wave originated, which gave us the redshift. The amplitude of the gravitational wave then gave us the first measurement of the Hubble constant with gravitational waves: $H_0 = 70.0^{+12.0}_{-8.0}$ km s^{-1} Mpc^{-1}. Cosmology with gravitational waves is already a reality.

But of course, with great power comes great responsibility. Gravitational-wave cosmology is challenging for a number of reasons. The main challenge probably is the identification of the host galaxy, which, as in the case of GW 170817, would yield the redshift, and thus, a measurement of the Hubble constant. A sufficiently close and well-localized source can have a clear galactic host, but more distant and/or less well-localized sources will pose greater challenges. How many such events will we detect in the future? This remains highly uncertain. It seems to have been a bit of a coincidence that the first gravitational-wave observation of merging neutron stars was accompanied by coincident electromagnetic observations. This occurred during the second observing run of LIGO and Virgo, which lasted roughly 9 months (with a $\sim$50% duty cycle—the fraction of time all interferometers were kept in lock). The third observing run lasted roughly 12 months with a slightly higher duty cycle, yet no electromagnetically coincident events were observed, although one other binary neutron star event was detected.

Another challenge is the binary's orientation. The inclination angle, the location of the source in the sky, and the polarization angle all affect the gravitational-wave amplitude, and none of them are known a priori. Indeed, what these quantities are matters a lot, because the amplitude of the gravitational radiation can be twice as large for systems that are "face on" (with the orbital angular momentum pointing along the line of sight) than "edge on" (with the orbital plane parallel to the line of sight). The polarization signature is very different in the two cases—face on gives circular polarization whereas edge on gives linear polarization—but with the current gravitational-wave network, polarization cannot be measured precisely. Without precise knowledge of these quantities, one must marginalize over them (see Chapter 4 and Appendix A), which then inflates the uncertainty in the measurement of the Hubble constant. The future situation will naturally be improved, but for now, electromagnetic counterparts help substantially to constrain the orbital inclination and the location of the source in the sky.

To get a sense for the challenge involved in tracking down a counterpart, suppose that we have a gravitational-wave event that has excellent localization: a solid angle of just $\Delta\Omega = 30\ \mathrm{deg}^2$ (or about 0.01 sr). Suppose also that we can measure its distance to be (200 ± 20) Mpc, which is also outstanding precision. Then, the uncertainty volume is $\Delta V = \Delta\Omega r^2 dr \approx 0.01(200\ \mathrm{Mpc})^2(40\ \mathrm{Mpc}) \approx 1.6 \times 10^4\ \mathrm{Mpc}^3$. The equivalent number density of galaxies like our own is $\sim 0.01\ \mathrm{Mpc}^{-3}$, so we'd expect more than a hundred galaxies in this quite favorable case, and many more in more typical cases; and that assumes that the source is near a galaxy. So where do we point our telescopes to in order to find the electromagnetic counterpart? Some telescopes do have a wide field of view, but the wider the field, the less far away you can see!

And even if the source was at exactly 200 Mpc and we knew where to look in the sky, there is no guarantee that we would be able to see the electromagnetic signal! Whether we can see it or not depends on how deep our telescopes can see, whether we can point the telescope to that patch of the sky in time to catch the signal, and whether when we finally get there, the signal is in the right wavelength range for our telescope (e.g., different phases of the postmerger will emit primarily in different frequency bands). Longer wavelengths emerge later, with radio waves being visible for years after, so the key is to be able to determine which object is the counterpart to the gravitational-wave signal. We do sometimes get lucky: GW 170817 was at just 40 Mpc, the counterpart was found rapidly, and then it seemed as if every telescope on the planet pointed to that patch in the sky. Double neutron star mergers might occur roughly yearly at distances $\lesssim 100$ Mpc, so simple patience may be all that is necessary. There are also discussions of how to use statistical correlations of gravitational-wave sources with large-scale maps of galaxies to obtain an estimate of the Hubble constant even without electromagnetic counterparts.

However, it is also important to determine what precision is needed so that gravitational-wave measurements of the Hubble constant can contribute significantly to our understanding of distances. If the purpose is to help judge between values in the $H_0 = (67\text{–}74)\ \mathrm{km\ s^{-1}\ Mpc^{-1}}$ range, then the answer is that a precision of $\sim 1\%$ is needed. The measurement of H_0 from GW170817 alone had a precision of $\sim 10\%$. It is therefore tempting to conclude that $\mathcal{O}(100)$ measurements are needed because the uncertainty in the population estimate should scale like the inverse square root of the number of measurements.

But this argument is flawed. It would be correct if every measurement involved only statistical uncertainties that were uncorrelated between different measurements. In reality, however, systematic errors cannot be treated in this way. Indeed, a lesson learned repeatedly by astronomers is that when greater precision is obtained, small errors that were previously ignorable become bottlenecks to further progress. An example of a possibility in gravitational-wave measurements is the calibration of the instruments. If, for example, the overall calibration is off by 2% (which is below the current level to which the calibration is known) then all amplitudes will be off by that same 2%. Thus, the reduction of systematic errors will be key in the future to do precision cosmology with gravitational waves and, for that matter, also to do precision tests of general relativity.

6.3.2 Probing Dark Energy Using Gravitational Waves

As we heard from Major Payne, unlike the situation at $z \ll 1$, for which the luminosity distance at a given redshift is determined solely by the Hubble constant via $D_L \approx z/H_0$, for larger redshifts, the luminosity distance depends on other cosmological parameters. Suppose that we use Ω_M to denote the fraction of the critical density in nonrelativistic matter (including dark matter), Ω_Λ, to indicate the fraction of critical density in a cosmological constant, and we ignore other contributions (e.g., from curvature, photons, neutrinos, and gravitational radiation). Then, $\Omega_M + \Omega_\Lambda = 1$ and defining

$$E(z) = \sqrt{\Omega_M(1+z)^3 + \Omega_\Lambda}, \tag{6.4}$$

the luminosity distance to a source at redshift z is

$$D_L = D_H(1+z)\int_0^z \frac{dz'}{E(z')}, \tag{6.5}$$

where $D_H \equiv c/H_0$.

The point is that if it is possible to measure both the luminosity distance and the redshift to sources with z not much less than unity, it is possible to infer Ω_M and Ω_Λ. Indeed, this is what led to the 1998 announcement that the expansion of the universe is accelerating (and to the 2011 Nobel Prize in Physics, which was awarded to Saul Perlmutter, Adam Riess, and Brian Schmidt). With more and better data, we can even go beyond this. If we suppose that there is some "fluid" that acts like dark energy, then for a cosmological constant, the fluid's equation of state is $p = -\rho$.

Dr. I. M. Wrong: Wait what? A constant is a constant that is a constant, so in particular, it is not a fluid! What are the authors doing!? They must have lost it...

Captain Obvious: Not so fast! Indeed, there is an analogy between the cosmological constant and a perfect fluid, and it goes like this. In quantum mechanics, we've learned that everything has a minimum energy state (if it is to be stable), and we called that minimum the "vacuum state." Spacetime, then, may also have a minimum energy density, which we will call "vacuum energy" for short. Moreover, spacetime can be thought of as an entity that can be continuously deformed by matter and that produces undulations that "flow" when this matter accelerates. If you think about it, this is essentially a fluid by definition!

Ok, so let's think of spacetime as a fluid whose minimum energy is this vacuum energy, but does it have a pressure? Obviously, not all fluids have a pressure; think of collisionless dust if you want, as an example. But many fluids are idealized as having "isotropic" pressure (i.e., pressure that is the same along any spatial axis) and as not having any other fancy properties, such as shear stresses, viscosity, or heat conduction. Let us then try to model spacetime as such a "perfect" fluid.

To figure out whether our "spacetime fluid" has a pressure, we can resort to our knowledge from thermodynamics. You must know from your earlier studies that the first law of thermodynamics is $dU = dQ - dW$, where dU is the change in the energy of the system, given an amount dQ of heat transferred, and an amount dW of work done. For our fluid, there is no heat transfer (it's perfect!), so the first law becomes $dU = -pdV$, which means that given a change in volume dV at a pressure p, the energy changes by dU. And here comes the kicker: how much vacuum energy there is in a given volume increases as the volume increases (simply because for the vacuum $U = \rho V$). Assuming that ρ is approximately constant as work is being done, then we have that $dU = \rho dV = -pdV$ and therefore $p = -\rho$! Incidentally, when the equation of state is $p = -\rho$, the Einstein equations then predict exponential expansion (which is indeed consistent with an accelerating universe). This is why people sometimes say that dark energy is a perfect fluid with a cosmological equation of state defined by the parameter $w \equiv p/\rho = -1$.

However, dark energy doesn't have to obey precisely this relation. An example generalization is $p = w\rho$, where $w = -1$ gives a cosmological constant and $w = 0$ would give a pressureless fluid, which is a good approximation to dark matter.

Second-generation gravitational-wave detectors such as LIGO, Virgo, and KAGRA are unlikely to be able to probe the universe to the depth required to constrain w. Third-generation detectors, which are anticipated to be 10–20 times more sensitive than second-generation detectors, might be able to reach far enough. At those distances, it might not be possible to identify the host galaxy electromagnetically, because there would be too many galaxies in the error volume, although methods have been proposed that take advantage of statistical associations or hoped-for features in the mass distributions of black holes or neutron stars. Moreover, at such distances, it is far from guaranteed that electromagnetic counterparts would be visible at all!

Most hope along these lines have therefore been pinned on observations with LISA of the coalescence of black hole binaries where both components have masses $\sim 10^5$–$10^7\ M_\odot$. Unlike stellar-mass black holes, these could carry with them substantial accretion disks that could light up in various ways as the holes approach each other. Numerous suggestions exist for what might be seen before, during, or after the merger, but the majority of them effectively ask the question "could we see those photons if we were looking directly at the source?" A potentially serious difficulty is that supermassive black holes with substantial accretion disks are active galactic nuclei (AGN), and AGN vary over a large range of frequencies, wavelengths, and amplitudes. Thus, rather than worrying about whether an associated electromagnetic signature is detectable, the concern could be that it is not distinguishable from the natural variation of AGN.

If conclusive association is possible, there is one more barrier to surmount. Gravitational waves, like light, can be gravitationally lensed—that is, the flux of gravitational waves that we receive could be greater than or less than the average value, depending on their specific path through and past galaxies and dark matter halos. This means that multiple otherwise identical events at identical redshifts would arrive at our detectors with a spread in fluxes, and thus, a spread in the inferred luminosity distance. Some work can be performed to model the likely effect of gravitational lensing along the path, but uncertainties in the masses of the intervening halos make this imperfect. If there were an enormous number of sources of this type, then it would be possible to perform a statistical analysis using the known probability distribution of magnification by gravitational lensing. However, the expected event rate is not large, so this might not be straightforward.

With luck, however, gravitational lensing could provide a tremendous benefit to gravitational-wave studies. On rare occasions (probability of a few percent at a redshift of a few and decreasing significantly at lower redshifts), a source will be multiply imaged by gravitational lensing. This means that we would get to see an event twice; this already happened electromagnetically, for example, with supernova Refsdal (named after Sjur Refsdal), which was observed first in 2014 and then a second time in 2015. The benefits of such a reobservation are that it would be possible to predict in advance where and when (for example) a binary coalescence would occur and possibly to stare deeply at that location at the appointed time to look for electromagnetic counterparts.

6.4 Exercises

1. In this problem, we will derive the evolution of the momentum of a freely streaming particle as the universe expands using the following hints:
   (a) Hubble's Law implies that for two points, both moving with the Hubble flow, separated by a small distance dr, the relative speed that they measure is
   $$dv = \left[\frac{(da/dt)}{a}\right]dr. \tag{6.6}$$
   (b) Because dr is small, we know that dv is small relative to the speed of light c.
   (c) The three-momentum p transforms with this small speed as
   $$p' = \gamma(dv)(p - E dv/c^2). \tag{6.7}$$
   (d) Because the Lorentz factor γ differs from 1 by $\mathcal{O}[(dv/c)^2]$ and dv/c is small, we set $\gamma = 1$.

(e) For massless particles $E = pc$, and for particles with rest mass m, $E = \sqrt{m^2c^4 + p^2c^2}$ and $p = \gamma_u mu$, where $\gamma_u = (1 - u^2/c^2)^{-1/2}$.

(f) Combined, all of this means that $dp \equiv p' - p$ can be written as

$$dp = -(E/c)(dv/c). \tag{6.8}$$

(g) Finally, we note that if our particle has a speed u relative to the Hubble flow ($u = c$ for a massless particle such as a photon or graviton, $u < c$ for a massive particle), then the elapsed time is $dt = dr/u$ and thus $dr = udt$ (we are assuming that $dv \ll u$).

Put all of this together to determine how the momentum p evolves with a for a massless particle and for a massive particle. What does this imply about how the temperatures of free-streaming (1) massless particles, and (2) very nonrelativistic particles evolve with the expanding universe?

2. Consider a merger of two black holes of arbitrary masses and spins. Suppose that the merger takes place at an unknown redshift z. Show that without knowing z, the waveform of the inspiral/merger/ringdown (meaning the observed frequencies, but not the amplitudes) is not sufficient to measure the rest-frame masses or angular momenta of the black holes uniquely. What combinations of masses, spins, and redshifts can be measured?

3. Suppose that the rate density of mergers between two 20 $M_\odot$ black holes is 20 $\text{Gpc}^{-3}\ \text{yr}^{-1}$.
   (a) Compute Ω_{GW} from these black holes as a function of frequency, from 10^{-4} Hz to 100 Hz.
   (b) Repeat this calculation for mergers of two 1.4 $M_\odot$ neutron stars, which we assume occur at a rate density of 500 $\text{Gpc}^{-3}\ \text{yr}^{-1}$.

Hint: it will be useful to know that when we apply Equation (2.21) to a population of merging binaries with chirp mass $\mathcal{M}$ and present-day comoving number density of remnants N_0, then

$$\Omega_{\text{GW}} = \frac{8\pi^{5/3}}{9}\frac{1}{c^2H_0^2}(G\mathcal{M})^{5/3}f^{2/3}N_0\langle(1+z)^{-1/3}\rangle, \tag{6.9}$$

where $\langle(1+z)^{-1/3}\rangle$ is the average, over the population, of $(1+z)^{-1/3}$, where z is the redshift of the mergers. This has a very weak dependence on the redshift evolution, so you can assume that the average is 1.

4. Very roughly, we can expect Milky Way–sized galaxies to have a major merger (i.e., a merger with a galaxy of comparable mass) once per 10^{10} years. Say that those galaxies contain black holes of mass $\approx 10^7\ M_\odot$ and have an effective number density of 0.01 Mpc^{-3}. Integrate out to a redshift $z = 1$ to estimate Ω_{GW} from these black holes at a gravitational-wave frequency $f_{\text{GW}} = 1/\text{year}$. Your answer should be well below the best estimate $\Omega_{\text{GW}} \approx 10^{-8}$. What additional considerations might increase the estimated background?

Useful Books

Choquet-Bruhat, Y. 2015, Introduction to General Relativity, Black Holes and Cosmology (Oxford: Oxford Univ. Press)

Freedman, W. 2004, Measuring and Modeling the Universe (Cambridge: Cambridge Univ. Press)

Hawking, S. W., Ellis, G. F. R., Landshoff, P. V., et al. 1975, The Large Scale Structure of Space-Time (Cambridge: Cambridge Univ. Press)

Hawley, J. F., & Holcomb, K. A. 2005, Foundations of Modern Cosmology (Oxford: Oxford Univ. Press)

Kolb, E., & Turner, M. 1994, The Early Universe (Boca Raton, FL: CRC Press)

Maggiore, M. 2018, Gravitational Waves: Volume 2: Astrophysics and Cosmology (Oxford Univ. Press: Oxford)

Peacock, J. A. 1998, Cosmological Physics (Cambridge: Cambridge Univ. Press)

Peebles, P. J. E. 1993, Principles of Physical Cosmology (Princeton, NJ: Princeton Univ. Press)

Ryden, B. 2016, Introduction to Cosmology (Cambridge: Cambridge Univ. Press)

Thorne, K. S., & Blandford, R. D. 2017, Modern Classical Physics: Optics, Fluids, Plasmas, Elasticity, Relativity, and Statistical Physics (Princeton, NJ: Princeton Univ. Press)

Weinberg, S. 1972, Gravitation and Cosmology: Principles and Applications of the General Theory of Relativity (New York: Wiley)

Chapter 7

Gravitational Waves and Nuclear Physics

Neutron stars are the compact remnants of massive stars whose cores have collapsed because they could no longer successfully oppose gravity. All neutron stars with precisely measured masses are in the range $\sim(1.2\text{–}2.1)\ M_\odot$; their radii have not yet been measured with much precision, with estimates of $\sim(10\text{–}15)$ km to be broad minded ((11–13) km is a more realistically probable range).

This combination means that the average density of a neutron star is greater than that of an atomic nucleus; for perspective, this means that at neutron star densities, you could cram all ~8 billion people on Earth into a teaspoon! The rotation frequencies of neutron stars are up to ~700 Hz (and down to once per several minutes or even less), which means that some spin faster than a kitchen blender. Neutron stars have the strongest persistent magnetic fields in the universe, with some estimated fields exceeding 10^{15} G. For comparison, Earth's magnetic field at its magnetic poles is about 0.5 G, the strength at the surface of a refrigerator magnet is about 10^2 G, and the strength of the field in an MRI machine is up to about 3×10^4 G.

All these extremes mean that neutron stars are attractive to study for people who want to push the envelope of fundamental theories of gravity, magnetic fields, and high-density matter. In this chapter, we discuss some of the basics of neutron stars (how do we know they exist? how are they formed?) and then talk about the contributions that can be made using gravitational-wave observations.

7.1 Basics of Neutron Stars

7.1.1 How Do We Know That Neutron Stars Exist?

As we mentioned, the extreme properties of neutron stars (gravity, density, and magnetic field) make them excellent objects for studies of how physics operates in unusual environments. However, until the 1960s it was not clear that they would ever be observed.

Then, Jocelyn Bell-Burnell, a graduate student at the University of Cambridge who was advised by Cambridge Professor Antony Hewish, constructed a radio array

doi:10.1088/2514-3433/ac2140ch7

with the help of several other members of Hewish's team. In the fall of 1967, Bell-Burnell saw a strange marking on one of the recording charts: a bit of "scruff" that was distinct from the usual types of noise that she had seen before. Hewish suggested that she put a high-speed recorder on the instrument, and when she did, incredibly, the scruff was resolved as a series of very regular pulses with an interval of about 1.3 s. After eliminating artificial sources, it was clear that this source and others like it were natural, coming from somewhere in the heavens. But what could cause this?

We know now that they are rotation-powered neutron stars, but it is instructive to go through the chain of logic leading to that conclusion. By 1968, there were four important facts established about pulsars:

1. The pulsation periods P were short: between 0.033 s and 3 s.[1]
2. Substructure within the pulses could be even shorter, <0.01 s in some cases.
3. They were extremely stable, with $P/\dot{P}$ equal to thousands to millions of years.[2]
4. Over time, their intrinsic periods always increased slightly and never decreased.

Let's consider what these facts imply, following a line of argument due to Thomas Gold, back then a professor at Cornell University. The fast periods and short pulses imply that whatever generated them must have been something small. Barring relativistic beaming, the size of the object would have to be $D < ct$, if t is the duration of the pulse. This is because the width of any given pulse cannot be larger than the light-travel time of the object that generated the pulse in the first place! That gives $D < 3000$ km for a pulse width of about $t \sim 0.01$ s. And because the mass of a star is roughly in the solar-mass range, this all demands a compact object (i.e., a white dwarf, a neutron star, or a black hole) instead of a normal star.

The next question is: What are ways in which one can get a nearly periodic signal from a star or stars? Generically, one might think of rotation, vibration, or a binary orbit (indeed, all three are seen in pulsars in various circumstances). Let's focus first on white dwarfs. Can they, by rotating, vibrating, or pulsating, give a period as low as 0.033 s? No: for all three processes, as we learned earlier, the minimum period is of the order of $2\pi(G\rho)^{-1/2}$, which is ~seconds at the $\rho \sim 10^8$ g cm^{-3} maximum densities of white dwarfs. Therefore, white dwarfs can't do it.

What about black holes? Can they produce periodic pulsations? No, but the answer isn't obvious. Getting a periodic signal out implies some fixed structure, and there just isn't any such place on the event horizon of a black hole. Of course, you can have structure farther out, such as accretion disks, and this structure can lead to quasi-periodic oscillations. But these oscillations are nothing as coherent as the pulses observed from pulsars.

That means we're down to neutron stars, but does the minimum period work out? Noting that the central density of neutron stars is a few times that of the maximum density reached in an atomic nucleus, 2.3×10^{14} g cm^{-3}, the minimum period is expected to be about $2\pi(G\rho)^{-1/2} \sim 10^{-3}$ s. But this is the minimum period. Nothing

[1] As mentioned above, today we know that the range is much greater, from about 0.0014 s to thousands of seconds or more

[2] Today we know that this range can be up to many billions of years!

says the period can't be longer than this, and how much longer will depend on what is causing it. Is it rotation, vibration, or a binary orbit? Starting with vibrations, the ultrahigh density means that the vibration period is of order $2\pi(G\rho)^{-1/2}$, which we already determined is milliseconds, but this is much too fast for the majority of pulsars.

What about binary orbits? There are two problems here. First, inspiral due to gravitational-wave emission would cause the period to decrease, not increase. Second, gravitational radiation from a binary orbit with a period of seconds would cause the two stars to spiral into each other in a few hours. Indeed, we saw back in Equation (5.1), in Chapter 5, that the amount of time to merge is about 1 day, if the initial semimajor axis is $a_0 \sim 2,000$ km for two $1.4M_\odot$ neutron stars, which corresponds to an orbital period of about 1 s. On the other hand, many pulsars have now been observed for decades, and at the time of Bell-Burnell and Hewish, the observations lasted for more than 1 day, so the periodic nature cannot be caused by binary motion.

As a result, the only possibility left is that pulsars are neutron stars whose periodic emission is related to their rotation. Indeed, rotational periods of a second, although fast, are not disallowed by any physical process we know of, as long as the body is sufficiently compact (like neutron stars are) to prevent it from flying apart. Moreover, if the star is slowed down by some process (such as "friction" due to interactions with their magnetic field), then its rotation period would increase, and this is consistent with the observations of 1968. And because neutron stars are so compact, it was reasonable to think that their rotational inertia must be huge, which in turn means they are hard to slow down, and thus, they are indeed very stable.

All of this put together then also implies that, remarkably, such rotating neutron stars are *very* observable. Today, not only can we detect them through observations of their radio emission, but also through radiation at other frequencies. For example, they have been detected through the X-rays emitted from cooling or accretion onto their surfaces, from various types of bursts, and from gravitational-wave events. There can be disagreement about whether they are genuinely "neutron stars" in the sense of having most of their mass be in neutrons, but that's just a name. What stands without doubt is that they certainly exist.

7.1.2 How Are Neutron Stars Formed?

The short answer is that neutron stars are one of the end points of the life of a star massive enough to undergo a core-collapse supernova. But to get better context we should review the evolution of the progenitors of supernovae.

Consider an isolated star that begins with at least 8 $M_\odot$. This star spends most of its fusion-powered life burning hydrogen to helium, which is when it is on the main sequence. When it gets through about 15% of its hydrogen (the part in the core that is dense and hot enough to fuse), it goes through a red giant phase, then burns helium to carbon in a "helium main sequence," then burns carbon to oxygen, oxygen to neon, neon to magnesium, and so on.

What prevents it from going on indefinitely? If you think about the nuclear binding energy per nucleon, you note that hydrogen to helium liberates about

7 MeV per nucleon, but that the pickings become slimmer as the nuclei increase in complexity. In fact, there is a maximum nuclear binding energy at ^{56}Fe.

Captain Obvious: The fundamental reason that there is a maximum nuclear binding energy per nucleon is that the nuclear strong force, which binds nucleons together, has a characteristic range of only $\sim 10^{-13}$ cm, which is roughly the size of a nucleon. In contrast, the electromagnetic force, which repels protons in a nucleus from each other, has unlimited range. Thus, although for small nuclei an increase in the number of nucleons increases the nuclear binding energy per nucleon (because the nucleons are all within the range of the strong force), for larger nuclei electromagnetic repulsion becomes more important. This is also why, for larger atomic numbers, stable nuclei have a larger ratio of neutrons to protons; nuclear binding is more effective for an equal number of neutrons and protons, but neutrons have no electrical repulsion from each other.

This nuclear binding energy per nucleon is just about 8.7 MeV, meaning that in the entire sequence from helium to iron-group elements, you only get about 1.7 MeV/(7 MeV) $\sim$ 25% of the energy you got during the main sequence. In addition, the nuclei have greater electric charge, meaning that the temperature has to be higher, and thus the burn rate greater, to fuse. When the last stage of fusion happens (silicon to iron, basically), the temperature is so high that photons can split up nuclei, and thus, decrease the net energy production even further. By the time the core is made of iron, it is held up against gravity not by generated energy but by gradients in the degeneracy pressure.

Major Payne: Hold on! This explanation omits details that can be important in some circumstances. In particular, the silicon-fusing core is extremely hot: $T \sim 3 \times 10^9$ K. Thus, the thermal energy per particle is $E \sim kT \sim 4 \times 10^{-7}$ erg. We can compare this with the gravitational binding energy per particle, which is $E_{\rm grav} = GMm/R$, which for $M = 1.4\ M_\odot$ and $R = 10^8$ cm (about right for a cold white dwarf at the Chandrasekhar mass limit), and $m = m_n \approx 1.7 \times 10^{-24}$ g is $E_{\rm grav} \sim 2 \times 10^{-6}$ erg. This means that the thermal energy is not a negligible fraction of the total; the core of a massive star just prior to core collapse is not identical to a cold, fully degenerate white dwarf, and thermal pressure gradients can help hold up the core. The net result is that the total mass of a core at collapse might be a few tenths of a solar mass higher than would be implied in the cold-core limit.

Even this degeneracy pressure, however, has its limits. We therefore need to discuss both degeneracy pressure and the Chandrasekhar mass, which is the limiting mass beyond which degeneracy pressure cannot hold up a star.

7.1.3 Degeneracy Pressure and the Chandrasekhar Mass

Let us now estimate the radius of a degenerate object as well as its maximum mass, which in the case of white dwarfs is called the Chandrasekhar mass.

To start, let's review degeneracy pressure. If you have taken quantum mechanics, you recall the uncertainty principle: $\Delta x \Delta p \gtrsim \hbar$, i.e., the product of the uncertainty in position with the uncertainty in momentum must be greater than or of order of $\hbar$ (actually greater than or equal to $\hbar/2$, but we're working to order of magnitude). For example, if you've ever done a square-well calculation in quantum mechanics, then you know that a particle that is confined within a well of size Δx actually does have a momentum of the order of $\hbar/\Delta x$.

Now, consider a white dwarf, which might typically have a radius of 10^3–10^4 km and a mass of 0.2–1.4 $M_\odot$. The electrons are all squeezed together much more tightly than in atoms (you can figure out the typical separations with a Fermi estimate if you'd like).

Captain Obvious: Challenge accepted! We know that in ordinary matter, almost all of the mass is in protons and neutrons rather than in electrons. The masses of neutrons and protons are about the same as each other; let's say the mass is $m_{\rm h}$, where "h" is for "heavy." Let's also say that there is one electron per heavy (okay, in a white dwarf it's typically one proton and one neutron per electron, so it would really be one electron per two heavies, but we're working in the $1 = 2$ approximation). Then, roughly, the mass of the white dwarf is $M_{\rm wd} = Nm_{\rm h}$, for a white dwarf with N electrons. The volume of the white dwarf is $V_{\rm wd} = (4\pi/3)R_{\rm wd}^3$ if its radius is $R_{\rm wd}$. If we assume that inside this ball there are N tiny electrons, which are typically separated by a distance r_e, we then have that $(4\pi/3)R_{\rm wd}^3 = (4\pi/3)r_e^3 N$, so $r_e = N^{-1/3}R_{\rm wd}$. Putting in $M_{\rm wd} = 0.5\ M_\odot$, $R_{\rm wd} = 10^4$ km, and $m_h \approx 1.7 \times 10^{-24}$ g gives $N \approx 6 \times 10^{56}$ and $r_e \approx 10^{-10}$ cm. In an atom, typical electron separations are $\sim$1 Å, or $\sim 10^{-8}$ cm, so in a white dwarf, they are packed much more closely than in atoms. As a result, they aren't in atoms; instead, they move freely without being bound to individual nuclei.

Like Captain Obvious said, the distance between electrons is really tiny, so we can model them as essentially free. Therefore, each electron feels the influence of neighboring electrons, and this acts as a confining square well. As a result, even at zero temperature, each electron has a Fermi momentum from the uncertainty principle of

$$p_F \sim \hbar/\Delta x \sim \hbar n^{1/3}, \tag{7.1}$$

where $n = 1/(\Delta x)^3$ is the number density for one particle inside a box of size Δx. There is a corresponding Fermi energy, which is $E_F \approx p_F^2/2m_e \sim \hbar^2 n^{2/3}/(2m_e)$ in the nonrelativistic limit and $E_F \approx p_F c \sim \hbar c n^{1/3}$ in the relativistic limit. Similarly, there are contributions to the energy density and pressure. This is the origin of degeneracy pressure.

Let us now figure out some properties of stars that are held up by degeneracy pressure. Let the star have a mass M and radius R. Assume, as Captain Obvious did, that the mass is made up of heavy particles of mass m_h, but that degeneracy pressure is supplied by possibly different particles of mass m_d, and that there are μ_d degenerate particles per heavy particle. For example, $\mu_d \approx 1$ for a neutron star (where most of the particles are neutrons, which are also the degenerate particles and thus $m_h \approx m_d$). In contrast, for a white dwarf, the electrons supply the degeneracy pressure and there is roughly one neutron and one proton per electron; both are heavy, so $\mu_d \approx 0.5$ for a white dwarf. Assume that the typical number density of the degenerate particles is $n \sim \mu_d(M/m_h)/R^3$, and say that the Fermi momentum is $p_F \sim \hbar n^{1/3}$; thus, we assume $p_F = f_1 \hbar(\mu_d M/m_h)^{1/3}/R$, where we are attempting to keep Major Payne a little happier by introducing a dimensionless order-unity factor f_1 to take into account the factors of π and 1/2 that we are missing, as well as the fact that the density in the star is not uniform. The total energy per degenerate particle is $E_{\rm tot} = \sqrt{m_d^2 c^4 + p_F^2 c^2} - f_2 GM(m_h/\mu_d)/R$, where we have also introduced an order-unity fudge factor f_2 to the Newtonian gravitational potential energy to take into account the unknown mass distribution in the star (e.g., for a uniform-density star $f_2 = 3/5$, but f_2 is larger than this for a realistic centrally condensed star).

The total energy per degenerate particle is then

$$E_{\rm tot} = \sqrt{m_d^2 c^4 + f_1^2 \hbar^2 c^2 (\mu_d M/m_h)^{2/3}/R^2} - f_2 GM(m_h/\mu_d)/R. \tag{7.2}$$

The condition for the equilibrium radius is simply $\partial E_{\rm tot}/\partial R = 0$, and solving this for R gives

$$R = \sqrt{\frac{f_1^2 \hbar^2 c^2 (\mu_d/m_h)^{2/3}}{m_d^2 c^4}\left(\frac{f_1^2 \hbar^2 c^2 (\mu_d/m_h)^{8/3}}{f_2^2 G^2 M^{2/3}} - M^{2/3}\right)}. \tag{7.3}$$

Because R obviously has to be a real number, this says that when the expression in the parentheses is negative, there is no equilibrium solution. The expression is zero at the Chandrasekhar mass

$$M_{\rm ch} = \left(\frac{f_1}{f_2}\frac{\hbar c}{G}\right)^{3/2} \frac{\mu_d^2}{m_h^2}, \tag{7.4}$$

which gives $M_{\rm ch} \approx 0.5(f_1/f_2)^{3/2}\ M_\odot$ for $\mu_d = 1/2$ (white dwarf); the correct value is about 1.35 $M_\odot$ for $\mu_d = 26/56$ (the value for pure iron). We see that this expression implies that for a neutron star ($\mu_d \approx 1$) the maximum mass should be about four times the maximum mass for a white dwarf, i.e., nearly 6 $M_\odot$. The actual maximum mass is unknown but is much less, in the 2 – 3 $M_\odot$ range. The primary reason is that for very compact objects such as neutron stars, general relativistic effects come into play. In particular, because all forms of energy gravitate, the high pressure adds gravitational mass and thus a neutron star collapses at a lower mass than you would otherwise have thought. Significant uncertainty in the maximum mass estimate is introduced because in neutron stars, the interactions and in particular the repulsion between nucleons and, even more fundamentally, their constituent quarks and gluons are important and extremely difficult to compute from first principles.

We can rewrite the radius now in terms of the Chandrasekhar mass to find

$$R = \left(\frac{f_1^3 \hbar^3}{f_2 Gc}\right)^{1/2}\left(\frac{\mu_d}{m_d\, m_h}\right)\left(\frac{M_{\rm ch}}{M}\right)^{1/3}\left[1 - \left(\frac{M}{M_{\rm ch}}\right)^{4/3}\right]^{1/2}. \tag{7.5}$$

For the radius, if $M \ll M_{\rm ch}$ we find from the above expression that $R \sim M^{-1/3}$. This is a general feature of matter supported by degeneracy pressure: the radius decreases with increasing mass, so the density increases rapidly. We also see that $R \sim m_d^{-1}$, so white dwarfs are much larger than neutron stars (by a factor of $\sim 10^3$, which is about right). Our formula would predict that $R \approx 3 \times 10^8 (f_1^2/f_2)$ cm for $M = 0.2\ M_\odot$ (remember that $M_{\rm ch}^{1/3} \propto (f_1/f_2)^{1/2}$). The actual radius is about 1.5×10^9 cm. Going farther than we have, i.e., effectively estimating f_1 and f_2, requires detailed calculations of the structure of compact objects of specified masses.

Captain Obvious: Look, I understand that the authors are trying to placate Major Payne by putting in all of these details, but in doing so they are obscuring what are really very simple principles that were first presented by the physicist Lev Landau. Because $p_F \sim n^{1/3}$ and $n \sim N/R^3 \sim (M/m_h)/R^3$, it means that $p_F \sim M^{1/3}R^{-1}$. If the degenerate particles are very nonrelativistic, then the Fermi energy is $E_F \sim p_F^2/(2m_d)$; this immediately tells us why we have to worry about the Fermi energy of electrons (with very small m_d) before we need to worry about the Fermi energy of the neutrons or protons (with m_d, that is $\sim$1800 times larger than the mass of the electron). Next, we note that the total energy per degenerate particle is the sum of the Fermi energy and the gravitational potential energy:

$$E_{\rm tot} = E_F + E_{\rm grav} \sim M^{2/3}/R^2 - GM/R. \tag{7.6}$$

Here we see that the Fermi energy term has a different dependence on R than the gravitational potential energy term, so there is a radius that minimizes the energy. Doing this shows that $R \sim M^{-1/3}$.

If we then jump to the relativistic limit, then again $p_F \sim M^{1/3}R^{-1}$, but $E_F \sim p_F c \sim M^{1/3}R^{-1}$. Now

$$E_{\rm tot} = E_F + E_{\rm grav} \sim M^{1/3}/R - GM/R. \tag{7.7}$$

But this is a different situation from what we had previously. Now the Fermi energy term and the gravitational potential energy term have the same radial dependence. If the total is positive, then the total energy is minimized by increasing R; eventually, because increasing R decreases the density and thus decreases p_F, the system is no longer fully relativistic and the star finds an equilibrium. In contrast, if the total is negative, then the energy is decreased indefinitely by lowering R; that is, it collapses. Because the M^1 dependence of the gravitational potential energy is steeper than the $\sim M^{1/3}$ dependence of the Fermi energy, this tells us that there is a mass $M_{\rm Ch}$ above which the star is unstable. This is the Chandrasekhar mass.

Major Payne: But don't you see that we have obtained additional insight because the authors properly put in all of the factors, including the to-be-determined factors f_1 and f_2? Not only do we have an estimate of $M_{\rm Ch}$, but we know how various quantities depend on the composition (via μ_d). We even know where we'd need to put our effort to make our calculations more precise, by doing the computations necessary to find f_1 and f_2. That's why I like to include these factors at the outset. I do admit that this approach complicates the algebra, but because I never make mistakes, that's not a problem.

Captain Obvious: Sure, whatever. Moving on, the authors motivated this subsection by talking about degeneracy pressure, but this concept never actually made it into their estimate of the Chandrasekhar mass. Let's then take a detour to present an equivalent derivation that, in the process, will provide a bit more insight. The general idea is that if you have a collection of fermions, then the Pauli exclusion principle induces an effective repulsion that we can quantify through a certain pressure that we call degeneracy pressure.

The derivation of this repulsive degeneracy pressure goes like this. Imagine that you have N fermions together that, because of quantum mechanics, must occupy different energy states. The highest energy state occupied is called the Fermi energy (not to be confused with a Fermi estimate!). If you think of these fermions as forming a kind of "gas," then you could attempt to associate an "effective" temperature T to them through the relation $E = k_B T$, where k_B is the Boltzmann constant and E is some average energy. Identifying E with the Fermi energy E_F, we can then define the Fermi temperature as $T_F = E_F/k_B$.

Now here comes the kicker. Continuing with our gas analogy, we can use the ideal gas law, $PV = Nk_B T$, to write a Fermi pressure, $P_F = Nk_B T_F/V = NE_F/V$, where V is the physical volume occupied by the gas. This Fermi pressure is what we call the degeneracy pressure. In the relativistic limit, this pressure becomes $P_F = Np_F c/V = N\hbar c N^{1/3}/(VR) \sim \hbar c(N/V)^{4/3}$, where we have used $V \sim R^3$. Notice that the Fermi temperature has nothing to do with the actual heat temperature of a star. Similarly, the degeneracy pressure has nothing to do with a classical force; its origin is purely quantum mechanical.

With that said, we can now play a "balance the pressures" game. What are the pressures at play? Well, degeneracy pressure is one of them, which we can rewrite as $P_F \sim \hbar c(N/V)^{4/3} \sim \hbar c(M_{\rm wd}/m_h)^{4/3}/R^4$, where we have used $M_{\rm wd} = Nm_h$ and $V \sim R^3$. The other pressure at play is that induced by the self-gravitational interaction of fermions. Using Newton's gravitation law, we then have $P_g = F_g/A = GM_{\rm wd}^2/R^4$. Setting these two pressures equal to each other, we then find the same Chandrasekhar mass estimate as the authors found in Equation (7.4). Just for kicks, one last fun fact: Chandrasekhar won the Nobel Prize in Physics in 1983 for a large body of work, including importantly this maximum mass limit and other work on white dwarfs.

7.1.4 What's Inside a Neutron Star?

Let's now switch gears from white dwarfs to neutron stars, but first, let's get some basic numbers. As we indicated above, if the energy and momentum are low, then the Fermi energy $E_F \sim p_F^2/(2m) \sim \hbar^2 n^{2/3}/(2m)$, where n is the number density. At some point, however, $E_F > mc^2$, and then, the Fermi energy $E_F \sim p_F c \sim \hbar c n^{1/3}$. For electrons, the crossover to the relativistic regime happens at a density $\rho \sim 10^6$ g cm^{-3}, assuming a fully ionized plasma with two nucleons per electron. For protons and neutrons, the crossover is at about 6×10^{15} g cm^{-3}. This is because this transition density scales as the particle's mass cubed (simply because the transition is where $p_F = m_d c$, or $\hbar n^{1/3} \sim m_d c$, or $n \sim m_d^3$ for a degenerate particle of mass m_d). The maximum density in neutron stars is less than this, so, for most of the mass, electrons are highly relativistic but neutrons and protons are at best mildly relativistic.

Let's now think about what that means. Suppose we have matter in which electrons, protons, and neutrons all have the same number density. For a low density, which has the highest Fermi energy? The electrons, because at low densities the Fermi energy goes like the inverse of the particle mass. When $\rho = 10^6$ g cm^{-3},

$E_F \approx m_e c^2 \approx 0.5$ MeV, so this is the density that marks the transition between nonrelativistic and relativistic electrons. Then, at 10^7 g cm^{-3}, the Fermi energy is about 1 MeV, and each factor of 10 increase in the density doubles the Fermi energy because $10^{1/3} \sim 2$ and $E_F \sim n^{1/3}$ in the relativistic regime. What that means is that the energetic "cost" of adding another electron to the system is not just $m_e c^2$, as it would be normally, but it is $m_e c^2 + E_F$. It therefore becomes less and less favorable to have electrons around as the density increases.

Captain Obvious: As I explained before, there is an analogy we can make between a collection of fermions and an ideal gas, and in this analogy, we can define a Fermi temperature via $T_F = E_F/k_B$, where we recall that k_B is the Boltzmann constant. When we do that at a Fermi energy of 1 MeV, we get a Fermi temperature of about 10^{10} K, which is already higher than the expected interior temperature of neutron stars more than a few thousand years old. The Fermi energies of electrons, protons, and neutrons in the cores of neutron stars are expected to be larger than about 100 MeV, which equates to Fermi temperatures of $\sim 10^{12}$ K. This Fermi temperature is vastly larger than the temperature of any neutron star beyond the first few seconds of its life. As a result, we sometimes say that, for many purposes, neutron stars might as well be at absolute zero temperature! What we mean really is that their actual temperatures have a negligible influence on their internal structures, which brings in a welcome simplification.

Back to the main story. In free space, neutrons are unstable. This is because the sum of the masses of an electron and a proton is about 1 MeV short of the mass of a neutron, so it is energetically favorable[3] for a neutron to decay to these two particles and an electron antineutrino, a process called beta decay. What happens, though, at high density? If $m_p + m_e + E_F > m_n$, then it is energetically favorable to combine a proton and an electron into a neutron; here we are assuming that the Fermi energies of the protons and electrons can be ignored, which is reasonably well below nuclear densities. Therefore, at higher densities, matter becomes more and more neutron rich. First, atoms get more neutrons, so you get nuclei such as ^{120}Rb, with 40 protons and 80 neutrons. Then, at about 4×10^{11} g cm^{-3}, it becomes favorable to have free neutrons floating around, along with some nuclei; this is called neutron drip because the effect is that neutrons "drip out" of the nuclei. When the average density reaches nuclear density $\sim 2.7 \times 10^{14}$ g cm^{-3} (a.k.a. "'nuclear saturation density," because this is the limiting density at the center of large nuclei), there aren't any individual nuclei,

[3] To forestall complaints from Major Payne: it's actually more nuanced than that. Of course, energy is conserved in microscopic interactions, so what do we mean when we say that something is "energetically favorable"? In more detail, the point is that a single particle (the neutron in this case) is replaced by a few particles (here, an electron, a proton, and an electron antineutrino). Those three particles can occupy many more quantum mechanical states (of position and momentum combined) than can the single neutron. Thus, it is more correct to say that the decay of a single particle into several particles is favored because there is more phase space that the multiple particles can occupy; in a sense, therefore, the decay is favored in free space because it increases entropy. Reverse interactions, which we're about to discuss (proton plus sufficiently energetic electron combine to make neutron plus electron neutrino), can in detail be justified in a similar manner, although it isn't as obvious because in that case we have two particles becoming two particles.

and thus, the nucleons form essentially a smooth distribution of neutrons plus a ~5%–10% smattering of protons and electrons.

At higher densities yet (here we're talking about nearly 10^{15} g cm^{-3}), the neutron Fermi energy could become high enough that it is favorable to have other particles appear. For example, baryons with at least one strange quark, such as hyperons, may emerge. Another possibility is that quarks could be deconfined, i.e., not bound in individual baryons, forming what is known as a quark–gluon plasma. Indeed, there is no reason that nuclei and deconfined quarks, or some blend including hyperons, could not all exist at the same density. It is currently unknown whether such particles will appear, and this is a focus of much present-day research. If they do, it means that the energetic "cost" of going to higher density is less than it would be otherwise, because energy is released by the appearance of other, exotic particles instead of more neutrons. In turn, this means that it is easier to compress the star: squashing it a bit doesn't raise the energy as much as you would have thought.

Another way of saying this is that when a density-induced phase transition occurs (here, a transition to other types of particles), the EOS is "soft." That means that the star can't support as much mass (although an important caveat is that if the new phase of matter is very hard, the net result can be that a star can support even more mass! Things can be complicated ...). As more mass is added, the star compresses more and more, so its gravitational compression increases. If pressure doesn't increase to compensate, in it goes and forms a black hole. What all this means is that by measuring the mass and radius of a neutron star, or by establishing the maximum mass of a neutron star, or by making other measurements of the structure, one gets valuable information about the equation of state (EOS), and hence, about nuclear physics at very high density. This is just one of many ways in which the study of neutron stars has direct implications for microphysics.

Captain Obvious: This "soft" and "hard/stiff" terminology pollutes the field of nuclear astrophysics to the point that it is sometimes hard to understand what people mean. So let's spend a few lines here to clarify the confusion. First, the definitions. A stiff EOS is one which as density increases, the pressure increases a lot, i.e., the tangent to the $p(\varepsilon)$ curve is steep. Similarly then, a soft EOS is one in which as density increases, the pressure does not increase a lot. This then means that stiff EOSs lead to stars that are hard to compress, and thus provide more support against gravity; conversely, soft EOSs lead to stars that are are easier to compress and thus, cannot support a lot of mass. Therefore, in terms of their mass–radius relation, stiff EOSs lead to neutron stars with larger radii for the same mass, compared to soft EOSs.

Sometimes astrophysicists like to use what is known as a polytropic EOS to talk about stiffness and softness. Such an EOS is $p = K\varepsilon^{\Gamma}$, where K is called the adiabatic constant, $\Gamma = 1 + 1/n$ is the adiabatic index and n is the polytropic index. A soft EOS corresponds to small Γ (or large n), while a stiff EOS is one with large Γ (or small n). Just for reference, an EOS with $n = 0$ corresponds to an incompressible fluid, meaning that its density is constant inside the star. This, as it turns out, is not a horrible approximation for the EOS inside a neutron star, although it is clearly unphysical, because the adiabatic speed of sound, $c_s = (dp/d\varepsilon)^{1/2}$, is infinite!

Captain Obvious has mentioned the EOS numerous times, and we have as well, so what is it really? In general, the EOS is a relation between the pressure and a bunch of other stuff, where "stuff" might include the energy density, temperature, composition, or other parameters. Captain Obvious argued above that the temperature doesn't matter for the bulk of the star, so that simplifies things. What about the composition? It would be lovely to be able to assume that in the core of a neutron star, which contains the vast majority of its mass, moment of inertia, etc., matter is in its energetic ground state. Then, even though we don't know what that ground state is, it would mean among other things that at those high densities the matter is in some fixed state rather than having its state depend on, for example, the history of the star. Is this reasonable?

At everyday densities, the ground-state assumption is quite definitively false. Remember that the nuclear binding energy per nucleon at low densities (such as ours) is maximized for ^{56}Fe, but most of us aren't solid iron! The basic reason for this is that although, indeed, ^{56}Fe is the ground state, the path to that ground state requires a huge amount of energy to get through the Coulomb potential barrier that keeps protons and other nuclei away from each other. At the temperatures and densities we encounter in everyday life, we simply don't have nearly that much energy available. Therefore, we have a wonderful diversity of nuclei and thus atoms. Like the Vulcans would say, infinite diversity in infinite combinations!

The same thing is expected to be true in the crust of the neutron star (which is the part below nuclear density). At such low densities, one can expect to see at least some diversity of nuclei, where as we mentioned before, the median nucleus is progressively more neutron rich at higher densities. But at the densities expected in neutron star cores, the potential barriers between different states of matter are much lower in energy than the Fermi energy. At such energies, then, it would be energetically favorable to decay to the ground state, whatever that may be. This is the basic argument of why we expect matter in neutron star cores to be in its energetic ground state at each density.

However, we should mention an important caveat: energy is not the only quantity to consider when we think about a transition from one state of matter to another. For example, if the original matter in a neutron star contained no strange quarks, then weak decays are needed to produce strange quarks if that leads to an energetically favorable state. Weak decays are slow by nuclear standards; for example, a neutron in free space takes $\sim 10^3$ s to beta decay into a proton, electron, and electron antineutrino. It is at least possible in theory that the particular pathways needed to reach an energetic ground state require billions of years. If this is true, then there might be some diversity of composition of neutron star cores, even when the density has been specified.

The usual assumption, which we will adopt here, is that matter inside the core of neutron stars is in its ground state. That, plus the irrelevance of temperature, means that neutron star EOSs are just the pressure as a function of energy density: $P = P(\varepsilon)$. When the pressure depends only on the density, the EOS is called barotropic. In the next section we will discuss how we can construct neutron stars given an EOS and central density, and how we can therefore infer some aspects of the EOS by macroscopic measurements of neutron stars.

7.2 How Can We Learn about the EOS?

There are a number of neutron star properties that, if measured precisely and accurately, could be used to constrain the EOS of matter beyond nuclear density. Examples of measured quantities include the gravitational mass, the circumferential radius (especially for a star of known mass), and the tidal deformability of a star of known mass. There are also potential future measurements that could be made that would have bearing on the EOS. Examples of potentially measurable quantities include the moment of inertia for a star of known mass, the rotational quadrupole moment for a star of known mass, the gravitational binding energy of a star of known mass, the oscillation frequencies of a neutron star, and the surface temperature of a cooling neutron star as a function of its age and mass.

As we indicated in the previous section, there is expected to be a unique EOS for "cold" matter (with temperatures sufficiently below the Fermi temperature) above nuclear density; any possible history dependence of the EOS can only occur at the lower densities in the crust, and this will have only a minor influence on measurable quantities. This has an important implication: all neutron stars should have the same EOS, even though only some may have quark–gluon plasmas in their cores because they have a sufficiently high central energy density. Given that all the quantities listed above can be computed using the EOS and an assumed central energy density, all such measurements can be used together to determine the high-density EOS, when put into the proper Bayesian statistical inference scheme.

In more detail, the basic procedure is as follows: pick many EOSs, determine their predictions for mass, radius, etc., and then compare those with the data to infer the mass, radius, and so on. Then, use Bayesian statistics to determine the relative weight of the EOSs in your sample and, from that, derive the distributions of other quantities based on the weighted EOSs. The wonderful benefit of this approach is that all data may be used, e.g., independent measurements of the same quantity (e.g., the mass of a given pulsar), or very different measurements (astronomical observations or laboratory experiments).

There are many ways to choose our EOSs, from using tables of specifically proposed EOSs, computed solving complicated equations in nuclear physics, to the use of different phenomenological parameterizations of the EOS, to the use of fully nonparametric phenomenological models. But in each case, given an EOS (i.e., a realization of the EOS with a choice of EOS parameters), we need to be able to compute macroscopic quantities, such as the mass, radius, moment of inertia, and tidal deformability, given a set of central densities. In this section, we discuss how to perform those calculations, with some details deferred to Appendix C. We assume that the star rotates slowly (where "slow" is compared with the rate of rotation that would cause the star to fly apart). Expressions exist for rapid rotation, but (1) those are a lot more complicated, and (2) even the most rapidly rotating neutron stars known to date can be approximated pretty well using the slow-rotation expressions.

7.2.1 Mass and Radius

To begin, we'll imagine a nonrotating neutron star. In equilibrium, such a star is spherical, which means that we can simplify our equations substantially compared with what would await us for an oblate or ellipsoidal star. Gravity is opposed by pressure gradients, and in Newtonian physics, this leads to the standard equation of hydrostatic equilibrium:

$$\nabla P = -\rho \mathbf{g}, \tag{7.8}$$

where $\mathbf{g}$ is the local acceleration of the gravity vector. Our assumption of spherical symmetry means that we can express this entirely in terms of the radius: $dP/dr = -\rho(r)g(r)$. If we use $M(r)$ at a radius r to represent the gravitational mass inside r, then because $g(r) = GM/r^2$ in Newtonian gravity, this becomes

$$\frac{dP}{dr} = -\rho(r)\frac{GM(r)}{r^2}. \tag{7.9}$$

Here ρ is the rest-mass or baryon density because in Newtonian gravity only the rest mass gravitates.

In 1939, Richard Tolman and, in a companion paper, Robert Oppenheimer and George Volkoff derived the equivalent equation (known as the TOV equation) when general relativity is taken into account:

$$\frac{dP}{dr} = -\varepsilon(r)\frac{GM(r)}{c^2r^2}\left[1 + \frac{P(r)}{\varepsilon(r)}\right]\left[1 + \frac{4\pi r^3 P(r)}{M(r)c^2}\right]\left[1 - \frac{2GM(r)}{c^2 r}\right]^{-1}, \tag{7.10}$$

where $\varepsilon(r)$ is the total mass–energy density as a function of radius (which reduces to $\varepsilon = \rho c^2$ in the Newtonian limit), because in general relativity all forms of mass–energy gravitate. We see that in the Newtonian limit (P small relative to ε and r large relative to $M(r)$) the TOV equation reduces to the Newtonian equation of hydrostatic equilibrium. The equation of continuity is

$$\frac{dM}{dr} = 4\pi r^2\frac{\varepsilon(r)}{c^2}, \tag{7.11}$$

which reduces to $dM/dr = 4\pi r^2\rho$ in the Newtonian limit. The system of equations then closes once we specify an EOS through a functional relation of the form $P = P(\varepsilon)$.

Major Payne: No, no, no! Well, yes: it's true that if all we want is general relativistic hydrostatic equilibrium in spherical symmetry, these equations are correct. But somebody has to explain, at the very least, how these equations are obtained. I'll do it here with enough generality that we can think about later calculating additional neutron star properties such as the moment of inertia, the rotational quadrupole, and the tidal deformability. The proper way to do this is to introduce the slow-rotation approximation, like Jim Hartle and Kip Thorne did back in the 1960s. Consider then a star that rotates *rigidly* with some angular frequency Ω_* that is much smaller than its Keplerian mass-shedding limit, i.e., $\Omega_* \ll (G\rho)^{1/2}$, or roughly $R_*\Omega_* \ll (GM_*/R_*)^{1/2} \ll c$, with R_* the stellar (equatorial) radius. One can then show that the line element can be written in Hartle-Thorne coordinates (t, r, θ, ϕ) as

$$ds^2 = -e^{\nu}[1 + 2\alpha h_2 Y_{2m}(\theta, \phi)]dt^2 + e^{\lambda}\left[1 + \frac{2\alpha m_2 Y_{2m}(\theta, \phi)}{r - 2M}\right]dr^2 \\ + r^2[1 + 2\alpha K_2 Y_{2m}(\theta, \phi)]\left\{d\theta^2 + \sin^2\theta[d\phi - (\Omega_* - \omega P_1'(\cos\theta)dt)]^2\right\}, \quad (7.12)$$

where (ν, λ) are unknown functions of radius at zeroth order in rotation, ω is an unknown function of radius at first order in rotation, (h_2, m_2, K_2) are unknown functions of radius at second order in rotation, while Y_{2m} and P_1 are spherical harmonics and Legendre polynomials (with the prime standing for a derivative), and $\alpha = \sqrt{\pi/15}$ is a normalization constant for later convenience. The quantity M is also an unknown function of radius, related to λ via $M = (1 - e^{-\lambda})(r/2)$, which can be thought of as the enclosed mass inside a radius r.

How should we now model the stress–energy tensor that describes the energy and matter content of the neutron star? As the authors argued earlier, to a very good approximation, isolated stars can be treated as objects at zero temperature with no viscosity or shear. If so, their stress–energy tensor can be modeled as that of a perfect fluid

$$T_{\mu\nu} = (\varepsilon + p)u_\mu u_\nu + p g_{\mu\nu}, \quad (7.13)$$

where ε and p are the energy density and pressure of the fluid, $g_{\mu\nu}$ is the metric tensor, and u^μ is the 4-velocity of the fluid. Because the star is rotating, we can model the latter via $u^\mu = u^0[1, 0, 0, \Omega_*]$, where u^0 is determined by requiring that the 4-velocity be a timelike vector, i.e., $u_\mu u^\mu = -1$.

With this at hand, now the calculation is "easy." All we have to do is insert this metric ansatz and this stress–energy tensor into the Einstein equations and expand in small rotation. At zeroth order in rotation, one can show that one indeed finds Equations (7.10) and (7.11). In addition, one also finds an equation for the unknown metric function ν, in terms of M and P. But because the metric function ν does not enter into Equations (7.10) and (7.11), one can solve the latter first for M and P, and then use these solutions in the differential equation for ν to find the full metric at zeroth order in rotation. As the authors will surely tell us below, we don't need ν to find astrophysical measurable quantities at zeroth order in rotation. But this metric function will be needed at higher orders.

It is instructive to write a computer code to solve both the Newtonian equation and the TOV equation for a simple EOS such as a polytrope (in which case $P(\rho) = K\rho^\Gamma$ or $P(\varepsilon) = K\varepsilon^\Gamma$, where we recall that K is the adiabatic constant and Γ is the adiabatic index). Because both are first-order differential equations, only one boundary condition is needed for each equation. Typically, we choose the central energy density ε_c to fix the boundary condition for the TOV equation, while the integration constant of the continuity equation is set to zero because $M(r) \to 0$ as $r \to 0$. What you should find is that for fixed K and Γ, a given choice of central density ε_c leads to a given neutron star mass M_* and a given neutron star radius R_*, where the former is simply $M_* = M(R_*)$ and the latter is defined as the radial coordinate where the pressure or energy density vanishes, i.e., R_* is such that $P(R_*) = 0 = \varepsilon(R_*)$ (in reality, we usually stop numerical integrations when the pressure drops several orders of magnitude below the central pressure).

You can now repeat the above integrations for a set of central densities, for a fixed EOS, and thus, generate a set of masses and corresponding radii that define the so-called mass–radius relation $M_* = M_*(R_*)$. What you should find is that the

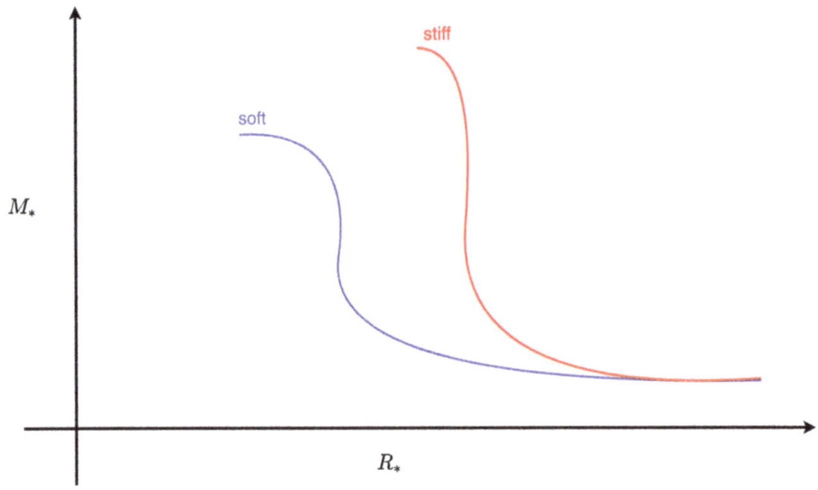

Figure 7.1. Cartoon showing a pair of mass–radius relations.

Newtonian equations give a mass M_* that increases monotonically with increasing central density. In contrast, the TOV equation gives a mass that increases with increasing central density, but reaches a maximum mass at a particular central density, after which the mass decreases with increasing central density. This maximum mass is one of the consequences of general relativity. Two cartoons of mass–radius curves are shown in Figure 7.1. Observe that softer EOSs lead to neutron stars with lower maximum mass and smaller radii, while stiffer EOSs lead to the opposite. Observe also that the mass–radius curves must approach each other at low mass because at low densities the EOS is better understood thanks to laboratory experiments.

As a side note, we see that even with the Newtonian equations one might worry about how to start the integration at $r = 0$ because the $1/r^2$ factor looks unfriendly. The proper way to deal with this is to "regularize," i.e., to use an analytical expression at $r \to 0$ that does not diverge. In this case, we know that at the center we have some density ρ_c or ε_c, and thus within a very small radius r_0 from the center, the mass is $M(r < r_0) = \frac{4}{3}\pi r_0^3 \rho_c$ in Newtonian gravity or $M(r < r_0) = \frac{4}{3}\pi r_0^3 (\varepsilon_c / c^2)$ in general relativity. Thus, $M(r < r_0)/r_0^2 \sim r_0$, which is therefore well behaved as $r_0 \to 0$.

Major Payne: Technically, that's not how you derive the boundary conditions for a system of differential equations. What you should really be doing is a local asymptotic analysis. This starts by making an ansatz for the behavior of all your functions near the boundary of interest. For example, for the problem at hand here, you could guess

$$M(r) = M_0 + M_1 r + M_2 r^2 + M_3 r^3 + \mathcal{O}(r^4), \tag{7.14}$$

$$P(r) = P_0 + P_1 r + P_2 r^2 + \mathcal{O}(r^3), \tag{7.15}$$

where we are considering the behavior of the function near $r \sim r_0 \ll R_*$. We could have selected a more complicated expansion of our functions near this boundary, but this ansatz suffices for most purposes. Note that we have not postulated an expansion for $\varepsilon(r)$, because the energy density as a function of the pressure can be found from the EOS $P(r) = P(\varepsilon(r))$, so we do not need to solve for it; in practice, the inversion to find $\varepsilon(P)$ is usually done numerically.

With these ansatzes in hand, all we must now do is determine the coefficients. To do so, we insert the ansatz into the differential equations (Equations (7.10) and (7.11)), and then expand them in $r \ll R_*$, matching coefficients order by order. When series-expanding about $r \ll R_*$, we must be careful because on physical grounds we want $M_0 = 0$, because $M(r) \to 0$ as $r \to 0$; in particular, terms of the form $1/M_0$ should not be allowed in the series expansion, which is most easily achieved by setting $M_0 = 0$ from the get go. With this in mind, one then finds that $M_1 = 0 = M_2$, $M_3 = 4\pi\varepsilon_0/3$ from the continuity equation, and $P_1 = 0$ and $P_2 = -2(P_0 + \varepsilon_0)(P_0 + \varepsilon_0/3)\pi$ from the TOV equation. The quantities ε_0 and P_0 are usually called the central energy density ε_c and the central pressure P_c, respectively (connected by the EOS via $P_c = P(\varepsilon_c)$). One can go to higher order in this expansion if one wishes, but this usually suffices for numerical integrations.

In any case, for a given EOS, there is a maximum gravitational mass for a slowly rotating star and there is a precise prediction for the circumferential radius of a star of a given gravitational mass. Numerous neutron star masses have been measured, particularly using radio observations; as we indicated earlier, the gravitational waveforms of coalescing neutron star binaries contain precise information about the chirp mass but it is significantly more challenging to obtain the individual masses. In principle, an EOS that predicts a lower maximum mass than the highest observed mass can be eliminated, although in practice mass measurements always have observational uncertainties. Thus, in reality, an EOS with a low maximum mass is merely strongly disfavored rather than being eliminated. Radius measurements are difficult to obtain without potentially large systematic errors, although this may not be an issue for phase-resolved spectroscopic measurements, such as those using NASA's Neutron star Interior Composition Explorer X-ray satellite.

But even more fundamentally, a given EOS determines all the major structural aspects of a neutron star: maximum mass, radius as a function of mass, moment of inertia, gravitational binding energy, and so on. And measurements of several of these quantities can provide complementary constraints on the EOS. We therefore continue by considering the moment of inertia for which the calculation given an EOS requires that we consider rotation to first order. We then consider the higher-order calculations of the rotational quadrupole moment and the tidal deformability. In this chapter, we give the essence of the ideas, but for those interested in the technical details of how to compute these quantities in general relativity, we give the details in Appendix C.

7.2.2 The Moment of Inertia

Because in general relativity a rotating object drags spacetime with it, a measurement of the frame-dragging effect, along with precise knowledge of the rotational frequency of a neutron star, would give its moment of inertia. This might be possible for some pulsars in double neutron star systems (which therefore also have very well-known masses) because the moment of inertia induces spin–orbit precession, altering the orbital trajectories, and therefore, the times of arrival of the pulses. The extraction of this effect, however, has turned out to be vastly more difficult than first envisioned. In principle, the gravitational waveform of a coalescing binary with at least one neutron star is also sensitive to frame-dragging, but the effect is of 1.5 post-Newtonian order and suppressed by the spin of the neutron star, so it is difficult to measure in practice with second-generation detectors. In any case, in order for a moment of inertia measurement to be used in EOS analyses, we need to be able to compute the moment of inertia for a given EOS and central density.

How do we do that? In Newtonian physics the moment of inertia for a spherical star is simply

$$I_{\text{Newt}} = \frac{8\pi}{3} \int_0^{R_*} r^4 \rho(r) dr, \tag{7.16}$$

where we recall that R_* is the (equatorial) radius of the star. One might hope that the general relativistic equivalent would be a slight modification of this expression (with, perhaps, a different expression for the volume element). However, as demonstrated by Hartle, the situation is far more complex in general relativity. One intuitive way to understand why this is so is to start with the relation $I = S/\Omega_*$, where S is the spin angular momentum and Ω_* is the stellar rotation frequency as seen from a large distance. As we said above, in general relativity, rotation introduces frame-dragging, which has no equivalent in Newtonian physics. For example, a fluid element moving with an angular velocity equal to the frame-dragging angular velocity at a given radius has zero angular momentum. Thus, instead of just being able to compute the angular momentum of a fluid element using $S = mvr$, this quantity depends on the spacetime of the star at that location. As a result, the angular momentum of the star, given an angular velocity, must be computed using a fully self-consistent solution of the star and of the spacetime. The details for how to perform this calculation are in Appendix C.1.

7.2.3 The Rotational Quadrupole Moment

At second order in perturbation theory, the equilibrium stellar shape is no longer spherical but is instead an oblate spheroid (at angular velocities close to the maximum, then as we mentioned in Section 2.2.1, the shape becomes a triaxial ellipsoid, but here we're considering slow rotation). Let us then consider the case of an isolated neutron star, like we've been doing all along, but to second order in the slow-rotation expansion. In Newtonian gravity, you expect the spheroid to be characterized by a quadrupole-moment tensor that will be diagonal (because the star rotates about a given axis, so the spacetime is axisymmetric), $Q_{ij} = \text{diag}(-Q/3, -Q/3, 2Q/3)$, where[4] the Newtonian expression of any of the diagonal components is related to

[4] We know, we know. We have changed notation by calling the quadrupole-moment tensor Q_{ij} instead of the I_{ij} we used in previous chapters. How dare we!? Well, for historical reasons, the quadrupole moment of a rotating star is typically called Q, so we'll stick with this notation in this chapter and in Appendix C. Sue us!

$$Q_{\rm Newt} = \int \varepsilon(r, \theta) r^2 P_2(\cos\theta) r^2 dr d\Omega, \tag{7.17}$$

and where the energy density at second order in rotation depends on both the radial coordinate and the polar angle.

The excitation of a quadrupole moment must enter at an even order in the slow-rotation expansion because we can consider rotation to have a sign; as we look at the star from one of the rotational poles, counterclockwise is positive spin and clockwise is negative spin, using standard conventions. If the deviation from sphericity were of odd order in rotation (e.g., first order), then when the spin angular momentum was in one direction the star would become oblate, whereas when the spin was in the other direction the star would become prolate! This is clearly absurd, because, for example, the sense of the spin reverses when we look at the star from the other rotational pole, yet the quadrupole should not depend on the observer. Therefore, the first deviation from equilibrium sphericity occurs at second order in rotation.

Of course, the naive Newtonian expression presented above is not enough to calculate the quadrupole moment in general relativity. As in the case of the moment of inertia, this is because the quadrupolar deformation of the star from sphericity affects the entire spacetime metric, inducing corrections both to the gravitational potential and to other diagonal components of the metric. In fact, unlike in the moment of inertia case, a closed-form integral expression for the quadrupole moment is horrendously complicated in general relativity, which is why this is not how we usually compute it. Instead, we have to solve the Einstein equations expanded to second order in rotation for all of the second-order-in-rotation metric functions; from that, we can then "read out" the quadrupole moment from the asymptotic form of the gravitational potential far from the star. The details of how to actually carry out this calculation are too much for this chapter, so you can find the details in Appendix C.2.

7.2.4 The Tidal Deformability

Let's now switch gears completely and consider what happens to a neutron star that is not rotating, but that is, instead, perturbed by some companion. A neutron star in a binary experiences tidal acceleration from its companion, and as a result its equilibrium shape is deformed away from a sphere. This leaves an imprint on the gravitational waveform, so it is important to understand what this deformation is. The deformation will be predominantly quadrupolar, with octupolar and higher multipolar corrections. Therefore, just as in the case of the rotational quadrupole moment, the tidal deformation will be governed by the Einstein equations, with perturbations to the metric in the diagonal sector (such that frame-dragging is not modified).

Major Payne: Technically, what the authors have described here are the so-called "electric-type" tidal deformabilities, because of how they transform under a parity transformation, similar to what we have seen in other contexts earlier in the book. It turns out that there are also "magnetic-type" tidal deformabilities, which do affect the off-diagonal components of the metric. These deformabilities, however, modify the gravitational binding energy (and thus the gravitational waves emitted) at higher post-Newtonian order (i.e., at higher powers of v/c) relative to the leading-order (the $\ell = 2$, electric-type) tidal deformability. I believe what the authors mean here when they say "tidal deformability" is this leading-order quantity.

To define the tidal deformability, suppose that a spherically symmetric star is put into a weak external quadrupolar tidal field $\mathcal{E}_{ij}$. Think of this as the quadrupole term in a multipole-moment expansion of the gravitational potential produced by some companion that is far from our star. In response to this field, the star develops a quadrupole-moment tensor Q_{ij}, and to leading order

$$Q_{ij} = -\lambda \mathcal{E}_{ij}. \tag{7.18}$$

The constant λ characterizes the tidal deformability of the star, i.e., how much the star deforms in response to a perturbation, and it is related very simply to the Love number, where the latter are a set of numbers introduced by Augustus Edward Hough Love (yes, you can call him "Dr. Love" and that wouldn't be inaccurate because he was a Professor of Mathematics at the University of Oxford in the early 1900s!) to describe tides on Earth. Of course, an external perturbation doesn't just induce a quadrupolar deformation; there will also be octupolar deformations, hexadecapolar deformations, etc. This is why λ is sometimes called the $\ell = 2$ tidal deformability. Moreover, as Major Payne hinted at, in general relativity the external tidal field comes in two flavors depending on their parity: electric type and magnetic type depending on whether, upon a parity transformation, they pick up a factor of $(-1)^{\ell}$ or $(-1)^{\ell+1}$, respectively. The tidal field $\mathcal{E}_{ij}$ is of electric-type parity, which is why λ is sometimes called the electric-type, $\ell = 2$ tidal deformability $\lambda^{E}_{\ell=2}$. We will only consider this quantity because it is the dominant one and just call it the tidal deformability λ. As was the case for the quadrupole moment, the calculation of the tidal deformability is quite involved, as it requires the solution of the Einstein equations for the metric functions; as before, we defer all details of how to compute the tidal deformability in general relativity to Appendix C.3.

But how does the tidal deformability affect the gravitational waves emitted in the coalescence of neutron stars? The dominant effect is not due to the change in spacetime around the deformed star.[5] Instead, we can understand the main change in the waveform by considering the energy of the system. We know that orbiting point masses lose energy to gravitational radiation, and this is obviously still true when the masses can no longer be modeled as points. When we think about tidal deformability, however, it is clear that the self-energy of a deformed star is greater (less negative) than the self-energy of a spherical star, because the equilibrium shape

[5] And before Major Payne complains, yes, it is true that the spacetime is modified because neutron stars have a rotation-induced quadrupole moment that is different from that of black holes, and this contributes to the binding energy of the orbit. This quadrupole, however, depends on the spins quadratically and it enters at second post-Newtonian order. It turns out that its contribution to the gravitational-wave signal is smaller than that of the tidal deformability, because the latter is enhanced by five powers of compactness, as we will see below.

of an isolated star is spherical. Therefore, tidal deformation requires energy, and that energy can only come from the orbit. In addition to this modification to the orbital energy of the binary, tidal deformations also modify the gravitational wave luminosity. This is because, as we saw in Chapter 3, the luminosity comes from derivatives of the multipole moments of the spacetime, and tidal deformations change the multipolar structure, leading to more gravitational wave dissipation than we would have had otherwise. Both effects (the change in the binding energy and the enhanced dissipation) tend to accelerate the inspiral, as compared to that of two point masses.

How exactly does this affect the waveform? Well, it introduces a correction to the orbital phase that enters at fifth post-Newtonian order. This is because, as we saw back in Equation (2.2), the binary's orbital energy is modified by a term proportional to $(k_1/C^5)(m/r_{12})^5$ relative to the Newtonian term, and by the virial theorem $(m/r_{12})^5 \sim (v/c)^{10}$. The quantity k_1 here is the tidal Love number, which is related to the tidal deformability via $k_1 = (3/2)(\lambda_1/R_1^5)$ for star 1 and similarly for star 2. In a binary with stars that have masses and tidal deformabilities of (m_1, λ_1) and (m_2, λ_2), it turns out that the dominant correction to the waveform depends primarily on the combination

$$\tilde{\Lambda} = \frac{16}{13}\frac{(m_1 + 12m_2)m_1^4\bar{\lambda}_1 + (m_2 + 12m_1)m_2^4\bar{\lambda}_2}{(m_1 + m_2)^5} = f(\eta)\bar{\lambda}_s + g(\eta)\bar{\lambda}_a, \tag{7.19}$$

often called the binary tidal deformability and sometimes the chirp tidal deformability, where we have defined the dimensionless tidal deformabilities $\bar{\lambda}_{1,2} = \lambda_{1,2}/m_{1,2}^5$ and the symmetric and antisymmetric tidal deformabilities $\bar{\lambda}_{s,a} \equiv (\bar{\lambda}_1 \pm \bar{\lambda}_2)/2$, while $f(\eta)$ and $g(\eta)$ are functions of the symmetric mass ratio, which can be inferred from the above equation. At leading post-Newtonian order, it is $\tilde{\Lambda}$ that can be measured by detectors, rather than the separate tidal deformabilities. At higher post-Newtonian orders (say at sixth post-Newtonian order), a different combination of tidal deformabilities enters the waveform; thus, with sufficiently sensitive detectors, one may be able to extract λ_1 and λ_2 independently, but with second-generation detectors, this is not possible without some additional help from universal relations (as we will see in Section 7.2.5).

Captain Obvious: Oh my stars! That's a lot of definitions, and to make matters worse, different researchers use different symbols to mean the same thing! At the cost of repeating a bit what the authors say, let's try to get all of this straight in the following summary:

- **The dimensionful tidal deformabilities**. For a relativist, this is the place to start. As the authors have mentioned, there are many such deformabilities depending on their parity properties ("electric" versus "magnetic") and the multipole-moment deformation they represent. The authors focus on the tidal deformability that matters the most for gravitational waves emitted in binary inspirals, the "electric-type," quadrupole ($\ell = 2$) tidal deformability, which they just label as λ (dropping the $\ell = 2$ and "electric" identifier). This quantity is defined via Equation (7.18), i.e., the ratio between the quadrupole moment and the quadrupolar external tidal field that induces the quadrupole moment, which then must have units of $(\text{length})^5$ because the induced quadrupole moment has units of $(\text{length})^3$ and the quadrupole tidal field has units of $(\text{length})^{-2}$. Because there are two stars in a binary, they use the notation λ_1 and λ_2 to refer to the tidal deformability of the two components in the binary.

- **The dimensionless tidal deformabilities**. These quantities are simply dimensionless versions of the tidal deformability λ. Because for a neutron star there are two natural length scales, its mass and its radius, one can in principle nondimensionalize λ with either. However, it has become customary in the gravitational-wave community to use the mass. There are two sets of notation that people use for these quantities: $\Lambda_A = \lambda_A/m_A^5 = \bar{\lambda}_A$, where the A subindex refers to the binary component with mass m_A.
- **The tidal Love number**. Dr. Love introduced this quantity in the 1900s when studying Earth's tides, which is why the tidal Love number is also sometimes called the tidal apsidal constant and is labeled through the letter k. As in the case of the tidal deformabilities, tidal deformations of different ℓ multipoles will induce multipole moments of different ℓ, so by k here the authors mean the "electric-type," $\ell = 2$ tidal Love number. As before, the authors drop the $\ell = 2$ subindex because they typically consider a binary system with two neutron stars, so k_1 and k_2 refer to the tidal Love numbers of stars 1 and 2 in the system. The tidal Love number of the Ath star with radius R_A in a binary system is related to the tidal deformabilities via $k_A = (3/2)\lambda_A/R_A^5$ (the factor of 3/2 is there for historical reasons). One sees then that the tidal Love number is generated by nondimensionalizing the tidal deformability with the "other" natural length scale in the problem, i.e., with the radius of the star instead of its mass.
- **The symmetric/antisymmetric tidal deformabilities**. The dimensionless tidal deformabilities ($\bar{\lambda}_A$ or Λ_A depending on your choice of notation) can be combined in a symmetric and an antisymmetric way via $\bar{\lambda}_s = (\bar{\lambda}_1 + \bar{\lambda}_2)/2$ or $\Lambda_s = (\Lambda_1 + \Lambda_2)/2$ and via $\bar{\lambda}_a = (\bar{\lambda}_1 - \bar{\lambda}_2)/2$ or $\Lambda_a = (\Lambda_1 - \Lambda_2)/2$. These combinations turn out to be convenient for a variety of applications, so they are commonly used in gravitational-wave analysis and in nuclear astrophysics. Clearly, because they are constructed from the dimensionless tidal deformabilities, they are themselves also dimensionless.
- **The binary or chirp tidal deformabilities**. These quantities are called this way because they are the certain combination of tidal deformabilities and masses that first enter into the Fourier phase of gravitational waves produced by a binary neutron star system. There are two such quantities that each have been used with a slightly different notation: $\bar{\Lambda}$ or $\tilde{\Lambda}$, and $\delta\bar{\Lambda}$ or $\delta\tilde{\Lambda}$. Both quantities are functions of $\bar{\lambda}_s$ or Λ_s, of $\bar{\lambda}_a$ or Λ_a and of the symmetric mass ratio η only (see Equation (7.19)), with $\bar{\Lambda}$ or $\tilde{\Lambda}$ entering at fifth post-Newtonian order and $\delta\bar{\Lambda}$ or $\delta\tilde{\Lambda}$ at sixth post-Newtonian order.

The lowest post-Newtonian order deformation to the waveform, and for that matter, to the spacetime, is static. Thus at frequencies significantly less than the oscillation frequencies of neutron stars ($\gtrsim$1500 Hz for the fundamental mode of oscillation, with somewhat different values for other modes), as the stars orbit each other, they essentially adjust smoothly to the increasing and slowly time-variable tidal field. At higher post-Newtonian orders, and at higher frequencies, oscillations can be induced that could induce dynamical tides, i.e., tides that cannot be modeled the way we've described them here. It may be possible in the future to measure some of these effects with third-generation gravitational-wave detectors.

7.2.5 Nearly EOS-independent Relations

We will now have a short discussion of a remarkable set of relations. As you know, black holes are extraordinarily simple objects. As we will see in the next chapter, in general relativity, a black hole in vacuum can be perfectly described by just three numbers (its mass, angular momentum, and electric charge), and because the net electric charge will be close to zero for astrophysical black holes, only two numbers suffice. In contrast, neutron stars are potentially very complicated, particularly given that we don't know the EOS. Indeed, the mass–radius curves, the moment of inertia–compactness curves (I–C), the quadrupole moment–compactness curves (Q–C), and the Love number–compactness curves (λ–C) curves all depend on the EOS, with EOS-induced variability of 50% or even higher. Recall here that the compactness is defined as the ratio of the neutron star mass to its radius.

It is therefore remarkable that there exists a set of relations between the moment of inertia, the tidal deformability, and the rotational quadrupole moment that are very nearly independent (at the level of ~1%) of the unknown high-density EOS. Based on the symbols used for these quantities, these have been dubbed the "I–Love–Q" relations. The simplest way to understand how these relations come about is to consider I, λ, and Q in the Newtonian limit.

> **Major Payne:** whoah, whoah, whoah ··· Newtonian limit!? What? Let's be a bit more precise here. It is obvious that in Newtonian gravity there aren't any neutron stars. As the authors pointed out before, neutron stars exist due to a delicate balance between gravitational attraction and degeneracy pressure, the latter of which necessarily requires quantum mechanics. But of course, when we construct models for neutron stars, we are never really solving the Schrödinger equation! That's because we are "hiding" most of the quantum subtleties inside of the EOS.
>
> Given an EOS, we are then free to solve the Newtonian equations of structure, or their relativistic counterparts, as the authors described in Section 7.2.1. And as the authors also pointed out in that chapter, if one takes the relativistic equations of structure and expands them in powers of $G/c^2 \ll 1$, the leading-order expressions reduce identically to the Newtonian equations of structure. Similarly, when we compute the moment of inertia, the quadrupole moment, or the tidal deformabilities, we can also do a second series expansion of the perturbed Einstein equations in powers of $G/c^2 \ll 1$ and keep only the leading-order terms. This is what the authors mean by taking the "Newtonian limit," and it is equivalent to expanding in the neutron star compactness $C = GM_*/(c^2 R_*) \ll 1$. Of course, this will never produce accurate models for neutron stars, because the typical compactness of neutron stars is in the (0.2–0.3) range. Yet, I can see how it may still provide some interesting insight, especially when trying to explain a complicated process, so I will allow it.

Taking the Newtonian limit, it turns out one can analytically solve for the metric perturbation, and thus for I, λ, and Q, for two different EOSs: an $n = 0$ ($\Gamma = \infty$) polytrope (also known as an incompressible fluid, or a constant-density star) and an $n = 1$ ($\Gamma = 2$) polytrope. And to placate Major Payne before we are interrupted again, yes, neither of these EOS models represents a realistic neutron star. Indeed, the speed of sound $c_s^2 = dp/d\varepsilon$ for a constant-density star is infinite, as we mentioned before. The calculations presented below can be done for more realistic EOSs, and even using the perturbed Einstein equations to higher or even all orders in G/c^2. However, doing so would require numerical calculations, and here we would like to gain some analytical insight.

Let us then plow forward and compute I, λ, and Q for these simple EOSs in the Newtonian limit. For an $n = 0$ polytrope, we have that

$$I_{\rm Newt,n=0} = \frac{2}{5} M_* R_*^2, \qquad \lambda_{\rm Newt,n=0} = \frac{1}{2} R_*^5, \qquad Q_{\rm Newt,n=0} = -\frac{1}{2} \Omega_*^2 R_*^5. \tag{7.20}$$

For an $n = 1$ polytrope, we have

$$I_{\rm Newt,n=1} = \frac{2(\pi^2 - 6)}{3\pi^2} M_* R_*^2, \qquad \lambda_{\rm Newt,n=1} = \frac{15 - \pi^2}{3\pi^2} R_*^5, \qquad Q_{\rm Newt,n=1} = -\frac{15 - \pi^2}{3\pi^2} \Omega_*^2 R_*^5. \tag{7.21}$$

When written in this way, it is clear that the EOS has an important effect on how I, λ, and Q scale with mass and radius. For example, $I_{\rm Newt}/(M_* R_*^2)$ is 0.4 for an $n = 0$ polytrope, while it is $2(\pi^2 - 6)/(3\pi^2) \approx 0.26$ for an $n = 1$ polytrope. Taking their fractional difference, we then find that $(1 - I_{\rm Newt}^{(n=0)}/I_{\rm Newt}^{(n=1)}) \sim 53\%$. There is clearly no EOS insensitivity here.

But now a miracle happens. The above expressions are relations between either I or λ or Q and some power of M_* and R_*. This means we can solve for R_* in terms of any one of I, λ, and Q, and then insert this relation into the other two quantities. Doing so and working with the nondimensionalized quantities

$$\bar{I} = \frac{c^4 I}{G^2 M_*^3}, \qquad \bar{Q} = \frac{c^4 Q}{G^2 M_*^3 \chi_*^2}, \qquad \bar{\lambda} = \frac{c^{10} \lambda}{G^5 M_*^5}. \tag{7.22}$$

where we have defined $\chi_* \equiv (c/G)(S/M_*^2)$, we then obtain relations directly between $\bar{I}$, $\bar{\lambda}$, and $\bar{Q}$, and these are the I–Love–Q relations. For both the $n = 0$ and the $n = 1$ polytrope cases in the Newtonian limit, we can write these relations as

$$\bar{I}_{\rm Newt} = C_{\bar{I}\bar{\lambda}}^{(n)} \bar{\lambda}_{\rm Newt}^{2/5}, \qquad \bar{I}_{\rm Newt} = C_{\bar{I}\bar{Q}}^{(n)} \bar{Q}_{\rm Newt}^{2}, \qquad \bar{Q}_{\rm Newt} = C_{\bar{Q}\bar{\lambda}}^{(n)} \bar{\lambda}_{\rm Newt}^{1/5}, \tag{7.23}$$

where the coefficients $(C_{\bar{I}\bar{\lambda}}^{(n)}, C_{\bar{I}\bar{Q}}^{(n)}, C_{\bar{Q}\bar{\lambda}}^{(n)})$ are the only quantities that depend on the EOS. Evaluating these quantities for our two test EOSs, one finds

$$\begin{aligned} C_{\bar{I}\bar{\lambda}}^{(n=0)} &= \frac{2^{7/5}}{5} \approx 0.528, & C_{\bar{I}\bar{\lambda}}^{(n=1)} &= \frac{2(\pi^2 - 6)}{3^{3/5}\pi^{6/5}(15 - \pi^2)^{2/5}} \approx 0.527, \\ C_{\bar{I}\bar{Q}}^{(n=0)} &= \frac{128}{3125} \approx 0.0410, & C_{\bar{I}\bar{Q}}^{(n=1)} &= \frac{32(\pi^2 - 6)^5}{27\pi^6(\pi^2 - 15)^2} \approx 0.0406, \\ C_{\bar{Q}\bar{\lambda}}^{(n=0)} &= \frac{25}{2^{14/5}} \approx 3.59, & C_{\bar{Q}\bar{\lambda}}^{(n=1)} &= \frac{3^{6/5}\pi^{12/5}(15 - \pi^2)^{4/5}}{4(\pi^2 - 6)^2} \approx 3.60. \end{aligned} \tag{7.24}$$

This time the relative fractional difference is very small: $(1 - C_{\bar{I}\bar{\lambda}}^{(n=0)}/C_{\bar{I}\bar{\lambda}}^{(n=1)}) = 0.2\%$, $(1 - C_{\bar{I}\bar{Q}}^{(n=0)}/C_{\bar{I}\bar{Q}}^{(n=1)}) = 1\%$, and $(1 - C_{\bar{Q}\bar{\lambda}}^{(n=0)}/C_{\bar{Q}\bar{\lambda}}^{(n=1)}) = 0.3\%$. We see that even though the EOSs are very different from each other, the coefficients are nearly independent of the EOS, and therefore, the *I*–Love–*Q* relations are nearly independent of the EOS.

The *I*–Love–*Q* relations are not just between the $n = 0$ and $n = 1$ polytropes in the Newtonian limit. The calculation above was just illustrative to explain the relations. One can compute the *I*–Love–*Q* relations in full general relativity, either in the slow-rotation approximation or even for rapidly rotating stars, and with any EOS one wishes to use (even tabulated ones). In all cases considered to date, the relations remain EOS insensitive, with variability introduced by the EOS at the 1% level or less. The only two ways known to date to break the EOS insensitivity are to either introduce very large magnetic fields or to consider newly born proto-neutron stars. In the first case, the breakage of EOS insensitivity is because a magnetic field also introduces a multipolar deformation, and this deformation depends on the strength and geometry of the magnetic fields. In the second case, the breakage is because neutron stars are born highly deformed and at very high temperature. In that case, however, EOS insensitivity is restored within a few milliseconds from birth, as the star cools down and settles to a stationary configuration.

Of course, the *I*–Love–*Q* relations are not the only ones that have been found to be EOS insensitive (although they do hold the record for EOS insensitivity to date!). Another set of EOS-insensitive relations are between the fundamental frequencies that are excited in the merger of a binary neutron star merger. The situation here is a bit murkier because binary neutron star mergers are extremely complicated, and full numerical relativity simulations that include all of this complicated physics (including magnetic fields, neutrino transport, and many other important nuclear processes) have not yet been completed. Another interesting set of EOS-insensitive relations is between the tidal deformabilities in a neutron star binary system. Indeed, it turns out that $\bar{\lambda}_1$ and $\bar{\lambda}_2$ satisfy EOS-insensitive relations when one also includes the mass ratio in the relations $\bar{\lambda}_1 = \bar{\lambda}_1(\bar{\lambda}_2, m_2/m_1)$; these are called "binary Love relations."

Because these binary Love relations have become quite useful in gravitational-wave data analysis (to, for example, use in conjunction with other EOS-insensitive relations to extract the neutron star radii), let us discuss them in a bit more detail. As before, let us consider the Newtonian limit for simplicity, and recall that $\lambda_{\text{Newt},(n)} = C_{\lambda}^{(n)}(R_*/M_*)^5$. If one has two neutron stars in a binary, there are then two tidal deformabilities, one for each star, and we can construct their symmetric and antisymmetric combinations $\bar{\lambda}_s = (\bar{\lambda}_1 + \bar{\lambda}_2)/2$ and $\bar{\lambda}_a = (\bar{\lambda}_1 - \bar{\lambda}_2)/2$; the convention here is that star 2 is the heavier one in the binary, so $M_{1,*}/M_{2,*} < 1$. Given the dependence of the tidal deformability on the compactness, we can then write

$$\bar{\lambda}_{a,s,(n)} = \frac{1}{2} C_{\lambda}^{(n)} \frac{1}{C_1^5} \left[1 \mp \left(\frac{C_1}{C_2} \right)^5 \right]. \tag{7.25}$$

Recall that $C_A = GM_{A,*}/(c^2 R_{A,*})$ is the gravitational compactness of neutron star *A*. It doesn't seem we have made much progress, until we realize that for any polytropic

EOS, the Newtonian continuity equation can be solved analytically to find that $M_{A,*} \propto C_A^{(3-n)/2}$. Using this, we can then rewrite the above expression as

$$\bar{\lambda}_{a,s,(n)} = \frac{1}{2} C_{\bar{\lambda}}^{(n)} \frac{1}{C_1^5} [1 \mp q^{10/(3-n)}], \tag{7.26}$$

where we have defined the mass ratio $q = M_{1,*}/M_{2,*} < 1$. Notice that the definition of the mass ratio is consistent with $q < 1$. As noted by Captain Obvious in Chapter 2, different people choose different conventions for the mass ratio, which is why using the symmetric mass ratio η avoids ambiguity. We will stick here with q as defined above, since this is customary in the literature. Taking the ratio of the symmetric and antisymmetric combinations of the tidal deformabilities, we then arrive at the binary Love relations in the Newtonian limit

$$\bar{\lambda}_a = F_{(n)}(q)\, \bar{\lambda}_s, \tag{7.27}$$

with $F_{(n)}(q) = (1 - q^{10/(3-n)})/(1 + q^{10/(3-n)})$. All of the EOS dependence is then encoded in the $F_{(n)}(q)$ function, which has a small EOS variability. Indeed, the relative fractional difference between an $n = 0$ and an $n = 1$ polytrope is only 12% for a mass ratio of $q = 1/2$. We see then that the EOS insensitivity is not as good as that of the I–Love–Q relations, but still not bad. A relativistic calculation, in fact, reveals that the compactness corrections to the above expression that we have ignored increase the degree of EOS insensitivity!

Let us close this section with some examples about why these relations are important. Consider first the binary Love relations and the Love–Q relations. These relations allow us to write $\bar{\lambda}_2$ as a function of only $\bar{\lambda}_1$ and q (from the binary Love relations), $\bar{Q}_1$ as a function of $\bar{\lambda}_1$, and $\bar{Q}_2$ as a function of $\bar{\lambda}_2$ (from the Love–Q relation), which in turn can also be written in terms of $\bar{\lambda}_1$ and q (from binary Love). Now, $\bar{Q}_1$, $\bar{Q}_2$, $\bar{\lambda}_1$, and $\bar{\lambda}_2$ all enter in the model of the gravitational waves emitted in the inspiral of binary neutron stars through various combinations, as we saw for example in Equation (7.19). Because we have these EOS-insensitive relations, we can replace $\bar{Q}_1$, $\bar{Q}_2$ and $\bar{\lambda}_2$ entirely in terms of $\bar{\lambda}_1$ and q, the latter of which is already present in the waveform model (through the symmetric mass ratio). We are therefore analytically breaking any degeneracies between these quantities and enabling the estimation of $\bar{\lambda}_1$ directly from the data. Once $\bar{\lambda}_1$ is extracted from the data, then one can also infer $\bar{\lambda}_2$ from the binary Love relation, as well as $\bar{Q}_1$, $\bar{I}_1$, $\bar{Q}_2$ and $\bar{I}_2$ from the I–Love–Q relations. This is important because without the EOS-insensitive relations, we would be forced to only estimate the $\tilde{\Lambda}$ combination of tidal deformabilities, which then requires the use of a given EOS parameterization to extract information about the radius. You may be worried about the systematic error one may incur due to the nonuniversality of the binary Love relations, but as long as this is smaller than the statistical uncertainty we are okay (which is the case for second-generation detectors, but may be not so for third-generation ones). In Appendix C we give details about the limits on the tidal deformability, and the implications for the I–Love–Q relations, that were obtained from the gravitational-wave observations of the double neutron star coalescence GW170817.

7.2.6 EOS Information from Coincident GW and Electromagnetic Observations

We conclude this chapter with a brief discussion of a promising, but complicated, path toward learning about the EOS of neutron star core matter when we have both gravitational-wave and electromagnetic observations of the same event.

At the wish-fulfillment level, we can certainly imagine a situation in which an electromagnetic observation could yield major information. Suppose, for example, that a binary coalescence were to be observed with such a high signal to noise that both masses, rather than just the chirp mass, can be determined with high precision. Suppose further that the masses are 2.5 $M_{\odot}$ and 2.7 $M_{\odot}$. If a strong electromagnetic signal were seen from such an event, then it would indicate that at least one of the two objects was a neutron star, and thus, that the maximum mass of a neutron star had to be at least 2.5 $M_{\odot}$. Sure, there would be some ambiguity depending on the spin of the neutron star, but this would be likely to show up in the gravitational waveform at such signal-to-noise ratios.

This would require a really high signal to noise, so it isn't likely. What about more probable paths toward an enhanced understanding of the EOS using both gravitational-wave and electromagnetic observations?

One approach that was proposed by a few groups prior to the direct detection of gravitational waves is related to the so-called short gamma-ray bursts. These are, well, short bursts of gamma-rays! They typically last for a few tenths of a second, and even before the advent of gravitational-wave astronomy, they were thought to be caused by the coalescence of two neutron stars, or a neutron star and a black hole. Further analysis suggested that the bursts occur after the coalesced object becomes a black hole and that the transition to a black hole has to happen in no more than a few tenths of a second. That's guaranteed if the original pair was a black hole and a neutron star, but if the pair was two neutron stars, then the sequence involves the two getting together and then collapsing to a black hole. To collapse into a black hole, the total mass has to be above the threshold that can be sustained for more than a few tenths of a second. Given that upon merger the now-single object has a lot of rotation, it can support itself against gravity a few tenths of a solar mass more than it could if it were not rotating, but the correction between rotation-supported objects and nonrotating objects is relatively well known.

However, turning this into a hard limit was difficult because we had no independent information about the masses of the neutron stars in short gamma-ray bursts, or even whether both objects were necessarily neutron stars; if one object was a black hole, this argument would not give us any limit on neutron star masses at all!

The double neutron star event GW170817, however, provided us with a precise chirp mass that was low enough to virtually guarantee that both objects were neutron stars. To make an estimate of the maximum mass of neutron stars we need the total mass $M_{\rm tot}$, not the chirp mass $\mathcal{M}$, but because $M_{\rm tot} = \eta^{-3/5}\mathcal{M}$, and because the symmetric mass ratio η is very close to 0.25 for neutron stars ($\eta = 0.25$ for equal masses and $\eta = 0.24$ even for an extreme (for double neutron stars) mass ratio of 1.5:1), knowledge of the chirp mass gives us the total mass to good precision. Based on this, several groups used GW170817 to derive an upper limit of $\sim 2.3\ M_{\odot}$ to the maximum mass of a nonrotating neutron star.

But there's more! A phenomenon predicted before GW170817, which might actually have been seen in events prior to GW170817 (but this is disputed) is variously called a "kilonova" or a "macronova" by different groups. The idea is that when neutron stars merge, it's a messy merger. Some of the matter becomes unbound and flies off; that matter is very rich in neutrons, which leads to the formation of many heavy elements. The radioactive decay of some of these elements ends up powering extended emission that was seen clearly in the intense observing campaigns following GW170817. The details of that emission (e.g., the total luminosity as a function of waveband, the time development of the emission, and so on) depend on how much stuff got thrown out and how fast it was ejected. That, in turn, depends on how big the stars were, and thus, by use of lots of numerical relativity simulations, people have used the kilonova/macronova emission to place constraints on the mass–radius relation for neutron stars and therefore on EOS of the dense matter in their cores.

Fantastic! Sounds like we're all set; we have upper limits on the maximum mass, radius measurements, and lots of other good input to our knowledge of the EOS. So what's with all the weasel-wording at the beginning, "promising, but complicated, path toward learning about the EOS"? Why not just say that we've got the information we need?

The answer is that a lot of the inferences depend on the accuracy of the models, and there aren't good independent checks for those. For example, let's think again about the way that we get an upper limit on the maximum mass. Recall that the idea is that to make short gamma-ray bursts, it is necessary that the objects form a back hole within at most a few tenths of a second. This is based on a very plausible and likely correct idea,[6] but the astrophysics involved is very complicated: jet formation, neutrinos, tangled magnetic fields, and all that. We can't be sure. Similarly, for kilonova/macronova emission, major-league numerical simulations are needed, and it's not easy to get all the physics at high-enough resolution to be certain of the results.

As a result, the most cautious analyses of the EOS do not put such electromagnetic-based inferences on the same footing as, say, measurements of neutron star masses or tidal deformabilities. Although these, too, have some uncertainties, they don't have the same potentially major systematic errors as more model-dependent calculations.

[6] For the interested: the idea is that if two neutron stars coalesce to form a rapidly rotating neutron star that persists for seconds or longer, then because all that matter is still around there will be a persistent wind of matter being blown off. That, in turn, would mean that the highly relativistic jet (Lorentz factors of hundreds!) that appears necessary to produce gamma-ray bursts would have to struggle through all that wind material, and this would lead to much longer events, more like tens of seconds, than a couple of seconds or less. It was also argued for GW170817 in particular that if there was a long-lived rapidly spinning neutron star remnant, then spindown of that remnant would have injected a large amount of energy into the afterglow. This was not seen, which either means that the remnant collapsed quickly to a black hole or that the spindown was too gradual to inject much energy.

7.3 Exercises

1. Consider a cold ($T = 0$) fluid of neutrons, protons, and electrons so dense that their Fermi energies are all in the ultrarelativistic limit, i.e., $E = pc$ for all three species. Assume that the reaction $n \leftrightarrow p + e^-$ is in equilibrium. Calculate the relative number densities of neutrons, protons, and electrons. Hint: ignore energies of interaction (e.g., binding energies between neutrons and protons that result from the strong force).

2. Like all great thinkers, Dr. Wrong has pursued many innovative ideas. Recently, Dr. Wrong has been dabbling in particle physics and nuclear physics and has come up with a model in which there is a previously unsuspected particle dubbed a "saneon" with a mass–energy of 8 GeV. In this model, the inner 0.5 $M_\odot$ sphere of slowly rotating neutron stars has a high-enough number density of neutrons that it is energetically favorable to convert them to saneons (that is, the total energetic cost of adding another neutron exceeds 8 GeV). Based on this idea, Dr. Wrong has applied for a physics faculty position at your institution. The chair of the physics department has been dubious about some of Dr. Wrong's previous ideas but is excited about this one. What report do you give to your chair? Note that the Fermi momentum at a neutron number density n_n is $p_F = (3h^3 n_n/(8\pi))^{1/3}$, where $h = 6.63 \times 10^{-27}$ cm^2 g s^{-1}. Hint: you may wish to consider the radius of a black hole in the context of this problem.

3. Suppose that the EOS is a polytrope, which means that the pressure P depends on the density ρ like $P = K\rho^\Gamma$, where K is a constant and Γ is the polytropic index. Now consider the Newtonian equation of hydrostatic equilibrium for a spherical star:

$$\frac{dP}{dr} = -\rho g. \tag{7.28}$$

We can write the acceleration of gravity at radius r as

$$g = \frac{GM(<r)}{r^2}, \tag{7.29}$$

where $M(<r)$ is the mass interior to the radius r. Say that the system is initially in equilibrium and consider a perturbation in which the entire star is shrunk by some amount; by "entire star" we mean that everywhere in the star the perturbation increases the density by the same factor (this is what would happen if the star were to contract homologously).

(a) Demonstrate that for $\Gamma > 4/3$ the star is stable against such a perturbation, whereas for $\Gamma < 4/3$ the star is unstable against the perturbation. Do this by showing that for $\Gamma > 4/3$, the magnitude of dP/dr increases by more than the magnitude of $-\rho g$, i.e., that for $\Gamma > 4/3$, a slight contraction of the star increases the pressure gradient more than it increases the gravitational force density. The reverse is true for $\Gamma < 4/3$.

(b) For the curious, determine what happens if the gravitational force law is steeper than $1/r^2$ (i.e., it can be locally represented as $1/r^{2+\varepsilon}$); this gives an idea of what happens to stability when general relativistic corrections to Newtonian gravity are important. Incidentally, pure radiation or degenerate matter in the ultrarelativistic limit tend to $\Gamma = 4/3$.

4. Suppose you were to observe the spin frequency of a pulsar with high precision over a long time. If you can measure the "braking index" $\Omega\ddot{\Omega}/\dot{\Omega}^2$, then you can in principle discriminate between spindown due to gravitational radiation and spindown due to magnetic torques. Recall that gravitational radiation gives $\dot{\Omega} \propto \Omega^5$, and magnetic dipole torques give $\dot{\Omega} \propto \Omega^3$. Use these to compute the braking indices for pure gravitational radiation and for pure magnetic dipole torques.

5. One burst source some people have proposed is pulsar glitches. In a glitch, the spin frequency of the pulsar changes suddenly, due (we think) to a sudden coupling between the crust and the underlying superfluid core. The energy release is $I\Omega\Delta\Omega$, but I is the moment of inertia of the crust, which is perhaps 10^{43} g cm^2, or 1% of the moment of the inertia of the star (because the crust exists only at low densities). In a really big glitch, astronomers observe that $\Delta\Omega \sim 10^{-6}\Omega$. Let's say that such a glitch happens to a star with $\Omega = 100$ rad s^{-1}, and that all the energy comes out in gravitational waves with frequency 2000 Hz (comparable to double the sound-crossing frequency), in a period of only 1 s. If this is a very close source, at 1 kpc (or about 3×10^{21} cm), could this be seen with advanced LIGO (sensitivity $\sim 2 \times 10^{-23}$ Hz $^{-1/2}$ at 2000 Hz)?

6. Accretion-induced collapse of a white dwarf to a neutron star has occasionally been proposed to explain various astrophysical phenomena. Just as the name would suggest, the idea is that enough matter falls onto a white dwarf that it goes over its Chandrasekhar mass limit and collapses to a neutron star. Such a white dwarf would have to be made of heavy-enough elements that nuclear fusion during the collapse would not produce a Type Ia supernova; maybe that condition is satisfied in some cases. Suppose that during the collapse the angular momentum of the star is preserved and so is the magnetic flux (when the magnetic flux is conserved, the surface magnetic field is proportional to $1/R^2$, where R is the radius of the object). For the purposes of this problem, we assume that we begin with an $M = 1.35\ M_\odot$ white dwarf with a radius of 10^8 cm, with a rotation period of 30 s (approximately the shortest known rotation period for a white dwarf) and a surface average magnetic field strength of 10^9 G (approximately the strongest known surface magnetic field for a white dwarf). Assume that both the white dwarf and the resulting neutron star, which we assume to have a radius of 12 km, rotate uniformly and have constant density, which means

that their moments of inertia are $\frac{2}{5}MR^2$ for mass M and radius R. What are the surface magnetic field strength and rotational period of the resulting neutron star? Do the implied field strengths and rotation stand out among the population of observed pulsars?

7. Could a neutron star rotate rapidly enough that it would not be able to collapse into a black hole? To investigate this, we will make rough assumptions and then ask what we would need to do to make the calculation more reliable. We will start by assuming that a neutron star is a uniform-density sphere no matter how rapidly it rotates, and that when it rotates, it does so uniformly (like a solid body). We will also use Newtonian physics, so that the moment of inertia of a neutron star of mass M and radius R is $I = \frac{2}{5}MR^2$ and its maximum angular velocity is $\Omega = (GM/R^3)^{1/2}$. To address the black hole question, we note that the dimensionless spin angular momentum of an object of spin angular momentum S and mass M is $\chi \equiv cS/(GM^2)$, and that black holes have $|\chi| \leqslant 1$. Your task is to compute $\chi_{\max}$ for a neutron star of mass M and radius R and to compute its value explicitly for a plausible mass and radius near maximum of $M = 2.3\ M_\odot$ and $R = 12$ km. If $\chi_{\max} > 1$, then the star would need to shed angular momentum before it collapsed to a black hole. After you obtain the answer, indicate a few improvements that you would suggest to the calculation.

Useful Books

Bertulani, C. A. 2007, Nuclear Physics in a Nutshell (Princeton, NJ: Princeton Univ. Press)

Crawford, N. 2019, Nuclear Physics: Concepts and Techniques (New York: Willford Press)

Glendenning, N. K. 2000, Compact Stars: Nuclear Physics, Particle Physics, and General Relativity (Berlin: Springer)

Lipunov, V. M. 2011, Astrophysics of Neutron Stars (Berlin: Springer)

Rezzolla, L., Pizzochero, P., Jones, D. I., Rea, N., & Vidana, I. 2018, The Physics and Astrophysics of Neutron Stars (Berlin: Springer)

Shapiro, S. L., & Teukolsky, S. A. 1983, Black Holes, White Dwarfs and Neutron Stars: The Physics of Compact Objects (New York: Wiley)

Thorne, K. S., & Blandford, R. D. 2017, Modern Classical Physics: Optics, Fluids, Plasmas, Elasticity, Relativity, and Statistical Physics (Princeton, NJ: Princeton University Press)

Chapter 8

Gravitational Waves and Fundamental Physics

8.1 What is Your Profession?!

So why do we test Einstein's theory of general relativity? After all, we are taught in school that this is **the** theory of gravity. The one that overthrew Newton's. The one that has been confirmed by every single test we have thrown at it. The one that predicts the most wonderfully bizarre phenomena in the cosmos, which in turn, one by one, have been observed and verified. Yet, we still doubt it and we still test it. Why?

There are many answers we can give to this question. The simplest is that testing hypotheses is what physicists do. We create a hypothesis and derive predictions from it, which we then contrast against data to either discard, amend, or confirm the hypothesis. Of course, this "scientific method" is not quite how physics is done nowadays, because modern physical theories are very complicated and often experiments or observations precede predictions. The essence of this philosophy, nonetheless, is still correct. In fact, this is in part how particle physics advanced in the 20th century, through predictions that were not confirmed by collider experiments, which then suggested that additions to the Standard Model of elementary particles were needed.

Ok, but beyond the "this is our duty" answer, are there any actual reasons to doubt Einstein's theory? One can answer this question from two different viewpoints: a theoretical one and an experimental one. The theoretical "problems," if we may call them that, start with the observation of Roger Penrose, Stephen Hawking, and others that singularities are ubiquitous and unavoidable predictions of Einstein's theory. The word "singularity" here means a place in spacetime where the curvature, as computed from some invariant, diverges. This, in turn, implies that other quantities that an observer could locally measure (or experience) would also diverge, such as the pressure and the energy density. Physicists abhor infinity (at least, some of us do!), and so instead of allowing for the concept of infinite physical observables, we simply argue that Einstein's theory must be incomplete. That there

doi:10.1088/2514-3433/ac2140ch8

must be some other, more fundamental theory, that "cures" these singularities by including quantum mechanics into the mix.

And this, the search for a quantum gravity theory, is one of the theoretical reasons we test Einstein's theory. As is well known in the popular psyche, quantum mechanics does not play well with general relativity (or is it the other way around?). What we mean is that when we take Einstein's theory (a so-called "classical field theory") and we attempt to quantize it, just like we do to produce quantum electrodynamics from classical electromagnetism, we run into some serious technical problems. When we quantize Maxwell's theory, we find that certain calculations diverge because they depend on (a finite set of) infinite integrals; think of, for example, the zero-point energy of vacuum. But these divergences can be "cured" through renormalization: you rewrite your Lagrangian in terms of measurable quantities (like the masses and charges of particles), instead of the "bare" quantities that you usually use for your Lagrangian, and this rewriting then generates a finite set of counter-terms that can be chosen so as to cancel all infinities. If one attempts the same trick with general relativity, one finds an infinite number of infinite integrals, and therefore, one needs an infinite number of counter-terms to renormalize it. This is why we say that general relativity is "perturbatively non-renormalizable." A possibility is then that standard perturbation theory is simply not valid when trying to quantize gravity, and instead, we need a fundamentally different description of gravity in the quantum regime.

Whatever the answer may be, wouldn't it be nice if experiment could guide theory? If we had an experimental inkling of a deviation from the predictions of general relativity, then maybe these data could light the way to a quantum theory, for example, ruling out some candidates or suggesting possibilities that have not yet been considered. Of course, any physicist well versed in dimensional analysis will immediately tell you that this is a pipe dream, that it's just not possible because one expects that quantum gravity effects should only be important at very short distances (maybe just a few Planck lengths away) from a singularity. At distances much farther from the singularity, for example, a macroscopic distance away from the horizon, one would then expect quantum gravity effects to be "Planck suppressed."

For the most part, dimensional analysis does tend to provide the right answer to most problems, but after we do a dimensional estimate, we always follow it up with a detailed calculation to see if our "naive" estimate was correct. This verification step is crucial because we have concrete examples in which dimensional analysis violently breaks down. Perhaps the most well-known example is the dimensional estimate for the cosmological constant Λ. Dimensionally, Λ has the same units as energy density, so one is tempted to associate it with the energy density of the vacuum via $\Lambda = 8\pi\rho_\Lambda$. The natural units for vacuum are Planck units, and because the Planck mass is $m_p = (\hbar G/c)^{1/2}$ and the Planck length is $\ell_p = (\hbar G/c^3)^{1/2}$, it then follows that $\Lambda \approx \rho_p \approx m_p/\ell_p^3 \approx \ell_p^{-2}$. However, when we rewrite the cosmological constant in terms of quantities that can be measured through cosmological observations, we find

$$\Lambda = 3\left(\frac{H_0}{c}\right)^2 \Omega_\Lambda, \tag{8.1}$$

where we recall from Chapter 6 that H_0 is the Hubble constant and Ω_Λ is the ratio of the energy density of the universe contained in the cosmological constant to the critical density that separates contraction from expansion. When one uses the measured values of these quantities, namely $H_0 \approx 70$ km (s Mpc)$^{-1}$ and $\Omega_\Lambda \approx 0.7$, one finds that $\Lambda \approx 10^{-122}\ell_p^{-2}$. This is 122 orders of magnitude smaller than the estimate obtained via dimensional analysis!

The failure of dimensional analysis in the case of the cosmological constant should not necessarily be used to argue that there must be quantum gravity modifications that are not Planck suppressed, for example, outside black hole horizons. Rather, the argument here is that, lacking a complete and reliable theory of quantum gravity, it may be safer to remain agnostic about the scale at which deviations from general relativity will be found. In an agnostic approach, one lets experiment and observation determine whether a deviation is present in the data. This has been the approach taken by many observers when analyzing a new astrophysical data set, and in some cases, it has led to anomalies, which have therefore provided the observational reason alluded to before to test or search for deviation of Einstein's predictions.

What is an anomaly? An anomaly is a feature in a data set that deviates from what is expected by the current broadly accepted physical model used to interpret the data set. Of course, if there is no broadly accepted model, then it is difficult to determine whether there is a true anomaly, or whether the model is incorrect. This can occur when dealing with data sets produced by the interplay of many complicated astrophysical processes that are difficult to model. For example, the modeling of accretion disks requires the study of fluids and electromagnetic fields around a rotating black hole, which in turn requires the solution to the Einstein equations, coupled to Maxwell's equations, coupled to the equations of relativistic hydrodynamics. Except in the most idealized scenarios, such solutions can only be obtained numerically, and in many instances, they require the modeling of instabilities and chaotic fluid motion.

But let us assume that one carries out an observation of a "simple" system, such that the data can be interpreted using a broadly accepted model. An anomaly can still arise if the broadly accepted model is not a sufficiently accurate representation of nature. This has happened many times before in the physical sciences, with perhaps the most relevant example being the anomalous precession of the perihelion of Mercury. After accounting for the perturbations of the planets and the oblateness of the Sun, Mercury's orbit is still observed to deviate from the Newtonian prediction by about 40″ per century. This is a classic example of an anomaly, whose resolution, in this case, would eventually be the overthrowing of Newton's theory in favor of Einstein's.

Several other anomalies have been discovered when observing astrophysical phenomena since the crowning of Einsteinian relativity, but all of them could be accommodated with a relatively innocent modification of general relativity. One of the best-known "anomalies" is the observation of the late-time acceleration of the universe, which we discussed already. One could of course try to concoct a modified theory of gravity to explain this observation or construct a dark-energy-like stress–energy tensor with negative pressure, but currently, the conservative modification is to add a cosmological constant to the Einstein equations and declare that this quantity is a new fundamental constant of nature that must be measured to be determined. Another example of an anomaly is the rotation curves of stars in galaxies, which one could attempt to explain with a modification of Newtonian gravity (and thus general relativity) at small accelerations. However, it is more common to argue that there is some "dark" component to the stress–energy tensor, dubbed dark matter, that provides the missing gravitation and fits the observations.

Given these theoretical and observational motivations, physicists and astronomers have put general relativity to the test over the last century, including the "classic" tests: the perihelion precession of Mercury, the bending of light by regions of large density, the gravitational redshift, and the Shapiro time delay. But given that there can be many modified theories of gravity that one could invent, how does one go about testing these theories in a unified and efficient way? In the 1970s, Clifford Will and Kenneth Nortdvedt developed a theory-agnostic formalism to connect and encapsulate all solar system tests: the so-called parameterized post-Newtonian (ppN) framework. In this scheme, one considers a family of metric tensors, whose members are labeled by certain post-Newtonian parameters. Each of these parameters has a physical interpretation in terms of different relativistic effects, such as a measure of how much spacetime is curved by a unit of mass. In general relativity, each of these parameters is a given real number, while in other theories of gravity, the parameters are functions of the coupling constants of the modified theory. With this ppN metric, one can then derive predictions for the classic tests, which will now be functions of the post-Newtonian parameters, and compare these predictions to data to measure the parameters directly. The difference between the measured value and the predictions of general relativity then allows for theory-agnostic constraints of generic deviations. This framework also allows for constraints on particular models once one employs the mapping between the post-Newtonian parameters and the coupling constants of a given theory.

A similar approach has been employed in tests of general relativity that use observations of pulsars in a binary with either a white dwarf or another neutron star. Binary pulsars emit radiation from one or both neutron stars that is collimated into a beam with a certain opening angle. When the radio beam crosses the line of sight to Earth, radio telescopes on Earth can detect a pulse of radiation. The times of arrival of these pulses therefore depend on both the rotational frequency and the orbital motion of the pulsar. The careful monitoring of these times of arrival can then be used to extract details about the orbital motion of the binary, including relativistic effects such as the pericenter precession of the orbit. In the 1990s, Thibaut Damour and Joe Taylor (yes, the same Taylor as in the Hulse–Taylor pulsar!) introduced a theory-agnostic formalism to connect and encapsulate all binary pulsar tests of general relativity: the so-called parameterized post-Keplerian (ppK) framework.

The idea here is to enhance the timing model that is fitted to the data with certain post-Keplerian parameters. Each of these parameters has a physical interpretation in terms of different relativistic effects, such as the rate of pericenter precession or the rate at which the orbital period decays. In general relativity, each of these parameters takes a particular functional form, which depends only on the unknown component masses and other Newtonian orbital parameters, such as the orbital period, that also enter the timing model. But in other theories of gravity, these parameters would not only depend on the unknown component masses but also on the coupling constants of the theory. Therefore, the measurement of at least two of these post-Keplerian parameters would reveal the component masses of the binary, but any additional post-Keplerian measurements would add redundancy and constitute model-independent tests of Einstein's theory. Through the mapping between these post-Keplerian parameters and their functional form in a given theory, one could also then find constraints on the coupling parameters of said theory.

To date, solar system and binary pulsar observations have severely constrained deviations from general relativity, and thus, many modified theories of gravity, which then raises the question: what is there left to test? Solar system observations are extremely precise, but they are only able to sample weak gravitational fields. If a modification of general relativity were to become important only when the gravitational field is strong, then solar system observations would not be able to constrain it. Binary pulsar observations, on the other hand, would be able to do so, provided the modification to general relativity is present around neutron stars. There exists an entire class of theories (so-called "quadratic gravity" theories) for which gravity is not modified much around neutron stars, and therefore, binary pulsar observations are not capable of constraining them.

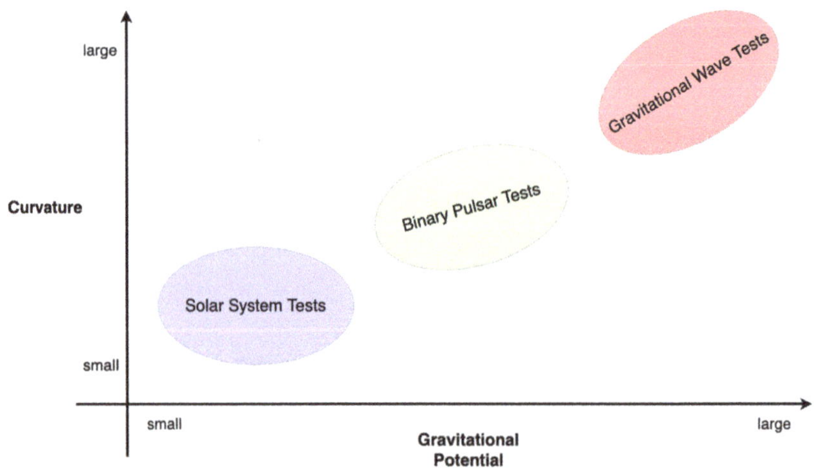

Figure 8.1. Cartoon of the different regimes probes by different tests of general relativity. The x-axis shows the magnitude of the gravitational potential, while the y-axis shows the measure of the curvature.

These thoughts are shown in Figure 8.1 as a schematic cartoon. The idea we want to convey here is that solar system tests typically probe weak gravitational fields and small curvatures, which is why we call them "weak-field" tests. Binary pulsar tests are able to probe stronger gravitational potentials and larger curvatures. Some observables, in fact, can probe the curvature right outside of a neutron star, for example, through measurements of the Shapiro time delay. The binary system in pulsar binaries, however, is very widely separated, so their orbital velocity is small, and a first-order post-Newtonian treatment suffices. Gravitational-wave tests, as we will see below, can probe a regime of larger gravitational potential and curvatures, because, in principle, they carry information about gravity when black holes collide at a good fraction of the speed of light.

Gravitational waves produced by binary systems are therefore ideal to probe gravity when the fields are very strong and rapidly changing. The question, of course, is how. To understand the answer to this question, we first need to understand how our gravitational-wave models are constructed. As we already saw in Chapter 3, gravitational waves are generated near the binary system, and then they propagate to the source over cosmological distances, until they hit the detector. One can then think separately of modifications to the generation of gravitational waves and to the propagation of gravitational waves. Before continuing to do so, a word of warning: both the generation and the propagation of gravitational waves are connected through the field equations of the theory one is considering. Typically, a modified theory of gravity will lead to modifications to both sectors simultaneously. There are some notable examples, however, in which only one of these two sectors is important, and we will discuss those below.

8.2 Generation of Gravitational Waves in Modified Gravity

But first let us talk about modifications to the generation of gravitational waves. In general relativity, gravitational waves are produced by the acceleration of matter sources as we saw back in Chapter 3. This is because the linearized Einstein equations reduce essentially to a source-wave equation.

Major Payne: Oh, come on! We can present that very important equation, can't we? We already saw back in Chapter 3 that the linearized Einstein equations take the form of Equation (3.8). Working to leading-order post-Newtonian order, this becomes simply

$$\Box_\eta h_{ij} = -16\pi \bar{T}_{ij}, \tag{8.2}$$

where $\bar{T}_{ij}$ is the trace-reversed stress–energy tensor of matter, and recall that $\Box_\eta$ is the D'Alembertian operator of flat spacetime. When one solves this equation by "inverting" the D'Alembertian, like Captain Obvious would say (but more properly, through the use of Green's functions), one finds the usual "quadrupole" formula

$$h_{ij}^{TT} \sim \frac{1}{r}\frac{\partial^2 I_{<ij>}}{\partial t^2}, \tag{8.3}$$

where we recall that h_{ij}^{TT} is the transverse-traceless metric perturbation, $I_{<ij>}$ is the transverse-traceless quadrupole moment of the binary (not to be confused with the moment of inertia), and r is the distance from the center of mass of the system to the detector. Indeed, we already saw a higher post-Newtonian order version of this equation back in Equation (3.8). Recall also that the (source) quadrupole moment for a binary system is $I_{ij} \sim m_1 x_1^i x_1^j + m_2 x_2^i x_2^j$, where $m_{1,2}$ are the component masses of the binary, and $x_{1,2}^i$ is the orbital trajectory (as measured from the center of mass). We see then explicitly how the generation of gravitational waves depends sensitively on the motion of dense objects.

Let us then begin by considering modifications to the motion of compact objects, and for simplicity, we consider the early inspiral phase only. Schematically, orbital motion is controlled by the binary's energy (or its Hamiltonian) and by any dissipative losses of energy and angular momentum that may be present. For a binary system of test particles in the center-of-mass frame (assuming this is conserved), the two-body problem maps to a one-body problem, where instead of having two objects of mass $m_{1,2}$, one has a single body of mass $\mu = m_1 m_2 / m$ in orbit around another body of mass $m = m_1 + m_2$ (as measured by distant observers). The proper way to write down the Hamiltonian in general relativity is then

$$\mathcal{H}(p_\mu, x^\nu, \tau) = \frac{1}{2} g^{\mu\nu} p_\mu p_\nu, \tag{8.4}$$

where $g^{\mu\nu}$ is the inverse of the metric tensor for the body of mass m, while $p_\alpha = \mu u^\alpha$ is the four-momentum of the particle with mass μ and 4-velocity u^α, and τ is the proper time. One can then consider deformations of the metric to find deformations of the Hamiltonian, which in turn, through the Hamilton equations, leads to modifications to the post-Newtonian equations of motion, i.e., modifications to Equation (3.10). Alternatively, one can consider deformations to the gravitational binding energy of the binary, which through its gradient would lead to modifications to the acceleration.

Major Payne: Let us see how these deformations are introduced in practice using the Hamiltonian formalism. First, we transform from a Hamiltonian that depends on 4-vectors of proper time $\mathcal{H}(p_\mu, x^\nu, \tau)$ to one that depends on 3-vectors of coordinate time $H(p_i, x^i, t)$. We do so by defining $H = -p_0$ and solving for p_0 from the normalization condition $p_\alpha p^\alpha = -\mu^2$ to obtain

$$H(p_i, x^i, t) = \left(\frac{\mu^2 + g^{ij} p_i p_j}{-g_{00}} \right)^{1/2}, \tag{8.5}$$

assuming the metric is diagonal. Assuming further that the metric of body m is spherically symmetric, its components can then be written as

$$g_{00} = -\left(1 - \frac{2m}{r}\varepsilon^2 + \alpha\delta g_{00}\right), \tag{8.6}$$

$$g_{rr} = -\left(1 - \frac{2m}{r}\varepsilon^2 + \alpha\delta g_{rr}\right)^{-1}, \tag{8.7}$$

where r is the field point distance (later to be set to the orbital separation r_{12}), while δg_{00} and δg_{rr} are functions of radius that characterize small deformations of the spherically symmetric metric. We have here also introduced the two bookkeeping parameters ε and α, which allow us to expand in weak fields and in small deformations. Restricting attention to the equatorial plane ($\theta = \pi/2$, $p_\theta = 0$) and noting that $|p_i p^i| = \mathcal{O}(\varepsilon^2)$, we then find

$$\begin{aligned} H(p_i, x^i, t) - \mu = &\frac{\varepsilon^2 p_r^2}{2\mu}\left(1 + \frac{1}{2}\alpha\delta g_{00} + \alpha\delta g_{rr}\right) + \frac{\varepsilon^2 p_\phi^2}{2\mu r^2}\left(1 + \frac{1}{2}\alpha\delta g_{00}\right) \\ &- \frac{\varepsilon^2 \mu m}{r} + \alpha\frac{\mu}{2}\delta g_{00} + \varepsilon^2\alpha\frac{\mu m}{2r}\delta g_{00} + \mathcal{O}(\varepsilon^3, \alpha^2). \end{aligned} \tag{8.8}$$

We recognize then that $H(p_i, x^i, t) - \mu$ is nothing but the orbital energy of the binary, which clearly holds for a system of comparable masses in the center-of-mass frame to leading order in a weak-field and small-deformation bivariate expansion.

Once one has a deformed Hamiltonian, one can now derive the equations of motion of the binary. The derivative with respect to p_ϕ is the angular velocity, i.e., $\omega \equiv \dot{\phi} = \partial H/\partial p_\phi$, because $p_\phi = L\mu$ with L the angular momentum per unit mass. The derivative with respect to p_r gives us $\dot{r}$, which can be solved for p_r and inserted into the Hamiltonian to obtain the effective potential $V_{\rm eff} \equiv \dot{r}^2/2$. Focusing on circular orbits, one can then solve for the gravitational binding energy E_b and the angular momentum of a circular orbit L_c by requiring that $V_{\rm eff} = 0 = dV_{\rm eff}/r$. When all of this is said and done, one finds an expression for the binding energy that is similar to what we presented in Chapter 3, namely

$$\frac{E_b}{\mu} = -\frac{m}{2r_{12}}c^2 + \frac{\alpha}{2\varepsilon^2}\left(\delta g_{00} + r\delta g_{00}'/2\right) + \mathcal{O}(c^3, \alpha^2/c^4), \tag{8.9}$$

where we have also required that the metric deformation be smaller than the weak-field expansion parameter. This requirement is not as restrictive as it may seem. The term of $\mathcal{O}(\varepsilon^2)$ in the expansion of g_{00} far from the system determines the total mass of the system (as measured by a far away observer). Therefore, if the $\mathcal{O}(\alpha)$ term in g_{00} were of the same size as the $\mathcal{O}(\varepsilon^2)$ term (and, in particular, if it also decayed as $1/r_{12}$), the metric deformation would change the mass of the system. But we started this whole calculation requiring that m be the observable mass, so then we must have that $\alpha \ll \varepsilon^2$. It may seem surprising that only the time–time component of the metric deformation affects the gravitational binding energy, but this is a well-known result of linearized gravity and post-Newtonian theory: the time–time component controls the leading-order Newtonian potential, while the space–space components only enter at higher order in perturbation theory.

The Hamiltonian picture above tells us the conservative side of the story. After all, given any Hamiltonian, orbital motion will be bounded, without any decay or losses. But the word "inspiral" implies that the orbital motion should not be fully conservative, but rather, that there ought to be radiative losses that force the orbital system to decay. These losses are due to the emission of waves and the emission of whatever fields are excited by the binary system and then leave the system, traveling at a given speed. In general relativity, the only field that could oscillate is the metric itself (the metric perturbation to be precise), and these waves leave the binary traveling at the speed of light and carrying with them energy, angular momentum, and linear momentum, as we saw in Chapter 3. But in modified theories of gravity, there could be other fields that are excited by the binary components, such as a scalar field or a vector field. And when the binary components move around each other, they will drag these fields, forcing them to oscillate and travel, usually as some type of wave, away from the binary system. These extra fields can also carry energy and angular and linear momentum away from the binary, therefore changing the radiative losses.

How can we model these additional radiative losses? Unfortunately, there is no generic way to do this, but there are some general statements we can make with our Fermi estimates. First, the field that is removing energy from the binary must be oscillating and traveling as a wave that reaches infinity with a $1/r$ fall off, as we saw in Chapter 1. Any faster than this, and the field would be negligible far away from the binary, and if it is not oscillating, well, it is then not traveling away from the system. These arguments suggest the field must take the form

$$\vartheta = \frac{\alpha}{r}\cos\phi, \tag{8.10}$$

where α is a parameter that encodes the magnitude of the ϑ field, and recall that r is the field point distance from the center of mass of the binary, and ϕ is the field's phase. We are using the symbol ϑ here to represent the field, but in practice, this field need not be a scalar, and instead, it could be a vector or a higher-rank tensor. As we also argued in Chapter 1, the energy carried away by this field per unit time (i.e., the luminosity) must scale as the square of the time derivative of the field times area, so including the general relativity (gravitational-wave) term we must have

$$\mathcal{L} \sim r^2\left(\dot{h}_{ij}\dot{h}^{ij} + \dot{\vartheta}^2\right). \tag{8.11}$$

In reality, one also has to average the above expression over several wavelengths, because wave-like perturbations are spread out over spacetime, and not localized in a finite region, but this is a minor detail.

And this is about as far as we can get with generic arguments because as should be obvious from the previous equation, we need to know the evolution equation of the extra field to estimate it. Back in Chapter 1, we were able to estimate the size of h because we knew that gravitational waves are sourced by the acceleration of masses, which mathematically boils down to Equation (8.2), which Major Payne presented. In a modified theory of gravity, the excitation of additional fields can be caused by a plethora of reasons, such as quadratic or higher-order curvature invariants, the stress–energy of matter sources or interactions among the multiple extra fields that may be present. Without knowing the field equations, it is impossible to estimate the magnitude of the extra fields.

But all is not lost. Because we expect the extra field to generate a wave induced by the motion of the binary system, it makes sense that it be sourced by derivatives of the orbital trajectory. We can then say that

$$\vartheta \sim \frac{\alpha}{r}\frac{\partial^n(\mathfrak{M}\mathfrak{L}^n)}{\partial t^n}. \tag{8.12}$$

Recall from Chapter 1 that $\mathfrak{M}$ and $\mathfrak{L}$ are the characteristic mass and length scale of the system (which in our case is the total mass and the orbital separation of the binary), while n represents the leading-order multipole moment of the field; for example, in general relativity $n = 2$ because gravitational waves are (to leading order in a far-field expansion) quadrupolar. Following the same arguments as before, it then follows that

$$\vartheta \sim \alpha\frac{m}{r}\eta^{n/2}a^n\omega^n, \tag{8.13}$$

where we recall that $\eta = \mu/m$ is the symmetric mass ratio, ω is the angular frequency, and a is the semimajor axis. From this, the luminosity can then be written as

$$\mathcal{L} \sim \mu^2a^4\omega^6 + \alpha^2m^2\eta^na^{2n}\omega^{2n+2}. \tag{8.14}$$

We can simplify this expression a bit by using Kepler's third law $\omega^2a^3 \sim m$ in the luminosity deformation to find

$$\delta\mathcal{L} \sim \alpha^2\eta^n(m\omega)^{2n/3+2}. \tag{8.15}$$

We see then that the luminosity deformation is suppressed quadratically by the deformation amplitude, but it can be enhanced over the general relativistic prediction at low frequencies, e.g., if the additional field has a dipole nature (i.e., if $n = 1$), then the luminosity deformation dominates at low frequencies (or large separations).

Understanding radiative losses is critical to understanding the inspiral because they control how rapidly the frequencies in the problem are changing with time. This is usually established through the balance law we saw back in Chapter 2 (see Equation (2.6)). In our context, this law has to be modified such that the energy carried away by all fields balances the rate of change of the gravitational binding energy of the system. This simple concept then implies that the rate of change of all quantities that depend on the radiation-reaction timescale must satisfy chain-rule laws of the form

$$\frac{d\omega}{dt} = \left(\frac{dE_b}{d\omega}\right)^{-1}\left(\frac{dE_b}{dt}\right) = -\left(\frac{dE_b}{d\omega}\right)^{-1}\mathcal{L}, \tag{8.16}$$

where we have used the orbital angular frequency here as an example (see also Equation (3.16) for a similar manipulation in general relativity). From our Hamiltonian analysis, we have seen that metric deformations will affect the gravitational binding energy, while from our radiative analysis we have seen that the sourcing of extra fields will affect the luminosity and the torque. Therefore, it is both the conservative and the dissipative part of the dynamics that affects the time evolution of the orbital phase and the orbital frequency!

Captain Obvious: But what happens if one of these modifications dominates over the other? Which one dominates depends on the functional form of the modifications one considers. In the inspiral, one uses the post-Newtonian approximation to solve the field equations in an expansion about small velocities and weak fields, resulting in series expansions that scale with powers of v and powers of m/a, which are connected by the virial theorem through $v^2 \sim m/a$. If a post-Newtonian expansion is admissible in the modified theory, then the modifications to the binding energy (through changes in the two-body metric) and to the luminosity and the torque (through activation of additional fields) will also scale with a given power of velocity. Indeed, we have seen that this is the case directly in the luminosity estimate of Equation (8.15). It is natural then to suggest the parameterization

$$\delta\mathcal{L} = A\ v^c, \qquad \delta E_b = B\ v^d, \tag{8.17}$$

where A and B control the magnitude of the deformation in the luminosity and the binding energy, while a and b are numbers that control the post-Newtonian order at which they first enter. Comparing these expressions to our expressions in Equations (8.9) and (8.15), we see that $A \sim \alpha^2\eta^n$ and $c = 2n + 6$, while $Bv^d \sim (\alpha/2)(\delta g_{00} + r\delta g_{00}'/2)$. We see then that whether the radiative correction dominates over the correction to the binding energy depends on whether $c > d$ (assuming obviously that both $A \neq 0$ and $B \neq 0$), and this depends on n and on the form of δg_{00} for our examples.

Given these modifications to the conservative and dissipative dynamics, what is the modification to the waveform? Once more, answering this question in general is not possible because the waveform depends on the equation of motion for the gravitational-wave metric perturbation, which in turn, depends on the modified gravity field equations. In particular, in modified gravity theories gravitational waves can have up to six polarizations, so four more beyond the two + and × polarizations of general relativity. Ignoring these other polarizations for now, one can show that deformations introduced to the gravitational binding energy and the luminosity will lead to modifications in the waveform of the form

$$\tilde{h} = \tilde{A}_{GR}(f)(1 + \alpha u^a)e^{i\Psi_{GR}(f)+i\beta u^b}, \tag{8.18}$$

to leading order in the deformation, where $\tilde{h}$ is the Fourier transform of the response function, $\tilde{A}_{GR}$ and Ψ_{GR} are the amplitude and phase of the Fourier transform of the response function in general relativity, while $u = (\pi\mathcal{M}f)^{1/3}$, and we recall that $\mathcal{M} = \eta^{3/5}m$ is the chirp mass and $\eta = m_1m_2/m^2$ is the symmetric mass ratio.

The waveform model presented above is the basis of the parameterized post-Einsteinian (ppE) framework for quasi-circular inspiral signals in modified gravity. The quantities α and β are post-Einsteinian "amplitudes" that depend on the coupling constant of the modified theory, as well as other dimensionless combinations of the binary parameters. If these amplitudes are set to zero, one recovers exactly the predictions of general relativity. The quantities a and b are post-Einsteinian "exponents," which are numbers that control the post-Newtonian order at which modifications to general relativity first hit the waveform. For example, when $b = -7$ the β term represents a -1PN deformation from general relativity, because the leading-order term in Ψ_{GR} goes as u^{-5} and $u = \mathcal{O}(v/c)$. One can think of a test that uses the ppE framework as a "ppN test", because in some cases (for some subset of values of a and b) the ppE corrections would enter at the same post-Newtonian orders as predicted in general relativity; e.g., when $b = -5$ the ppE correction enters at the same order as the "Newtonian" term of general relativity, when $b = -3$ it enters at 1PN order, etc. However, in the ppE framework, a and b are also allowed to take values that are not predicted in general relativity, such as the value $b = -7$, which represents the activation of dipole radiation.

One can then, in principle, use such a waveform meta-model to filter gravitational-wave data and place constraints on the magnitudes of α and β deformations from general relativity. Because the model does not refer to a specific theory of gravity in particular, such constraints are said to be model independent or agnostic. Given a modified theory of gravity, one can also in principle calculate that theory's values of (α, a, β, b) at leading post-Newtonian order and to leading order in a deformation from general relativity. The construction of such a dictionary is crucial if one wishes to infer constraints on modified theory parameters from constraints on α and β. In this way, one can turn model-independent constraints to constraints on particular models, just as is done in the ppN and the ppK frameworks.

The ppE formalism, however, is not the "be-all and end-all" of tests of general relativity, and like the ppN and ppK formalism, it has its weaknesses. For one, the formalism, as described above, requires that the signal observed contain enough cycles of inspiral. If the only part of the signal that can be detected by a given instrument is the merger and ringdown, then the ppE inspiral waveform described above cannot be used to test general relativity. This, in turn, depends on the instrument used to observe gravitational waves and on the total mass of the system. Why does it depend on the instrument? Because interferometric detectors are not "wide-band" antennas, but rather operate in a given frequency range. Ground-based detectors, for example, are limited at low frequency (at about 10 Hz) by ground motions.

Captain Obvious: But why does it depend on the total mass of the system? Because this quantity determines the gravitational-wave frequency at merger. One can realize that this had to be the case using our Fermi instincts: the frequency has units of inverse seconds and, in general relativity, a mass can be converted into a quantity with units of seconds by multiplying with G/c^3, so therefore $f_{\rm merger} \sim c^3/(Gm)$, with m the total mass of the binary. We can check whether this dimensional argument is valid with our Newtonian intuition because we know from Kepler's third law that $2\pi f = (Gm/r_{12}^3)^{1/2}$, where f and r_{12} are the orbital frequency and the orbital separation respectively. Setting the orbital separation to twice the radius, which for a nonrotating black hole is twice its individual mass, $r_{12} = 2Gm/c^2$, and setting the gravitational-wave frequency to twice the orbital frequency, $f_{\rm GW} = 2f$, we then have $f_{\rm merger} = (\sqrt{2}/8)\pi^{-1}[c^3/(Gm)]$ and thus $f_{\rm GW} \sim 10^{-1}[c^3/(Gm)]$. We see then that our Fermi estimates were not too far off! Plugging in numbers, we find that $f_{\rm GW} \sim 375$ Hz if $m = 60M_\odot$, while $f_{\rm merger} \sim 10$ Hz if $m = 2 \times 10^3 M_\odot$. Of course, these numbers are not quite right, because we are still using Newtonian formulae, but the scaling with m^{-1} is indeed correct.

Another weakness of the ppE formalism is that it considers only the leading-order modification to the general relativity waveform in a series expansion in both small velocities (i.e., in a post-Newtonian expansion) and small deformations from general relativity (i.e., small post-Einsteinian amplitudes). If the modification to general relativity does not admit a post-Newtonian expansion or a small-deformation analysis, then this particular theory will not fit in the ppE framework. Are there theories like that? Although the majority of the theories considered are not like that (and thus, they do fit in the ppE framework), one can concoct theories that do not admit a post-Newtonian expansion. There are two general classes of such theories that have been discovered.

The easiest ones to understand are those that add a new massive field (such as a scalar field or a vector field) to the description of gravity. The mass of the new field introduces a new length scale to the problem through its Compton wavelength ($\lambda_C = 2\pi\hbar/(m_{\rm field}c)$); inside this length scale, the field is active and can have a large effect, but outside, the field is typically exponentially suppressed. If the mass of the extra field is just right, one can have a "Goldilocks" situation in which the orbital separation starts out much larger than the Compton wavelength so initially there is no effect of the new field, and the inspiral proceeds as in general relativity. But then, some time later during the inspiral observation, the orbital separation crosses the Compton length scale and then the field can have a large effect on the orbital evolution (for example, a speeding up of the inspiral because of the sudden activation of dipole emission); for LIGO this happens for fields with mass $m_{\rm field} \approx (10^{-11}, 10^{-13})$ eV for binaries with mass $m \approx (3\text{–}100)M_\odot$. As you can expect, a theory like this will break the post-Newtonian expansion because, somewhere in the middle of the inspiral, the scalar field will "turn on" (on a scale comparable to the orbital one), and the dynamics will be greatly altered, leading to an evolution that cannot be well represented by a finite series in powers of the velocity.

Another class of theories for which a post-Newtonian expansion may break down is that which predicts the phenomenon of "dynamical scalarization." These theories are somewhat similar to the ones described above in that they contain an additional field (a vector or a scalar), but this time the field is massless. Nonetheless, the theories are such that the additional field can be effectively inactive when the binary is widely separated, only to then turn on rapidly when the binary has shrunk and the binding energy has exceeded a given threshold that depends on the coupling parameters of the theory. This "dynamical scalarization," the turning on of the scalar field due to the dynamics of the binary, is quite similar to the activation of a massive field described above. And just as in the previous case, when the additional field turns on, the binary dynamics can be greatly modified, for example by speeding up the inspiral due to the extra field emitting dipole radiation. When this is the case, the orbital dynamics cannot be modeled as a power series in velocity, meaning that the post-Newtonian approximation is not well equipped to provide an accurate model.

So is it all lost then? Can we just not apply the ppE tests to these theories? In principle, that's right, but in practice, it seems like we can get away with it! Researchers have found that if the true theory of nature was a modification to general relativity that contained one of these non-post-Newtonian effects, a ppE analysis would still reveal that a deviation from general relativity is present, provided the signal-to-noise ratio is high enough. Of course, the ppE analysis would not be able to pinpoint what non-Einsteinian physical effect caused the deviation, because the ppE model would not be able to faithfully recover the non-post-Newtonian signal. However, a ppE analysis would be able to identify that an anomaly was present in the data.

8.3 Propagation of Gravitational Waves in Modified Gravity

We have considered above modifications to gravitational waves due to corrections in the way the waves are generated by a binary system, but what about corrections that are introduced as the waves propagate from the source to us? In fact, one may expect these propagation effects to dominate over generation effects because, even if the propagation modification is small, it could build up over cosmological distances, leading to an enormous modification at the detector. Notice that we are not here talking about a constant change in the speed of propagation of gravitational waves. Such an effect would be 100% degenerate with the time of coalescence unless you have an electromagnetic counterpart (as we discuss below). Changing the speed of gravity by a constant would "look" like the inspiral took longer than it really did, thus changing the time of coalescence in the waveform model. Instead, we are here talking about changing the dispersion relation of gravitational waves, or more precisely, its phase and group velocity. Such a modification would mean that waves of different frequencies propagate at different speeds, so one expects a bunching up of the wave train, as measured at the detector. This effect is, in principle, not degenerate with anything else introduced by the general relativity model.

How would such a modification scale? Let us try to use dimensional analysis and our Fermi instincts to "predict" what the modification should be. We have already argued that the modification should scale with distance, but distance has units of length, so let us nondimensionalize this quantity using the only other natural length scale of the problem, the total mass of the system, (D/m), where we recall that D is the distance to the source. What else should the modification depend on? It must depend on the coupling constants of the modified theory, let us call them all ζ, such that when these are taken to zero, one recovers general relativity. If ζ is not dimensionless, then it must define a Compton-like scale we will call λ_ζ, which must be made dimensionless again via normalization with the total mass, so the correction must be proportional to (m/λ_ζ). Notice here that we've put λ_ζ in the denominator, because we are demanding that as the Compton-like scale goes to infinity, then this is equivalent to $\zeta \to 0$ and one recovers general relativity. We will also raise the power of the combination (m/λ_ζ) to the $p \in \mathbb{N}_{>0}$ because we don't know a priori how the modification should scale with λ_ζ, while we do know that it will scale linearly with distance. Finally, we know that the modification induces a bunching up of the wave train, because the wave speeds will be a function of the frequency, so the correction to the gravitational-wave phase should scale as (mf) to a given power $q \in \mathbb{N}_{>0}$. Putting all of this together, our Fermi estimates suggest the modification to the gravitational-wave phase should scale as

$$\delta\Psi(f) \sim \left(\frac{m}{\lambda_\zeta}\right)^p \left(\frac{D}{m}\right)(mf)^q. \tag{8.19}$$

Great! So we have an estimate, but is it correct? It certainly has the right units and it has the right scaling. To figure out whether we got it right, and also to understand any order-unity numerical coefficients we may have left out, we need to do this calculation carefully. We begin with the dispersion relation in general relativity. In Einstein's theory, gravitational waves travel at the speed of light, so their phase velocity $v_p = \omega_{\rm GW}/k$ and group velocity $v_g = d\omega_{\rm GW}/dk$ equal unity, where $\omega_{\rm GW}$ and k are the frequency and wavenumber of the wave. If we think of gravitational waves as a train of particles we will call "gravitons," then assuming that $E_g = \hbar\omega_{\rm GW}$ and $p = \hbar k$, we must have that $E^2 = p^2c^2$, the mass–energy relation of special relativity for a massless particle. Adding a mass to this relation is simple, $E^2 = p^2c^2 + m_g^2c^4$, as we know from special relativity for a particle with mass m_g. If one wished to deform this relation further, one could for instance write

$$E^2 = p^2c^2 + m_g^2c^4 + \mathbb{A}E^\alpha, \tag{8.20}$$

where $\mathbb{A}$ is an amplitude that controls the magnitude of the deformation and α is a number that controls the type of deformation.[1] From this dispersion relation, one finds that such a wave would have

[1] This α has nothing to do, in principle, with the α deformation parameter we introduced in the previous section when talking about the generation of gravitational waves.

$$v_{\text{phase}} \equiv \frac{\omega_{\text{GW}}}{k} \sim 1 + \frac{1}{2}\frac{m_g^2}{k^2} + \frac{1}{2}\mathbb{A}k^{\alpha-2} + \mathcal{O}(m_g^4/k^4, \mathbb{A}^2), \tag{8.21}$$

$$v_{\text{group}} \equiv \frac{d\omega_{\text{GW}}}{dk} = 1 - \frac{1}{2}\frac{m_g^2}{k^2} - \frac{1}{2}\mathbb{A}(1-\alpha)k^{\alpha-2} + \mathcal{O}(m_g^4/k^4, \mathbb{A}^2), \tag{8.22}$$

$$v \equiv \frac{p}{E} = 1 - \frac{1}{2}\frac{m_g^2}{E^2} - \frac{1}{2}\mathbb{A}E^{\alpha-2} + \mathcal{O}(m_g^4/E^4, \mathbb{A}^2). \tag{8.23}$$

We see then that although the phase velocity v_{phase} can exceed unity, the group velocity v_{group} and the particle velocity v are smaller than unity provided $\alpha \leqslant 1$. Of course, one can still have $\alpha > 1$, provided one makes $\mathbb{A}$ sufficiently small such that the $\mathcal{O}(m_g^2/k^2)$ term is larger. We also see that when $\alpha = 0$, the $\mathcal{O}(\mathbb{A})$ deformation becomes equivalent to a mass term, while when $\alpha = 2$, the $\mathcal{O}(\mathbb{A})$ deformation becomes a constant and is degenerate with a change in the constant speed of gravity.

Consider now two gravitons emitted from the same source but at two different times, say t_e and t'_e, with corresponding energies $E_e = hf_e$ and $E'_e = hf'_e$, and received at a detector very far away from the source, at arrival times t_a and t'_a. Assuming that the difference in emission time is much smaller than the expansion rate of the universe and that the gravitons are traveling at the group velocity of Equation (8.22), one can show that

$$\Delta t_a = (1+z)\left[\Delta t_e + \frac{D_0}{2\lambda_g^2}\left(\frac{1}{f_e^2} - \frac{1}{f_e'^2}\right) + \frac{D_\alpha}{2\lambda_{\mathbb{A}}^{2-\alpha}}\left(\frac{1}{f_e^{2-\alpha}} - \frac{1}{f_e'^{2-\alpha}}\right)\right], \tag{8.24}$$

where $\Delta t_a \equiv t'_a - t_a$ and $\Delta t_e \equiv t'_e - t_e$. Recall that z stands for cosmological redshift, and we have defined the length scale $\lambda_{\mathbb{A}} = h(1-\alpha)\mathbb{A}^{-1/(2-\alpha)}$. The quantities D_α and $D_0 = D_{\alpha=0}$ are distance measures, both with units of length, but slightly different dependence on redshift:

$$D_\alpha \equiv \frac{c(1+z)^{1-\alpha}}{H_0}\int_0^z \frac{(1+z')^{\alpha-2}dz'}{\sqrt{\Omega_M(1+z')^3 + \Omega_\Lambda}}. \tag{8.25}$$

Recall that H_0 is the Hubble constant and $\Omega_{\Lambda,M}$ are the fractional energy density parameters (see Chapter 6). Equation (8.24) shows precisely the frequency-dependent bunching up of the signal: when $\alpha < 2$, waves emitted at lower frequencies travel more slowly than waves emitted at higher frequencies (and vice versa when $\alpha > 2$), leading to a bunching up the wave train toward merger.

Captain Obvious: Let's pause for a second and consider what the authors have presented. Clearly, the second and third terms in Equation (8.24) must have units of length, because they are to be added to the first term in that equation, which is a time difference. The second term in this equation makes sense, because D_0 is a distance and λ_g is a wavelength. The third term, however, is a bit more complicated. Because we know that D_α is a distance measure, it must have units of length, so this means the constant $\mathbb{A}$ cannot be dimensionless. Indeed, $\mathbb{A}$ must have units of $(\text{Length})^{2-\alpha}$, so that $\lambda_{\mathbb{A}}$ has units of $(\text{Length})^1$, and the units of the third term in Equation (8.24) work out.

But what is this distance measure D_α? It is not the same as the luminosity distance. We can see that this is the case by comparing Equation (8.25) to Equation (6.5). Alternatively, we can also show this clearly by taking the small redshift limit, $D_\alpha/D_L \sim 1 - (1 + \alpha/2)z + \mathcal{O}(z^2)$ when $z \ll 1$. We see then that $D_\alpha < D_L$ in general, because typically one considers cases where $\alpha > 0$. But because $D_L \sim z/H_0$ for $z \ll 1$, we also see that the difference between D_α and D_L is really proportional to z^2, and thus, in the nearby universe, they are nearly the same.

How does this change in the propagation affect the gravitational waves observed at the detector? To answer this question one must propagate the correction found above through the calculation of a gravitational wave in the stationary phase approximation. The calculation is not very complicated (and is in fact ideal for a homework problem...), as it really boils down to the following integral

$$\Psi(f) = \int_{f_c}^{f} (t - t_c) df, \tag{8.26}$$

where $\Psi(f)$ is the Fourier phase of the gravitational-wave response function, f_c and t_c are the frequency of the wave and the time at coalescence, and t is the function of frequency derived above. Carrying out this elementary integral, we find

$$\tilde{h} = \tilde{A}(f)e^{i\Psi(f)}, \tag{8.27}$$

where $\tilde{A}$ is the Fourier amplitude of the response function as calculated in general relativity, $\Psi(f) = \Psi_{\rm GR}(f) + \delta\Psi_{\rm prop}$ with $\Psi_{\rm GR}(f)$ the Fourier phase as calculated in general relativity. When $\alpha \neq 1$ we have

$$\delta\Psi(f) = -\frac{\pi^2}{1+z}\left(\frac{\mathcal{M}}{\lambda_g}\right)^2\left(\frac{D_0}{\mathcal{M}}\right)u^{-3} - \frac{\pi^{2-\alpha}}{(1-\alpha)(1+z)^{1-\alpha}}\left(\frac{\mathcal{M}}{\lambda_{\mathbb{A}}}\right)^{2-\alpha}\left(\frac{D_\alpha}{\mathcal{M}}\right)u^{3\alpha-3}, \tag{8.28}$$

while when $\alpha = 1$ we have

$$\delta\Psi(f) = -\frac{\pi^2}{1+z}\left(\frac{\mathcal{M}}{\lambda_g}\right)^2\left(\frac{D_0}{\mathcal{M}}\right)u^{-3} - 3\pi\left(\frac{\mathcal{M}}{\lambda_{\mathbb{A}}}\right)\left(\frac{D_1}{\mathcal{M}}\right)\ln u, \tag{8.29}$$

where as you recall $u = (\pi\mathcal{M}f)^{1/3}$. We recognize the term proportional to λ_g^{-2} as a 1PN correction to the leading-order term in $\Psi_{\rm GR}$, while the term proportional to $\lambda_{\mathbb{A}}^{-2+\alpha}$ is a $(1 + 3\alpha/2)$ correction and the term proportional to $\lambda_{\mathbb{A}}$ is a 2.5 PN order correction.

How did our Fermi estimate of Equation (8.19) compare to the actual calculation above? We see that indeed the structure above is exactly what our Fermi estimates predicted, with the Fermi exponent p related to α and the scale $\lambda_{\mathbb{A}}$ related to λ_ζ. Equation (8.28), however, contains two terms (one proportional to λ_g^{-2} and one proportional to $\lambda_{\mathbb{A}}^{-2+\alpha}$), while the Fermi estimate contained a single term. Why is this? It's because we have separated the massive graviton modification from the generic dispersion relation modification. Typically, one would consider only one such modification at a time, and we see that the Fermi estimate can recover either. What the Fermi estimate cannot recover, however, is the logarithmic dependence in Equation (8.29) when $\alpha = 1$. This dependence comes about because the integral of a polynomial is always a polynomial *except* when the power is exactly -1! In that case, it becomes a log, and this is not something our Fermi estimate predicted.

Incidentally, we also see that the above modification has exactly the form of a ppE modification. The main difference is that the propagation correction to the wave is here tagged onto the entire inspiral–merger–ringdown signal, while the inspiral ppE modification could only be applied to the inspiral phase. Either way, once a constraint on the post-Einsteinian parameter β is obtained, one can in principle connect it to a constraint on λ_g or $\lambda_{\mathbb{A}}$, and from there constrain particular modified gravity theories for which Equation (8.20) is known.

8.4 The Nature of Black Holes

We have so far considered tests of gravity in the generation of gravitational waves or in their propagation, but we are missing one big elephant in the room: tests of the very nature of a black hole. What do we mean by this? We mean tests that try to determine whether the compact object observed is truly a black hole or whether it is some other exotic compact object, or "ECO" as they are sometimes called (not to be confused with the word "echo," which we will also discuss later in this section).

What determines whether an object is a black hole or not? That's a deep philosophical question, but in general relativity its answer is almost by definition: a black hole is an object with a curvature singularity that is hidden by an event horizon. There are a lot of technical words here, so let's unpack them. "Curvature singularity" here means a region in spacetime where some curvature invariant becomes infinite. Indeed, for a nonrotating and isolated black hole in most standard coordinate systems (like in the usual Schwarzschild coordinates), any scalar you can construct from the curvature tensor (i.e., the Riemann tensor) diverges at the origin (except for the Ricci tensor, which obviously vanishes). As we mentioned earlier in this chapter, this is a problem to physicists because as observers approach this region of spacetime, they would also measure (and experience!) infinite distortions of spacetime and of themselves!

But such an observer would not be able to communicate this singular behavior to observers outside the black hole because she would be inside the black hole "event horizon." We have mentioned this term before in this book, and you may have encountered it in your other studies as well, so we won't dwell on it much here. Just think of an event horizon as a one-way membrane, a boundary in spacetime from inside of which nothing moving at or slower than the speed of light can escape. In particular, any message sent at the speed of light from inside the event horizon will never reach observers sitting outside the event horizon. In many ways, the event horizon therefore defines the "surface" of the black hole, although this is a very special surface. It is a "political" surface, like the boundary between countries. There is no matter, radiation, or anything else at the event horizon that an observer could encounter as she crosses it (at least in standard general relativity). In fact, for a sufficiently massive black hole, an observer may not even notice crossing the horizon at all!

The vacuum Einstein equations seem to abhor curvature singularities, so all black hole solutions to the equations of general relativity that we know of are hidden behind an event horizon. Indeed, in 1969 Roger Penrose proposed the hypothesis that in fact all singularities that come about from solutions to the Einstein equations are hidden behind an event horizon, a statement that came to be known as the Cosmic Censorship Hypothesis. There has been a long mathematical history related to this hypothesis and about whether one can even formalize it enough to prove it mathematically, but we will not go into such details here. The key point is that if nature is described by general relativity, then naked singularities (i.e., those that are not hidden by an event horizon) do not exist. Of course, if nature were not described by general relativity, but perhaps by some other quantum gravitational theory that plays well with quantum mechanics, then it could be possible that the singularities that arise in the classical solutions to the Einstein equations would be "cured" by the quantum modifications introduced. And if this is the case, then these new non-singular solutions would not need an event horizon to hide anything, because, well, there wouldn't be anything to hide!

Therefore, determining whether the compact objects in nature are black holes and whether they are the black holes of general relativity are two very important questions, but notice that they are distinct. The first question is about finding compact objects in nature that look like black holes (because they are very compact), but are not black holes (because they do not have an event horizon). Such objects are sometimes called black hole mimickers, and finding one would be revolutionary, as it would necessarily indicate the need to do some violence to Einstein's theory (either in the form of introducing exotic matter that probably violates the energy conditions, or modifying the way matter curves geometry). The second question is about finding a compact object in nature that is a black hole, but that is not the Kerr (electrically uncharged) or the Kerr–Newman (electrically charged) solutions for isolated and rotating black holes. Could it be the case in nature that black holes are not those of Kerr?

The no-hair theorems would suggest this not to be so. In the 1970s, several physicists (including Stephen Hawking, Werner Israel, Ivor Robinson, Brandon Carter, and others) developed a series of mathematical theorems that, when put together, essentially state that, given a series of conditions, then the black holes of general relativity are the Kerr (or Kerr–Newman) spacetime. What are these conditions? They assumed (i) that general relativity is valid, (ii) that the black holes are isolated, meaning they are vacuum solutions to the Einstein equations without any matter in the spacetime, (iii) that the black holes have settled down to their final, relaxed state, so they are stationary, (iv) that spacetime is four dimensional, and (v) that a given black hole is rotating about an axis, and so the spacetime is axisymmetric. If these conditions hold, then the spacetime is that of the Kerr (or

Kerr–Newman) spacetime, which means it can be fully characterized by only three numbers: their mass, their spin, and their charge. Astrophysically, if black holes formed with any charge, they are expected to neutralize very fast due to the ambient plasma that surrounds it right after formation. Thus, we typically just say that "black holes have two hairs," their mass and their spin, following the terminology of John Wheeler.

Dr. I. M. Wrong: Yes, I've got it! What we must do then is test these no-hair theorems with observations. If we take some data, be it from gravitational-wave or electromagnetic observations, and we can show that the black hole observed is not described by the Kerr metric, because we need more than just these two hairs to describe it, then we've proven that these "luminaries" were all wrong and the no-hair theorems are just poppycock! I'll jump right in and analyze the Event Horizon Telescope data …

Major Payne: Wait a minute! A theorem cannot be poppycock, unless there is a mathematical flaw in the logic used to prove it. Given a set of assumptions, sometimes called premises, the mathematician uses logical arguments to prove a conclusion. Therefore, it makes no sense whatsoever to talk about "testing the no-hair theorems." That's like trying to use data to prove or disprove that $1 + 1 = 2$.

What you can do, however, is test whether the premises on which the theorems rely hold in nature. This, of course, need not be true, and therefore, there are various ways in which the no-hair theorems could not apply. For example, if nature were not described by the Einstein equations, then one could find black hole solutions that differ from the Kerr metric in modified gravity. This has indeed been done in a variety of modified theories, although there are also some modified theories in which the Kerr spacetime is also a solution.

Another example is to imagine that nature is not described by a four-dimensional spacetime but rather by a higher-dimensional spacetime. If so, then again the no-hair theorems would not apply, and one could find higher-dimensional black hole solutions of the higher-dimensional version of the Einstein equations that are not described by the Kerr geometry. This was done in the early 2000s, by for instance Roberto Emparan, Harvey Real, and collaborators, who found black hole solutions that look like black rings. These solutions were later generalized to other cases, perhaps most interestingly into solutions that look like the planet Saturn, with a spherical black hole surrounded by a black ring.

But the most obvious example is to imagine a black hole surrounded by an accretion disk, and we don't have to stretch our imaginations to believe such objects exist in nature! In astrophysics, when we consider a black hole with an accretion disk, we are typically interested in the behavior of the gas, the magnetic fields, and the radiation emitted as the gas spirals into the black hole. These systems are usually just modeled using the Kerr spacetime, but this is technically not true. A black hole with an accretion disk is not a vacuum scenario, so the no-hair theorems do not apply, and the Kerr spacetime will not be the correct solution to the Einstein equations. Rather, one must include the stress–energy tensor that represents the accretion disk, and then solve the nonvacuum Einstein equations. It turns out, however, that because the mass of accretion disk material near the black hole is truly tiny relative to the mass of astrophysical black holes, the Kerr geometry is an excellent approximation. With that said, though, technically speaking, astrophysical black holes do not satisfy the premises of the no-hair theorems, and thus, are not exactly described by the Kerr spacetime.

Although we cannot test the no-hair theorems, we can test the premises these theorems rely on, and so people have focused on testing the so-called Kerr hypothesis, that all astrophysical black holes are described by the Kerr geometry. Certainly, we expect violations of the Kerr hypothesis because black holes in nature are never exactly isolated. However, we also expect such deviations from the Kerr spacetime to be tiny (and probably unobservable with current or near-future instruments). Therefore, what one is testing when confronting the Kerr hypothesis is the possibility that astrophysical black holes are not described by the Kerr geometry because they are described by black hole solutions in modified gravity (be it due to modifications to the way matter curves geometry, or through the inclusion of scalar or vector fields that couple to gravity), or in higher-dimensional theories.

Ok, so how do we go about testing the Kerr hypothesis or searching for black hole mimickers? Let's discuss the Kerr hypothesis first. The Kerr spacetime can be expanded through an infinite sum of multipole moments. But because the metric only depends on the mass and the spin of the black holes, all of the multipole moments must depend on only these two quantities. Indeed, this is the case, as R. O. Hansen and Robert Geroch showed in 1974:

$$M_\ell + iS_\ell = M(ia)^\ell, \tag{8.30}$$

where M_ℓ are called mass-multipole moments and S_ℓ are current-multipole moments, because of how and where they enter into the far-field expansion of the Kerr spacetime. For instance, $M_0 = M$ is the mass of the black hole, while S_1 is the magnitude of its spin angular momentum. Therefore, if one could go about measuring the multipole moments of the spacetime, one could check whether the equation above is satisfied, and therefore test the Kerr hypothesis.

How would one go about doing so? One way is through extreme mass-ratio inspirals (EMRIs). As we discussed in Chapter 3, EMRIs encode information about the spacetime background of the supermassive black hole around which the small object is moving. This is because their trajectory depends sensitively on the background. In some sense, they act as "tracers" of spacetime through the gravitational waves they emit, and thus, they are able to reveal the multipole structure of the supermassive black hole spacetime. This is very much akin to the science of geodesy, in which one tries to understand, among other things, the gravitational field of Earth (for example, through observations by the GRACE and LAGEOS satellites). As Fintan Ryan showed in the 1990s, EMRIs would then allow for spacetime geodesy, and therefore, for tests of the Kerr hypothesis through the verification of Equation (8.30).

Another way to test the Kerr hypothesis is through the gravitational waves emitted during the ringdown phase of a black hole merger. As we saw in Chapter 3, the collision of two black holes results in a distorted object that relaxes down through the emission of gravitational waves that can be described as exponentially damped sinusoids, or QNMs. The damping time and the real oscillation frequency of these waves carry a signature of the gravitational theory in play. This is because, as discussed above, in general relativity, the Kerr geometry is fully described just by the mass and spin of the black hole. Therefore, if one is able to measure the first N complex quasi-normal frequencies of the gravitational waves emitted during the ringdown phase, then one could use one of them to infer the mass and spin of the final black hole and the remaining $N - 1$ frequencies to carry out $N - 1$ tests of general relativity.

How would one carry out such a test in practice? One way to do this is to take a page from the binary pulsar community and the ppK formalism. Imagine the M–a plane, where M and a stand for the mass and spin parameter of the final black hole the spacetime is relaxing to. In this plane, the real and the imaginary parts of each quasi-normal modes define a curve, whose shape depends on the theory of gravity. For example, imagine you have detected the real part of the $(\ell, m, n) = (2, 2, 0)$ ringdown mode to be some number $\hat{\omega}_{220}$ with a 1σ uncertainty of $\delta\omega_{220}$; then, the equation $\omega_{220}(M, a) = \hat{\omega}_{220} \pm \delta\omega_{220}$ (with ω_{220} the fit presented in Equation (3.36) if one assumes general relativity is correct) defines a thick curve in the M–a plane, as shown in Figure 8.2. Each new measurement of the real or of the imaginary part of a QNM frequency defines a new thick curve in the same way. In general relativity, two such thick curves will intersect in a region of the M–a plane, revealing the best-fit values of the mass and spin of the final black hole remnant (with the area of this

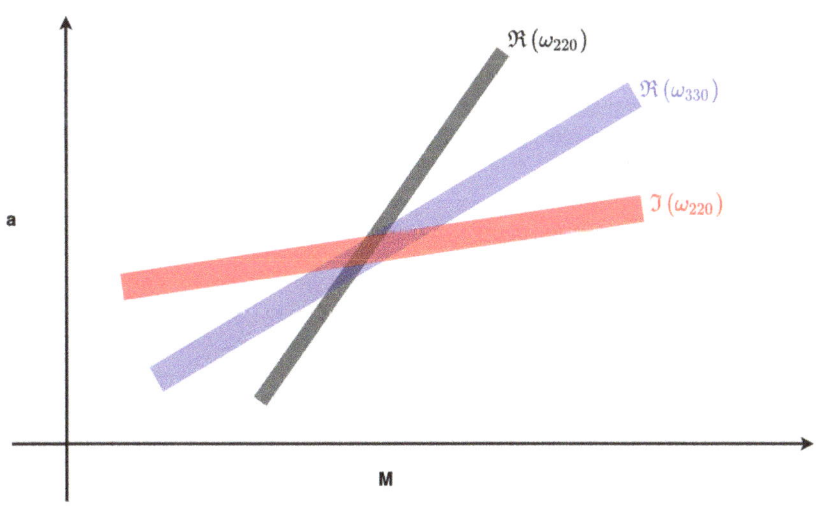

Figure 8.2. Cartoon of how to carry out a test of general relativity with measurements of the quasi-normal frequencies. Each region in the figure corresponds to a measurement of either the real or imaginary parts of the $(\ell, m, n) = (2, 2, 0)$ mode or the $(\ell, m, n) = (3, 3, 0)$ mode, picked arbitrarily here for illustrative purposes. The reason each measurement does not lead to a thin line, but rather to a thick line, is the statistical uncertainty inherent in each measurement. If general relativity is correct and the object is a Kerr black hole, then all thick lines should overlap somewhere in the M–a plane.

intersecting region providing the uncertainty in the inference). If general relativity is correct, any additional thick curve must also cross through the same intersecting region of the M–a plane. If this is the case, then the test is passed; if this is not the case, then the functional forms used for the real and imaginary parts of the QNM frequencies were wrong, implying a deviation from general relativity.

Beyond these model-independent tests, there are other more model-dependent exercises one could perform. One way to proceed is to construct a straw man and to use the data to refute it. In the context of tests of the Kerr hypothesis, the straw man is a particular deviation from the Kerr spacetime. Such a parametric deviation is called a bumpy black hole, mostly because of the idea that any deviation introduces bumps on the otherwise nicely ellipsoidal Kerr geometry. How would you construct a bumpy black hole metric? Well, because you have to deform the Kerr spacetime through functions, and there are 10 components of the Kerr metric, then you have at least 10 functions of 4 coordinates that you could use. If you wish to restrict attention to stationary and axisymmetric spacetimes, then you need only modify four components of the Kerr metric through functions that depend only on two coordinates. But still, these are functional degrees of freedom, and because there is an infinite number of functions to choose from, there is an infinite number of bumpy black holes you can construct.

This has, somewhat unfortunately, led to a proliferation of bumpy black hole metrics, not all of which are created equal. The first bumpy metrics were constructed by Scott Hughes and collaborators, who chose functions by requiring that the bumpy spacetime satisfy the vacuum Einstein equations. But because of the no-hair theorems, such metrics unavoidably introduce naked singularities.

Captain Obvious: Can we rule out naked singularities or ECOs based on (lack of) observations or based on simple logical arguments? Sometimes you see arguments that we can, but those arguments have loopholes and truly rigorous disproofs have eluded us thus far.

We'll start with ECOs. Suppose for simplicity that you have a spherically symmetric, uncharged ECO with a surface at a radius just slightly larger than the radius of a black hole with the same mass as the ECO. Suppose also that the ECO accretes mass at a rate $\dot{M}$ that we can estimate. If the accreting matter started far from the ECO with a speed much less than the speed of light, then the total energy per mass released by the time the matter settles on the ECO surface is $\approx c^2$. Based on this, an attempted "quick kill" of ECOs is that if the luminosity from a dark object is much less than $\dot{M}c^2$, it can't be an ECO. This inequality does hold for numerous specific dark objects, including the $\sim 4 \times 10^6\ M_\odot$ object at the center of our Galaxy. Voila! ECOs have been disproved!

But it's not that simple. An initial attempt to kill the quick kill noted that the closer the radius of the surface is to what would be the event horizon radius, the smaller the range of photon propagation directions away from radially outward that allow the photons to escape. For many directions, the photons would just curve back around and hit the surface, without escaping to the observer at infinity. Thus, for a given surface temperature, the luminosity is less than it would be if all photons could escape. However, although this statement is true as it stands, the argument is incomplete. The radiation from an ECO that comes back to the surface heats the surface, so given enough time the surface will heat up enough that the fraction of the radiation that does escape will carry away $\approx \dot{M}c^2$ in luminosity. This further thinking leads us finally to what are two possible loopholes in the $\dot{M}c^2$ argument. The first possibility comes from the "given enough time" caveat above: if the ECO surface is close enough to the radius of a horizon, then it might be that even in the age of the universe, the surface doesn't heat up enough that the small fraction of radiation that escapes can carry anywhere close to $\dot{M}c^2$. The second possibility comes from noting that as the temperature increases, the neutrino emissivity increases dramatically, and because neutrinos have much weaker interactions than photons, those could have evaded our detectors. No quick kill here.

What about naked singularities? Here, again, you sometimes see attempts at a stake through the heart. For example, perhaps we can argue that if the singularity is a point, then gravitational accelerations diverge near the point, and thus (for example) if we drop an electron toward our singularity, then the luminosity in electromagnetic radiation will diverge. However, here again it is not clear that the argument holds up: for example, speeds near the speed of light lead to beaming of radiation, in this case toward the singularity, which would therefore capture a larger fraction of the radiation as the beaming increases and the electron gets closer to the singularity. Is it a given that the amount of radiation that escapes will diverge? There seems to be general agreement in the community that under certain perfectly symmetric conditions naked singularities could exist, but that they would be unstable to any perturbations and thus could not form in reality; that is, in the specific cases that have been examined with mathematical rigor, naked singularities are unstable. There are also intriguing conjectures related to other hypotheses, such as the suggestion that in any universe where gravity is the weakest of the fundamental forces, naked singularities will always be clothed by event horizons. Naked singularities probably don't exist, but we're not yet at the level of a mathematical proof.

Another class of bumpy metrics was constructed by Dimitrios Psaltis and collaborators, mostly based on simplicity, by adding the least number of simple functions possible to modify the quadrupole moment. Such modifications, unfortunately, typically also introduce naked singularities or other pathologies. A third class of bumpy metrics was constructed by requiring that the spacetime retain certain symmetries (like stationarity, axisymmetry, and the existence of a Carter constant, a concept we encountered back in Chapter 3), without requiring that the bumps satisfy the Einstein equations. Non-Kerr black holes in modified gravity theories have been shown recently to typically not have a Carter constant to begin with, so requiring the opposite is not justified any longer. Finally, a new set of bumpy metrics has recently been introduced by Luciano Rezzolla and collaborators, where the bumps are modeled through something called continued fractions (a fraction, of a fraction, of a fraction, etc). Such bumpy metrics are not required to be solutions to the Einstein equations, and they were built to guarantee there are no naked singularities or other pathologies outside their event horizon. Unfortunately, such bumpy metrics typically require that many bumpy parameters are different from zero to recover non-Kerr solutions in modified gravity. Parameter degeneracies may make it difficult to place constraints on more than one such bumpy parameter from the data.

Major Payne: Ok, ok, ok, I must interrupt now. There is a serious problem with this kind of straw-man testing that must be addressed. Usually, in these types of tests, the reasoning goes as follows. One first constructs a straw-man hypothesis that is different from the canonical set of beliefs one is working with. In our case, that would be the construction of a bumpy black hole spacetime that differs from the Kerr spacetime. Then, one compares this hypothesis with the data, and although the experimenter wouldn't know it, let's assume for one moment the data are actually consistent with a Kerr black hole. The comparison of the straw-man hypothesis with the data then tells us that if this hypothesis is correct, then the data must have been highly unlikely given your choice of prior beliefs.

But one has to be careful! In reality, all one can say is that the particular bumpy metric considered is not favored by the data given the priors one brought into the analysis. To be extreme or absurd, one could for example imagine constructing a metric that represents a Kerr black hole with fire-resistant elephants stuck right outside the event horizon. One could compare data to this hypothesis and find that the data do not support it. All one has then shown is that fire-resistant elephants were probably not outside of the event horizon for the data analyzed. One cannot, however, conclude that the data were probably those of the Kerr geometry! There are many, many other possibilities, beyond fire-resistant elephants, that one has not actually considered.

And while we are on this topic, a final word of caution: it is not entirely clear what one learns by showing that there are no fire-resistant elephants outside of an event horizon. Sure, it's always good to gain information from data. But did we actually believe that this could be the case in the first place? Does the existence of such elephants resolve a major problem in physics, such as the existence of singularities? And if not, what good does it do us to know such elephants do not exist? What theoretical knowledge have we gained? Of course, if the data support the existence of such elephants, then we surely start packing our bags to go to Stockholm and pick up the Nobel Prize … I mean, it would be revolutionary (though more likely there is something wrong with your data or in your analysis). The probability of this being true a priori is so insignificant, that one may wonder whether it's worth the effort.

The moral of the story then is that one must be careful about the conclusions one extracts from straw-man hypothesis testing, and one should always strive to construct hypotheses that, once rejected, inform theory in some way. An even better way to proceed is to construct many models (ones that are consistent with general relativity, and many others that are not but that are theoretically interesting), and then let the data decide which ones are better supported. This is much more along the lines of proper Bayesian model selection, and its outputs are probabilistic statements about the support of competing hypotheses (through the Bayes factor we discussed in Chapter 4). This, of course, can and is being done with bumpy metrics, but a warning and a reminder are never out of place.

Let us now discuss tests for the existence of black hole mimickers. The idea here is to use data to determine whether the compact objects involved could be devoid of an event horizon, i.e., if it could be an ECO. If such is the case, then the compact object must have a surface, and this may be completely reflective (i.e., when things fall on it, they just bounce off) or partially absorbing/partially reflective (i.e., when things fall on it, there is some chance that the material will go through the surface, and some chance it will bounce off). In contrast, an event horizon can be thought of as a perfectly absorbing surface, so when material falls onto it, then it goes right through and nothing escapes. But if the ECO has a surface, then gravitational waves that are generated in its vicinity need not be completely absorbed, and rather, they could be reflected from its surface. You then have a situation in which gravitational waves are first generated, with some of them escaping to infinity and some heading toward the ECO, but then, these ingoing waves could get reflected and escape back to infinity! As measured by a detector very far away from the ECO, you would then detect the first burst of gravitational waves, and then, some time later, you would hear the same burst again but with a lower amplitude. Such a repetition is essentially an echo of the original signal, an "echo produced by an ECO." In fact, you can have several echoes, because once the first reflection occurs, it is possible for that first outgoing wave to be reflected back to the ECO surface by the curvature of spacetime; when this second ingoing wave hits the ECO surface it would be reflected again, and again some of it would escape to infinity, producing a second echo of lower amplitude than the first.

The search for echoes is then a model-independent test for the existence of black hole mimickers that can be carried out on gravitational-wave data. From a data analysis point of view, the main characteristics of the model are the delay time and the amplitude decrease between echoes, which, in turn, depend on the location of the surface relative to the would-be event horizon and on how reflective the surface is. This echo idea was initially proposed by Vitor Cardoso, Paolo Pani, and others in the mid-2000s, and then, after the first few detections of gravitational waves, Niayesh Afshordi and others claimed a detection of echoes in the LIGO data. Since then, the LIGO/Virgo collaboration has looked into this in detail, as have several other data analysis groups. The current consensus is that there is no unambiguous (higher than about $(3\text{–}4)\sigma$) detection of echoes in the data. But because such an echo idea is quite general, because it does not need a precise model of quantum gravity, and because a detection would be revolutionary, the search for echoes continues.

Before we conclude this section, let us briefly touch on what it would mean to detect echoes in a signal. As we argued above, an echo could be associated with a compact object devoid of an event horizon; therefore, lacking other explanations for the echo (e.g., if strong lensing is too improbable), an echo would prove the existence of an ECO. How could such an ECO been formed, what field equations do they solve, and are they even stable? If the ECO is to exist within general relativity, then by the no-hair theorems, it must be a nonvacuum solution. But the matter content (be it a scalar field or vector field) would probably violate the strong energy conditions, which has never been observed in nature before. Of course, the ECO could be a solution to a modified gravity theory, but which one? As of the writing of this book, we are not aware of an action or set of field equations whose solutions lead to an ECO. And even if such a theory existed, it is not guaranteed that generic gravitational collapse would lead to ECOs. And even if the collapse does form ECOs, it is not clear that spinning ECOs would be stable against a variety of instabilities that plague extremely compact objects with a surface. And even if they are stable, it is likely that a modified theory would also predict modifications to the generation or the propagation of gravitational waves (and for that matter, maybe also modifications to solar system, binary pulsar, and cosmological observables), as we saw in the previous sections, which can be constrained or ruled out independently of the existence of echoes. Needless to say, then, there is much work that remains to be done on the theory side to better understand whether ECOs are completely impossible, nearly impossible, or plausible compact objects.

Captain Obvious: There have been many ECOs discussed in the literature, but perhaps the simplest and one of the most popular are boson stars. A boson star is a compact object formed from the condensation of bosons with a repulsive interaction. You need the latter because otherwise there is nothing to prevent the boson condensate to continue to collapse and form a black hole. Such a star can then be extremely compact, yet it would be invisible, like a black hole, because these bosons wouldn't shine. Numerous studies on the linear and nonlinear stability of these objects have been performed. The problem with them is that the more compact you make them, the more unstable they become to black hole formation or complete dispersal. Therefore, if you start with two boson stars compact enough to be black hole mimickers, then when they merge, they will form a black hole and not another boson star. If that's the case, then there will not be any echoes in the merger, because the remnant will be a black hole. Obviously, if you start with two black holes in a binary, then you would also form a black hole upon merger, which again would be devoid of echoes. It is thus not clear what initial conditions would allow the formation of a boson star ECO upon merger that would then allow for gravitational-wave echoes. But this is a very active area of research, and the theoretical search continues!

8.5 Other Tests with Gravitational Waves

We have so far discussed tests that probe the propagation of gravitational waves or their generation in the inspiral phase of the orbit, but what about other phases? Carrying out a test of gravity during the merger is extremely interesting, but this is easier said than done. To construct a model for the merger, first one needs to destroy the four-dimensional nature of the theory by decomposing spacetime into three spatial dimensions and one temporal dimension, a so-called "3+1" decomposition.

Major Payne: Why is this actually necessary? Is it a practical matter, to make things easier, or is it actually impossible to do the evolution without introducing a 3+1 decomposition?

Great question, me! Let's think about this more carefully. What does the "evolution" of a system of ordinary differential equations mean? It means that given some initial conditions on the functions you are solving for (or "initial data," as it's called), you will then use the differential equations to figure out what the values of the functions are at a future time. With partial differential equations, the setup is similar, except that there, you are dealing with functions of space and time or "fields," instead of functions of time only.

Now a key concept in the above discussion is that encoded in the words initial and evolution. To define what "initial" and "evolution" mean, you need to have a concept of what time is, or better yet, a concept of the direction in spacetime in which you wish to learn about the behavior of your fields. Remember that in general relativity, space and time coordinates can mix through coordinate transformations, so the meaning of a coordinate v may not mean "time," even if it appears in the zeroth slot of a 4-vector. This is indeed the case in so-called ingoing Eddington–Finkelstein coordinates of a Schwarzschild black hole spacetime, in which the "time" coordinate v is related to the Schwarzschild coordinate t and r via $v = t - r + \mathcal{O}(M/r)$ far from the black hole.

But once you've picked a "time" direction in which to push your fields from some initial state to some future state, you are automatically defining a three-dimensional "hypersurface" orthogonal to this direction. That is, by choosing the time evolution vector, you are introducing a 3+1 split of spacetime: one dimension that describes the flow of your time evolution vector, and three dimensions that describe the space orthogonal to this time evolution vector. So is the 3+1 formulation the only way to evolve the Einstein equations (or any other set of gravitational field equations) numerically? Certainly not, if we are being rigorous (and I am!), there are other formulations, such as a "2+2" framework, which uses two-dimensional null hypersurfaces instead that are well adapted to radiation in general relativity. However, by far, the most well-adopted method is that of the 3+1 decomposition.

After a 3+1 split, one needs to show that this decomposition of the field equations leads to a "well-posed initial value problem." The latter is a very technical set of requirements, but it essentially boils down to this. First, choose "initial conditions" for the fields in your theory that satisfy the field ("constraint") equations at some instant in time. Then, show that there is a unique evolution of these initial fields into new fields that now satisfy the field equations at a future instant in time. This was proven by Yvonne Choquet-Bruhat back in the 1950 s for the Einstein equations, but outside general relativity, well posedness of the initial value problem has only been proven for the simplest of theories (although this field is rapidly advancing as well!). Needless to say, without such a formulation, it is very hard (if not impossible) to understand the merger of binaries, and the associated merger gravitational waves, in modified gravity.

After the binary components have merged, however, the resulting object will tend to settle down by relaxing into a stationary and stable configuration. For example, when two black holes collide in general relativity, the result is a highly distorted black hole that relaxes through the emission of gravitational waves into a spinning (i.e., Kerr) spacetime, as we discussed in Chapter 3. Indeed, during this phase one can carry out a test of the Kerr hypothesis as discussed in the previous section.

There are other consistency tests that can of course also be performed. A classic one is a residual test. The idea here is quite simple: take your best-fit model and subtract it from the data. Because there are statistical uncertainties (because the data are noisy, and the noise is not stationary and Gaussian), and possibly also some small systematic errors (because the model is not an exact solution to the Einstein equations, but rather an approximate one), the subtraction will not be perfect, and there will always be something left behind. The question is whether this leftover something is noise or whether it contains some amount of modified gravity signal that was not removed by the Einsteinian model. One can check this by calculating the signal-to-noise ratio on the residual and ask whether this is statistically consistent with the signal-to-noise ratio in other stretches of data where there is no signal. This is sometimes called a residual test.

A perhaps more interesting test is one in which one searches for the polarization content of the signal. As we discussed in Chapter 1, gravitational waves have two polarizations in general relativity (corresponding to what we call transverse and traceless modes). In a modified theory of gravity, however, gravitational waves can have up to six polarizations! The four missing ones correspond to two "scalar" polarizations and two "vector" polarizations, and the best way to understand their difference is to imagine how they affect a circle of test particles as the waves go through. In the transverse-traceless case, we call the wave polarizations + and × because this is the shape a circle of particles would deform to if a wave were traveling right through it. If the wave had scalar polarizations, it would either "inflate" and "deflate" the circle (a so-called breathing mode), or it would drag the sides as it goes by. If the wave had vector polarizations, it would elongate and rotate the circle as it goes by; in this sense, vector polarizations are longitudinal, just like one of the scalar polarizations. Figure 8.3 shows how different polarizations affect a ring of particles.

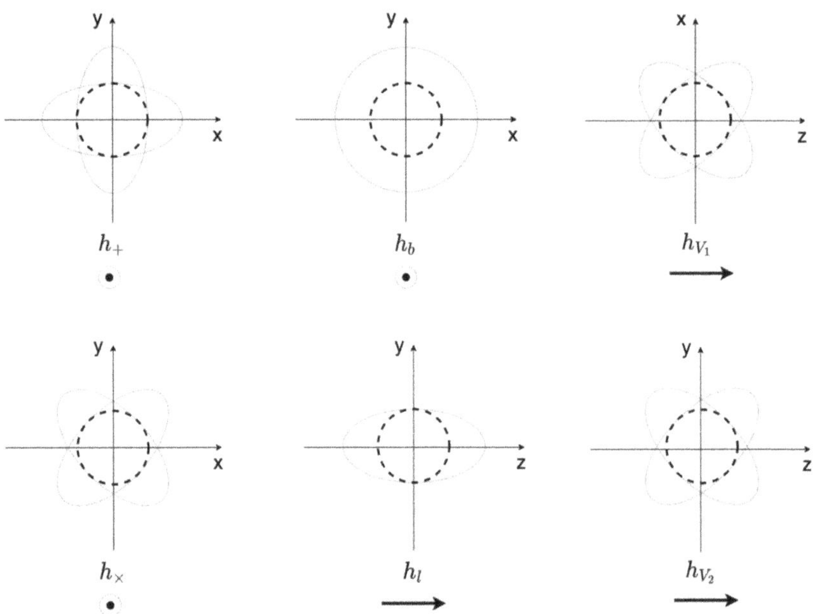

Figure 8.3. Cartoon of how a ring of particles would be affected by different gravitational-wave polarizations traveling in different directions (indicated right below the polarization label). In general relativity, gravitational waves only have two polarizations (the + and × modes). But in modified gravity, they can have up to six polarizations.

Major Payne: Wait, wait, wait, wait. Surely, the response of a detector will depend on the properties and shape of the detector, no? I mean, it can't be that all detectors just always respond in this way, because then L-shaped detectors would never be sensitive to some modes.

Another great question, me! And of course, I am totally right. The picture above is highly idealized, and it is just the response of a circle of test particles to an impinging wave. If you want to be superprecise (and I do!), then you would also need to take into account the fact that these particles should have a mass, and that they are probably connected to each other through some mechanical system. The reality is that the shape and the details of the interferometer greatly affect the way they respond to a gravitational wave. For example, an L-shape detector, such as the laser interferometers used in LIGO, Virgo, and KAGRA, are greatly sensitive to the $+/\times$ transverse-traceless polarizations, and not so much to a breathing mode. However, a spherical bar detector would be highly sensitive to a breathing mode. In fact, spherical bar detectors would be sensitive to all polarizations. What you gain in breadth, however, you lose in sensitivity, because bar detectors are notorious for not being "broad band." What I mean here is that they are more sensitive to waves in a given frequency band (near where they are resonant) and not very sensitive elsewhere. In contrast, L-shaped interferometers are sensitive to a broad range of frequencies in some sensitivity bucket.

As you can imagine, if different polarizations have such a drastically different effects on a circle of particles, they will also have a similarly different effect on a laser interferometer. Unfortunately, the effect of the different polarizations is degenerate with the direction in which the wave is traveling, which, in turn, depends on the location of the source in the sky and its polarization angle. In order to break this degeneracy, you must observe a given gravitational wave with more than one instrument. For example, if you have two instruments that are sufficiently far away from each other and pointing in different directions (on the surface of Earth), then you could use the observation of a wave by these two instruments to separate the $+$ and $\times$ polarizations, assuming general relativity is correct. Then, if you could observe the same wave with more instruments, then you could in principle extract additional polarizations. The LIGO instruments, unfortunately, have similar orientations on Earth's surface, and they are not that far away from each other, so they cannot really be used as independent instruments to extract the $+$ and $\times$ polarizations.

Dr. I. M. Wrong: Wait, what? That's crazy, I mean, totally bananas! Who would be so dumb?! Obviously, the placement of the detectors should be such that all polarizations can be observed. After all, we know that Einstein is wrong, and I have several ideas about how to fix his theory.

Captain Obvious: Actually, it's not insane at all! The detectors were coaligned deliberately so that if a signal with a $+/\times$ polarization were to be observed, it would be seen in the two detectors with similar features. This would add confidence that what was observed was a real signal and not an artifact of the noise. You need to remember that prior to 2015, many astronomers (and also some physicists) still doubted that nature would provide enough sources in the LIGO band for LIGO to detect, or that LIGO would actually achieve the required sensitivity to detect such waves. Weber's announcements starting in the 1960s of direct detection, which were later demonstrated to be artifacts of the noise and insufficiently detailed analysis of the data, did not help the LIGO case either. Understanding this historical context, it is then clear that physicists would choose a detector configuration that would enhance their confidence of an observation over the benefits of extracting multiple polarizations. Now that gravitational waves have been detected, we can definitely use the multiple detectors that have been built to constrain the existence of Einstein's theory, or, as you said, to search for a deviation.

As the network of detectors grows from the two LIGO ones, to include Virgo in Italy, KAGRA in Japan, and LIGO-India in India, there will be enough detectors to, in principle, extract different polarizations and carry out a polarization test of general relativity. This, of course, requires that the signal observed be strong enough and that it come from a favorable orientation. This is because the sensitivities of Virgo and KAGRA are not as great as that of the LIGO detectors, so tests of general relativity, and, in fact, all gravitational-wave observations are dominated by the signal-to-noise ratio collected in the LIGO instruments. A nondetection, however, in this case can be crucially important. This is because the response of L-shaped detectors to gravitational waves has nodes along which the detector would sense very little power if a wave were propagating from that direction. These nodes depend on the polarization of the signal, so even the lack of power detected in a lower-sensitivity instrument can help localize a source or even constrain the existence of additional polarizations.

How would one go about carrying out such a test? A clever way to do this is through the construction of null streams. The idea is quite simple. Imagine that one has detected, with three independent instruments, the gravitational waves emitted by some source. One can then use the data streams from two of them to extract the $+$ and $\times$ polarizations, assuming general relativity is correct. Then, one can linearly combine the three data streams to construct a new combined stream that should have no signal, a so-called "null stream," provided that gravitational waves only have the two $+$ and $\times$ polarizations. One can check whether this is the case in the same way as in the residual test, by calculating the signal-to-noise ratio of the null stream and then checking whether this is statistically consistent with the signal-to-noise ratio when there is no signal. Given then N linearly independent data streams, one can construct $N - 2$ null streams. With current plans to have five second-generation interferometers, we will be able to construct two null streams (because the two LIGO ones are not linearly independent).

A final test one can carry out involves the simultaneous observation of gravitational and electromagnetic waves from the same source. Such coincident events, as you may have guessed, are not the most common ones in nature. On general grounds, what would be the ideal conditions for the observation of coincidences? Let's use our intuition to answer this question. Clearly, you want the binary to emit light, and light is only emitted by matter (not by black holes, which, by definition, are black). This means we must have either binary neutron stars or a mixed neutron star/black hole binary for electromagnetic coincidence with ground-based detectors.

Captain Obvious: Wow, wow, wow … wait a minute. Accretion disks shine, and they shine a lot! That's why we can see AGNs, active galactic nuclei in which a supermassive black hole is accreting large quantities of matter, to very high redshift. So why can't we use a black hole binary with an accretion disk as a source that could produce coincident gravitational-wave and electromagnetic signals?

Let's consider this question because I guess the authors are being a bit loose here. If you read carefully, the authors say that they want a mixed binary or a neutron star binary to have electromagnetic observations with ground-based detectors. The words "ground based" are here crucial because these detectors are not capable of observing gravitational waves below roughly 10 Hz. Recall from our previous discussions that the gravitational-wave frequency scales inversely with the total mass. So if a binary black hole system has a mass much larger than $10^3 M_\odot$, then its gravitational waves will fall outside the sensitivity bucket of ground-based instruments.

If one could use space-based instruments, however, the story would be different. In fact, LISA is expected to be sensitive to gravitational waves between 10^{-5} to 10^{-1} Hz, which means it could see gravitational waves from black hole binaries with masses between about 10^3 and $10^7 M_\odot$. Some of these sources will surely have a circumbinary accretion disk, so there is great expectation about their observation both with gravitational waves and with electromagnetic waves.

In fact, circumbinary accretion disks may produce light before the merger occurs! This is because as the inspiral proceeds, different parts of the disk crash into each other, form particular shapes, and heat up the nearby gas, therefore emitting radiation. After the merger, the gravitational-wave burst emitted will take mass–energy away from the system, thus forcing the disk to readjust and emit light again. The observation of such an event would teach us loads about astrophysics, provided of course that the electromagnetic signal is bright enough to be seen on Earth in the first place.

But it's not likely that we will see the light emitted by the individual neutron stars in the binary before they merge; they would have to be extremely close to Earth to allow for this and we would have to know where to look ahead of time!

Captain Obvious: What about using the gravitational-wave data prior to the merger to find a likely source of the event in the sky, and then pointing a wide field-of-view telescope at it? Maybe some gamma-ray burst precursors could be observed as the neutron star crust cracks before merger.

That's a good idea, and there have been some papers that have explored this possibility. The problem is that the inspiral itself only lasts at most tens of minutes in the sensitivity band of second-generation ground-based detectors, like advanced LIGO and Virgo. Initially, the analysis of this data took too long, on the order of weeks or months for a detailed data analysis investigation. At present, and in the near future, such an analysis is being accelerated with a variety of techniques, such as reduced-order models. The more detailed the analysis, the better the localization of the source in the sky. With an initial analysis, it is typically hard to pinpoint where the source is coming from.

And because there can be hundreds if not thousands of galaxies in the initial localization box, it is hard to find the source electromagnetically, unless it is very close by. This, of course, should not prevent us from searching for such precursors, as their observation could teach us a lot about the equation of state of nuclear matter in the crust.

Therefore, we want to make sure that enough of a catastrophic collision occurs that matter is spewed in all directions after the merger, possibly leading to a gamma-ray burst. Such a gamma-ray burst would be ideal, because, when their jets are pointed at us, their fluxes are large, so we can see such events from far away. A binary neutron star merger would do that, but if it is a mixed binary, then we must have the neutron star disrupt before the black hole swallows it whole. This only happens if the black hole is very light (less than about 10 $M_\odot$, as we mentioned in Chapter 2), although the details also depend on the radius of the neutron star, its equation of state, and the spin parameter of the black hole, as well as the direction of the encounter relative to the spin.

So let's assume we have observed the gravitational and electromagnetic waves from such a merger event, now what? The electromagnetic waves from the collision will not be emitted until the collision happens, while gravitational waves are being emitted during the entire process, including during the inspiral. Therefore, if gravitational waves travel at the speed of light, we expect to see electromagnetic waves after we have begun to see gravitational waves. Precisely how much of a delay there is depends on the details of the process that emits the gamma-ray burst shortly after the collision. This involves "dirty gastrophysics" that is not fully understood, though the expectation is that the gamma-ray burst will be emitted within some seconds of the merger. On general principles then, we know that if gravitational waves travel at the speed of light, then we will see those first, and then within a few seconds of the peak in the gravitational-wave signal (associated with the time of coalescence), we should see the associated gamma-ray burst. A coincident event then allows for a natural way to measure the (constant) speed of gravitational waves and to constrain modified theories of gravity that predict speeds different from that of light. You may remember that we already discussed tests of the propagation of gravitational waves, but back then we pointed out that a constant shift in the speed of light would not be measurable with gravitational waves alone (because the associated modification in the gravitational wave would be degenerate with the time of coalescence). This is precisely why we need a coincident event: the electromagnetic signal is, in effect, acting as a "yardstick" against which we can measure the speed of gravity.

Can we make a Fermi estimate of how powerful these tests could be? Of course we can! The speed of gravitational waves is a quantity with units of km s^{-1}. The information we have at hand is the time delay in the observation of gravitational waves and electromagnetic waves $\Delta t_{\rm obs}$ and the distance to the source D_L. We may be tempted to then say that the gravitational wave had to travel a distance $D_L = v_g \Delta t_{\rm obs}$, but this would not be right. The quantity $\Delta t_{\rm obs}$ is the difference in the time of arrival between gravitational and electromagnetic waves, not the travel time of either wave T_γ or $T_{\rm GW}$. A better estimate is to argue that the difference between the speed of gravity and the speed of light Δv_g must be proportional to the time delay $\Delta t_{\rm obs}$. This must clearly be so, because the longer the delay, the more different the speed of gravitational waves must be from the speed of light. But if we say that $(\Delta v_g)/c \sim \Delta t_{\rm obs}$, the units don't work out, because the left-hand side is dimensionless, while the right-hand side has units of seconds. We can fix this unit discrepancy through our only other observable quantity, D_L. The only combination that works is $(\Delta v_g)/c \sim c(\Delta t_{\rm obs})/D_L$. A Fermi estimate of the strength of this test suggests then that we should be able to constrain the speed of gravity from the speed of light to $c(\Delta t_{\rm obs})/D_L \approx 10^{-16}$ for a time delay of 1 s and a distance of 100 Mpc. Astonishing!

In 2017, the first coincident event was observed, which we think probably consisted of two merging neutron stars. The two LIGO interferometers detected gravitational waves, and roughly 1.74 s after the peak of the gravitational-wave signal, the Fermi telescope detected a gamma-ray burst, correcting of course for the altitude and location of the satellite relative to the LIGO instruments. This then established that there was a delay of approximately 1.7 s between the compact binary merger and the time when electromagnetic waves started to be emitted. With this information, one can then construct an upper limit and a lower limit on the speed of gravity relative to the speed of light. You get the upper limit as follows. Assume first that the gamma-rays were emitted at exactly the same time as the peak of the gravitational-wave signal. If this is the case, then the delay in the arrival of the gamma-rays had to be completely because of the gravitational waves moving faster than light, and we have

$$\Delta t_{\rm obs} = T_\gamma - T_{\rm GW} \geqslant \frac{D_L}{c} - \frac{D_L}{v_g}, \tag{8.31}$$

where we recall that $\Delta t_{\rm obs}$ is the delay in the observing time between gamma-rays and gravitational waves, T_γ and $T_{\rm GW}$ are the travel times of the gamma-rays and the gravitational waves, respectively, D_L is the luminosity distance, and v_g is the speed of gravitational waves. Solving for v_g from the above equation, one finds

$$\frac{v_g}{c} \leqslant 1 + \frac{c\Delta t_{\rm obs}}{D_L}. \tag{8.32}$$

For the event observed by LIGO and Fermi, this translates into an upper bound of about $v_g/c \leqslant 1 + 7 \times 10^{-16}$.

You can now obtain a lower bound in a similar way. Instead of assuming that the gravitational waves and gamma-rays were emitted at the same time, let us assume that the gamma-rays were emitted 10 s after the gravitational waves. If this is what happened, then it must have been the case that the gravitational waves were traveling slower than the gamma-rays, so that the gamma-rays could catch up to the gravitational waves and only be observed with 1.7 s of difference in times of arrival. If this is the case, we then have

$$\Delta t_{\rm obs} = T_\gamma + \tau_{\rm int} - T_{\rm GW} \leqslant \frac{D_L}{c} + \tau_{\rm int} - \frac{D_L}{v_g}, \tag{8.33}$$

where $\tau_{\rm int}$ is the intrinsic delay between the peak of the gravitational-wave signal and the emission of the first gamma-rays. Solving for v_g, we then have

$$\frac{v_g}{c} \geqslant 1 - \frac{c(\tau_{\rm int} - \Delta t_{\rm obs})}{D_L}. \tag{8.34}$$

For the event observed by LIGO and Fermi, this translates into a lower bound of $v_g/c \geqslant 1 - 3 \times 10^{-15}$ if we assume that $\tau_{\rm int} = 10$ s.

Such an agnostic measurement of the speed of gravity is phenomenal, but what is even more impressive is that this single measurement ruled out a broad class of modified gravity theories that attempted to explain the late-time acceleration of the universe without a cosmological constant. This is because these theories predicted that gravitational waves would not travel simply at the speed of light, but rather they would satisfy a modified wave equation. Generically, when considering gravitational waves propagating on a cosmological (Friedman–Robertson–Walker) background, one finds that

$$\ddot{h}_{ij} + (3 + \alpha_M)H\dot{h}_{ij} + (1 + \alpha_T)\nabla^2 h_{ij} = 0, \tag{8.35}$$

where we recall from Chapter 6 that $H = \dot{a}/a$ is the Hubble parameter with the scale factor $a(t)$, ∇^2 is the Laplacian operator of flat space and the overhead dot here stands for a partial time derivative. The quantities α_M and α_T are called the running of the effective Planck mass and the tensor speed excess respectively, both of which are zero in general relativity. One can see easily that the first modification changes the "friction" created by the expansion of the universe, while the second term changes the group velocity of gravitational waves. It is precisely this second quantity, α_T, that the coincident gravitational-wave and gamma-ray observation was able to constrain. Because a plethora of modified theories required $\alpha_T \neq 0$ in order to explain the late-time acceleration of the universe, the single LIGO–Fermi event allowed us to throw a big heap of them into the trash bin.

8.6 Exercises

1. We have argued in this chapter that gravitational-wave tests may allow us to place stringent constraints on a regime where gravity is extreme, meaning strong and highly dynamical. Figure 8.1 attempted to quantify this in schematic form. In this problem, we will evaluate a few examples of solar system tests, binary pulsar tests, and gravitational-wave tests to fill out Figure 8.1.
    a) Find the location of the classic general relativity test of the perihelion precession of Mercury, the location of the orbital period decay test using the double binary pulsar J0737–3039, and the location of the GW 150914 event to construct Figure 8.1 for these experiments.
    b) In the gravitational-wave case, do you find a single point in this diagram or a curve? Explain why.

Hint: For the curvature, use the inverse square root of the Kretschmann scalar, $\mathcal{R} \sim (\mathfrak{M}/\mathfrak{L}^3)^{-1/2}$, while for the potential use $\Phi \sim \mathfrak{M}/\mathfrak{L}$, where $\mathfrak{M}$ and $\mathfrak{L}$ are the characteristic mass and length scales probed by the experiment.

2. In this problem, we will investigate orbital motion in modified gravity.
   a) Start from the Hamiltonian in Equation (8.5) and the deformed Schwarzschild metric in Equations (8.6) and (8.7) to derive the Hamiltonian in Equation (8.8).
   b) Use this to derive the Hamilton equations and find the binding energy and the angular moment of circular orbits to leading order in α. Don't forget that the metric deformations are functions of radius!
   c) From this, find the modified version of Kepler's third law.
   d) Show that the only way to have the orbital frequency not increase with decreasing orbital separation is for $\delta g_{00} \gtrsim +4\, m/r_{12}$.
   e) Is such a choice possible? If you had an isolated black hole with a metric given by Equations (8.6) and (8.7) with this δg_{00}, what mass would an observer measure this black hole to have by measuring, e.g., the orbital motion of a test particle around it in a circular orbit?

3. In this problem we will explore antichirping and floating orbits. The idea here is that in general relativity, compact binaries inspiral, and therefore, the orbital frequency increases with time (and thus, the wave "chirps"). But could you have a general relativity modification that leads to the opposite behavior? Or better yet, could you have a general relativity modification such that the frequency doesn't evolve at all, leading to a "floating orbit"? Consider the rate of change of the orbital angular frequency in Equation (8.16), and assume the modified theory only introduces modifications to the gravitational-wave luminosity through a propagating scalar field, as in Equation (8.14). Compute what n and α must be so that $\dot{\omega} = 0$ and you have a floating orbit. Why is this physically impossible in this case? (Floating orbits can exist in modified gravity, but not through the mechanism described in this problem).

4. In this problem, we study modifications to the propagation of gravitational waves.
   a) Take the dispersion relation in Equation (8.20) to compute the phase, group, and particle speeds of gravitational waves given in Equations (8.23) and (8.21).
   b) Now imagine that $\mathbb{A} = 0$ and that you have just detected the GW 170817 event and Fermi just detected its counterpart. Use the fact that the gravitational-wave event (the peak of the strain) was detected 1.7 s before Fermi detected the short gamma-ray burst to place a constraint on the mass of the graviton.
   c) Given that detecting another coincident gravitational-wave/electromagnetic event at larger distances is difficult, what do you conclude about the ability of ground-based detectors to place more stringent constraints on the mass of the graviton?

5. Let's continue our investigation of propagation effects and ask whether we can constrain them without an electromagnetic counterpart. In this chapter

we saw how a modification to the dispersion relation due to a massive graviton alters the gravitational-wave phase.

a) Consider then the Fourier phase given in Equation (8.28) for the case of a massive graviton only and estimate the accuracy to which the mass of the graviton can be constrained using the Fermi estimate method of Equation (4.16). Hint: Be careful here to apply the estimate to the parameter m_g^2 (instead of m_g) and then take the square root!
b) Evaluate this estimate for the GW 150914 event, which had a signal-to-noise ratio of 24, was at a luminosity distance of about 430 Mpc (corresponding to a redshift of $z \sim 0.1$), and merged around 200–300 Hz.
c) Is this constraint more stringent than what one obtains from a coincident gravitational0wave/electromagnetic observation (see the previous problem)?

When answering this question, take into account the limitations of the Fermi estimate.

Useful Books

Blanchet, L., Spallici, A., & Whiting, B. 2011, Mass and Motion in General Relativity (Berlin: Springer)

Chandrasekhar, S. 1998, The Mathematical Theory of Black Holes (Oxford: Clarendon)

Choquet-Bruhat, Y. 2009, General Relativity and the Einstein equations (Oxford: Oxford Univ. Press)

Choquet-Bruhat, Y. 2015, Introduction to General Relativity, Black Holes and Cosmology (Oxford: Oxford Univ. Press)

Maggiore, M. 2007, Gravitational Waves: Volume 1: Theory and Experiments (Oxford: Oxford Univ. Press)

Maggiore, M. 2018, Gravitational Waves: Volume 2: Astrophysics and Cosmology (Oxford: Oxford Univ. Press)

Petrov, A. 2020, Introduction to Modified Gravity (Berlin: Springer)

Poisson, E. 2007, A Relativist's Toolkit (The Mathematics of Black-Hole Mechanics) (Cambridge: Cambridge Univ. Press)

Poisson, E., & Will, C. M. 2014, Gravity: Newtonian, Post-Newtonian, Relativistic (Cambridge: Cambridge Univ. Press)

Will, C. M., & Yunes, N. 2020, Is Einstein Still Right? Black Holes, Gravitational Waves, and the Quest to Verify Einstein's Greatest Creation (Oxford: Oxford Univ. Press)

Appendix A

A Primer on Bayesian Statistics

Statistics is crucial to science. Indeed, a decent understanding of probability and statistics is somewhere between useful and essential for anyone who wants to participate knowledgeably in society. And yet you wouldn't have to look through many research papers to find one that misused statistics. Some of those misuses might be minor and benign, but others can really confuse a field. It is your responsibility to learn statistics well so that you don't participate in the confusion!

As part of that, we urge you to ask yourself two questions before you perform any statistical analysis. First:

How would I perform this analysis if I had unlimited time and resources?

and then second:

How can I perform my analysis, with the least loss of accuracy and precision, given my finite time and resources?

In particular, you need to make any approximations or compromises consciously, and you need to understand their consequences. Most people don't do that. This is largely understandable; because of our finite time we often offload parts of a project onto a collaborator or onto some prewritten analysis package; we can't do everything ourselves. But with statistics, many people go way too far in that direction. They don't ask themselves simple questions about the data, or about the results of their analysis, and thus, their conclusions can end up being completely wrong.

With that admonition in mind, we'll start with a brief overview of some statistical sins that we all commit (or thought about committing at some point), but that ought to be avoided. Then, we discuss the barest essence of Bayesian statistics, but we strongly encourage you to read other sources as well (see, for example, some suggestions at the end of this appendix). This book is not going to be enough for you to master the subject, but at least, it should give you a sense of the basics, so that you can then study this in more detail by yourself in the future.

doi:10.1088/2514-3433/ac2140ch9

A.1 The Seven Sins of Sad Statistical Analysis

The First Sin: Ignoring Systematics

There's a saying that in astronomy 3σ happens half the time. That's a little tongue in cheek because really 3σ should happen 0.3% of the time, assuming a normal distribution. The reason this is said is that it is very rare indeed that we understand our instruments and contaminating effects perfectly. Maybe that short-duration excess amplitude was a massive binary, but maybe it was just a truck passing by. Maybe the detector had a nonlinear response to some photons, or perhaps its calibration isn't perfectly understood. Maybe that gravitational-wave signal was not of astrophysical origin, but rather the signal was because of a glitch somewhere in the detector, or maybe a plane just happen to fly over it. There are also cases in which contaminating astrophysical sources can intervene.

A perhaps more insidious systematic is one that plagues analyses that rely on a template or model. Any statistical inference of the parameters of a given model relies implicitly on the assumption that the model is a good approximation to the signal contained in the data. Many times we don't know if this is the case; e.g., are the photons we detected from the accretion disk of this black hole coming from the innermost stable circular orbit of a perfectly thin (Novikov–Thorne) disk? Many times we think our model is highly accurate, but in truth, the model is constructed from perturbative solutions to some field equations (such as in post-Newtonian theory). For this reason, it is imperative that we analyze the data with more than one model, so that model systematics can be accounted for, especially when trying to measure a small effect in the signal. Think twice before you rewrite the laws of physics and open a bottle of champagne⋯

The Second Sin: Not Estimating "Trials" Correctly

In an otherwise uninteresting stretch of LIGO data, you see an intriguing bump, which you excitedly calculate to have a false-alarm rate of 10^{-3}. Wow! Did that record the production of a quark star? Maybe, but did you take into account that in your data you have 1000 time intervals similar to what you examined and thus that there were 1000 chances to have a bump that is improbable at the 10^{-3} level? Many times people will not account correctly for the number of "trials" they perform, sometimes called the "background,"' and thus they overestimate the significance of the effect. This can be insidious, in the sense that it may not be obvious how many trials are being performed. For example, sometimes the sequence in research is (1) see something that looks interesting, then (2) calculate the probability that exactly that thing should happen. This is Feynman's "'license-plate fallacy": isn't it remarkable that yesterday the car parked next to mine had a license plate that read HSX 495? That exact license plate, out of all possibilities!

You therefore have to ensure that the signal is strong enough to be noteworthy despite the many trials. This is why gravitational-wave scientists spend so much time calculating noise samples through time shifts to estimate their background, as we discussed in Chapter 4. Only once you understand your background can you estimate a false-alarm probability and therefore your confidence that the event you observed is real.

The Third Sin: Thinking That You Need to Bin

It is very common in statistical analyses to assume a Gaussian distribution for some quantity. Many tools require that assumption (e.g., this underlies the calculation of χ^2). But people usually understand that when one has a small number of points, the distribution will typically not be Gaussian. So they take their data and group it so that there are a larger number of data points per group, so that the statistics are closer to Gaussian. For example, if you are analyzing the distribution of black hole masses inferred from gravitational-wave events, this philosophy would dictate that you set up mass bins that each have many events. We have, incredibly, had PhD scientists tell us that this improves the precision of the resulting statistical inference. No, no, no! By grouping data you lose track of where in the group the data originated, so you are guaranteed to lose information. Now, it could be that the information you lose is of negligible importance, or that the data come in binned form (e.g., there are enormous numbers of gravitons or radio waves in observations, so we don't count each one separately), or that it is computationally infeasible to use all the data in their original form. But if you are somehow forced to bin, you should do so with eyes wide open.

The Fourth Sin: Confirmation Bias and the Elimination of "Outliers"

It's easy to want certain results from an analysis. But because we do know that glitches occur, sometimes an observation or a point in that observation might not really be representative of the source. As a result, we can be tempted to try to identify those "outliers" and eliminate them, to get "clean" data. But beware! This leads to a statistics version of confirmation bias, by which we reinforce our prejudices when we see something we like and dismiss evidence that contradicts our prior conclusions.

Consider the case of the first gravitational-wave observation. This event consisted of two black holes merging with masses around $30M_\odot$ to produce a remnant with mass of about $60M_\odot$. Before this observation, it was not exactly clear that such heavy black holes merged frequently enough to be detectable—or even existed in the first place! Fortunately, the LIGO collaboration had in place algorithms to detect burst-like signals with excess power tools and with wavelet tools. But had they not had this, then matched filtering searches would have missed the first event if the template bank did not include black holes heavier than $25M_\odot$! This is a concrete example of how confirmation bias could have affected gravitational-wave astrophysics, but fortunately, it didn't in this case.

The Fifth Sin: Subtracting a Background or a Glitch Rather than Modeling It

Suppose that you're looking for an excess above a background, which is often the case in gravitational-wave data analysis. Why is subtracting the background wrong? Consider electromagnetic astronomy as an example and remember that fluctuations in the data depend on the total intensity or number of counts. Suppose, for example, that we're in the Gaussian regime, where if the average number of counts in some interval is N, we'd expect $N \pm \sqrt{N}$ in a particular observation. Then if (for instance) we use χ^2 statistics, $\sqrt{N}$ is what we use for the standard deviation of the data. If the background has 99% of the counts, then if we subtract the background, we erroneously conclude that the fluctuation level (and the standard deviation we use in our χ^2 analysis) is $\sqrt{0.01N}$, or only 1/10 of the correct value. The right procedure is to include a model of the background as part of the overall modeling of your data.

Consider now glitch subtraction in gravitational-wave data. Spikes or "lines" in the signal power spectrum are commonly eliminated from the data because they are of known origin (e.g., the 60 Hz line associated with electrical currents in North America). But glitches also occur regularly. Eliminating the glitches can be dangerous in this case. Instead, a better approach is to follow Cornish's motto: "model everything and let the data sort it out." What can be done to achieve this is to construct a glitch model, a Gaussian noise model, and a signal model, and then to carry out model selection with a given stretch of data.

The Sixth Sin: Using a Black Box Code

As discussed above, we have finite time and thus we naturally focus our personal resources on a limited set of things. But when we do statistical analyses, this can come back to bite us. Someone points us to a particular statistical package, which is used for our type of analysis. Yay! These can save us a lot of effort; for example, who wants to spend a lot of time writing their own code from scratch to interpret data from a particular instrument? But the analysis performed by the package will usually make certain assumptions, and those won't always be valid. It is your responsibility to determine the assumptions used in any package you employ and to understand the consequences of those assumptions.

Let us give you a dangerous example. Imagine that you want to analyze a stretch of LIGO data because you want to use it to constrain a deviation from general relativity. You decide to use one of LIGO's software tools to do this calculation (at the time of the writing of this book, that's something like PyCBC, LALInference, or Bilby). You don't have time to learn Bayesian statistics carefully, so you figure out how to make the software use your modified gravity template, and you just run it against the data, using a Markov Chain Monte Carlo (MCMC) approach. After a day or so, you look at the run and ask the software to construct for you the posterior probability distribution for your modified gravity parameter (properly marginalized over all other parameters). If you weren't careful, it is entirely possible the posterior you got will be completely wrong because the MCMC exploration has not yet converged after one day. Understanding when to stop the software, how to check whether the exploration has converged, and how many truly independent steps have been taken requires some nontrivial knowledge of statistics. Ignoring all of this and just pressing "Shift–Enter" without understanding what the code does dramatically increases the chances your conclusions will be wrong.

The Seventh Sin: Not Thinking about Whether Your Answers Make Sense

Try actually looking at your data and at the way your exploration of the parameter space is being conducted! Did you truly explore all of the parameter space, or did you get stuck somewhere? Do the conclusions you drew from your analysis pass the gut check test? If not, think again. For example, it can easily be that you do an analysis, estimate parameters, and end up with some clear conclusions, but actually your model doesn't fit the data. Or, you can do something in a formally right way that leads to an answer that is actually absurd. Remember, you are the master of the statistics; statistics shouldn't boss you around!

What are some gut checks that you could conduct? Well, that depends strongly on what calculation you are actually trying to do. If you are running a parameter estimation study, a simple thing you can do is simply run the exploration for longer to see if your probability distributions have converged or not. Another thing you can do is plot the parameters that are being explored at each step and check whether you are just locked into a tiny region. You can also calculate (well, estimate really) the number of independent steps you've taken in the parameter exploration (through something called the autocorrelation length) and see if you have enough independent samples. You can raise your likelihood to the power of zero and check that your posterior distributions return your priors. But to do any of this, you need to understand what the likelihood is, what the prior is and what the posterior is. So let's jump right in!

A.2 Bayes' Theorem

If you have encountered statistics in a class, odds are that it has been "frequentist" statistics. Frequentist statistics often involves null hypothesis testing and uses tools such as chi-squared (χ^2), f-tests, t-tests, and so on. However, over the last few decades, a different type of statistics called Bayesian statistics (after the Reverend Thomas Bayes) has become more and more popular in physics and astronomy and is now an essential part of gravitational-wave data analysis. Philosophical battles abound. For example, given the critical role of prior probability distributions in Bayesian statistics, various jokes can be made.[1] However, from our perspective, the philosophy is unimportant; what matters are the results! Indeed, the practical utility and reliability of Bayesian statistics, when coupled with algorithms and computers fast enough to perform the calculations, are what have caused the ascendency of these methods.

In this section, we set things up by deriving and discussing Bayes' theorem. In the following sections we talk about the role of (and the necessity for!) prior probability distributions, and then parameter estimation and model comparison. We assume throughout that you have encountered the fundamental concepts of probability theory (if not, please consult the suggestions at the end of this appendix.)

Suppose that we have two events, A and B, each with some probability. Let us consider the probability that both happen: $P(A \text{ and } B)$. This is equal to the probability of B by itself times the probability that A happens given that B happened:

$$P(A \text{ and } B) = P(A|B)P(B), \tag{A.1}$$

[1] A Bayesian is someone who expects to see a horse, actually sees a donkey, and concludes that they have seen a mule!

where $P(A|B)$ is the conditional probability that A happens if B happens. Note that A isn't required to depend on B; for example, if A and B are completely independent, then $P(A|B) = P(A)$, but that isn't the general form. We can also write this the other way around:

$$P(A \text{ and } B) = P(B|A)P(A). \tag{A.2}$$

Therefore, $P(A|B)P(B) = P(B|A)P(A)$, which also means that we can write

$$P(A|B) = \frac{P(B|A)P(A)}{P(B)}, \tag{A.3}$$

as long as $P(B) \neq 0$. This is Bayes' theorem.

The power of this for statistical analysis comes from replacing A with a particular hypothesis (e.g., that the mass of a black hole is 31.2 $M_{\odot}$) and B with the data you have in hand. Then the factors in this equation may be interpreted as follows:

- $P(A|B) \to P(\text{hyp}|\text{data})$ is the probability of the hypothesis given the data and prior information. This is called the posterior probability.
- $P(B|A) \to \mathcal{L}(\text{data}|\text{hyp})$ is the probability that the data would be observed if the hypothesis were true. This is usually called "the likelihood of the data given the model."
- $P(A) \to q(\text{hyp})$ is the prior probability of the hypothesis being true; in other words, the probability you assigned to the hypothesis before you took the data.
- $P(B) \to P(\text{data})$ can be considered as a normalizing constant, given that probabilities must integrate to unity. This is called the evidence.

The job of a "Bayesian" is then clear. Given some model, some data, and some priors, map out the posterior by evaluating or "sampling" the likelihood over the entire parameter space. Once your map is complete enough, then look at what the (marginalized) posterior distribution is for your set of parameters (which we typically do through something called a "corner plot," a figure that contains many panels with one-dimensional and two-dimensional marginalized posterior distributions) and how this has updated your prior beliefs. In practice, likelihood analyses usually use the natural log of the likelihood (which we creatively call the "log likelihood") rather than the likelihood itself. That's because products of exponentials and powers can often lead to values that are huge or tiny, which makes them difficult to use on a computer. Logs are better behaved.

As a concrete example, suppose that we are presented with a coin that might or might not be fair, and say that our model is that the probability of heads in a given flip is a. If we genuinely don't know a, our prior probability distribution for a might be $q(a) = 1$ from $a = 0$ to $a = 1$; note that this satisfies the requirement for probability distributions that the integral of $q(a)$ over all possible values of a is 1. Now say that we flip the coin one time and it turns out to be heads; thus our data set B is one heads. If our model is that the probability of heads is a, then the likelihood of a heads in one flip is $\mathcal{L}(B|a) = a$. Therefore, the posterior probability of the model given the data and our prior is

$$P(a|B) \propto a. \tag{A.4}$$

But the posterior probability distribution must integrate to unity, so we need to determine the normalizing constant. We do this by dividing the unnormalized $P(a|B)$ by its integral over all a. Given that $\int_0^1 a\, da = 1/2$, we get finally

$$P(a|B) = 2a. \tag{A.5}$$

Note that the normalizing constant $P(B) = \int \mathcal{L}(B|a)q(a)\, da$. This is generally true, and the integral of the prior times the likelihood over all allowed values of the parameters (for any number of parameters) is called the evidence, as noted above.

With this introduction, we now discuss priors, marginalization, parameter estimation, and model comparison in Bayesian statistics.

A.3 Priors

The use of priors is a common sticking point when people first encounter Bayesian statistics. The normal statistics you use give the impression that they'll just tell you the answer, with no subjective priors. So why do we need them and, if we do need them, how should we pick them?

To address that, let's begin with another simple example. Suppose that you have a coin. The coin appears and feels completely ordinary, and we assume that you acquired it in an ordinary way, e.g., maybe you got it in change from a store. You idly flip the coin 10 times and get 8 heads and 2 tails. Your friend watches this and then, being a betting person, puts down \$1 to bet that in the next 100 flips there will be at least 50 tails. You don't have anything else to do, and \$1 isn't too bad, so you agree to the bet and put down \$1 of your own. But your friend objects. Because the first 10 flips got eight heads, it is clear that this coin is biased toward heads. Indeed, when the two of you do a careful calculation,[2] you find that based on the data the probability that flips with this coin give tails at least half the time is only 0.0327. Thus your friend demands that you put down $(1 - 0.0327)/0.0327 \approx \$30$ to make the bet fair.

Do you take the bet?

Most astronomical problems are like this to some degree. Even if you think you are just analyzing data, you have to make some prior assumptions to make progress. This is why you need to make prior assumptions, and why it is a good thing that in Bayesian statistics you are required to specify your priors explicitly. If you don't, then people can't reproduce your work.

[2] For the curious: say that we begin with a prior that the probability that the fraction of heads is a is given by a flat distribution: $q(a) = 1$. The likelihood of getting 8 heads and 2 tails in 10 flips is $\mathcal{L} \propto a^8(1 - a)^2$, so because the prior is $q(a) = 1$, the posterior is also $P(a) \propto a^8(1 - a)^2$. When normalized so that $\int_0^1 P(a)da = 1$, we find $P(a) = 495a^8(1 - a)^2$. Integrating from $a = 0$ to $a = 0.5$ gives 0.0327.

But you might now be settling into despair. How are you supposed to choose your priors? After all, if you choose your priors narrowly enough, you'll get any answer you like. For example, in the coin-flipping example that started us off, if you have an absolutely unshakable faith that the probability of tails is 0.5 to an arbitrary number of significant figures, no number of flips would suffice to change your mind. Someone could flip a coin 1000 times, getting all heads, and it wouldn't matter to you, in spite of the data, which is sad.

Despair, however, is not necessary. We can divide the choice of priors into two broad categories:

- Cases in which you have specific prior information. For example, we know from other data and theoretical understanding that we can't have neutron stars with negative masses, and to a lesser (but still very high) confidence, that their mass cannot be more than five times the mass of the Sun.
- Cases in which we don't have a lot of prior information. For this case, it is appropriate to use uninformative priors, which we'll now discuss.

Clearly, unless we really do have a lot of prior information, we shouldn't use priors so narrow that they force a final answer identical to our priors. Thus, we'd like to use a broad prior, so that if we end up with a narrow posterior, it's because the data demand it, rather than because our prior beliefs biased the result.

What the right broad prior is can depend a bit on the problem. If you're interested in the fraction of times a weirdly shaped object lands on "H" rather than "T," then maybe a uniform prior probability from 0 to 1 would be appropriate. If you don't know the scale of the problem, then a logarithmic prior could be appropriate. For example, if we are interested in the distance to an object and have no idea how far it is, you might think that you should have a uniform prior over the distance, e.g., from 0 distance to 10 billion light-years. But in doing that you would be unintentionally setting the prior so that large distances are preferred; after all, with that prior, there is a 99.99% prior probability that the distance is more than a million light-years; this is simply because a million light-years is 10^{-4} of the total distance of 10 billion light-years, so the probability with this prior that your distance is more than a million light-years is $1 - 10^{-4} = 0.9999$! To be scale independent, you would want a logarithmic prior, which would mean that there is an equal prior probability that the distance is between 1 and 10, or 10 and 100, or 100 and 1000, etc., light-years.

Major Payne: Of course the authors are not being thorough in their discussion of priors, and they seem to have forgotten that they are writing a book on gravitational waves! A prior often used in gravitational-wave analyses (e.g., inference of rates from a sample of events) is the Jeffreys prior, in which the prior probability density is proportional to the square root of the Fisher information matrix. In the case of an event rate R, this turns out to mean that $q(R) \propto R^{-1/2}$.

One reason to use a Jeffreys prior is that then the relative probability of a portion of parameter space is invariant to reparameterizations. The lack of invariance to reparameterizations was one of the main points raised against Bayesian statistics in the early

1900s by other statisticians, so you would think that the authors would have featured this prior! Thus, although there is no single correct prior for all situations, the Jeffreys prior should be considered.

One can say with some justification that if you try several reasonable priors and these give you wildly different answers, your data didn't contain enough information to judge between them, so you can't say much. It is often appropriate to try to select priors that are as uninformative as possible so that the data speak for themselves. But you need to specify your prior and indicate what it is so that others know how you performed your analysis!

It's also always a good idea to think about plotting your prior together with your posterior so that you can visually assess how much information you have learned from the experiment or observation. There's even a way to compute quantitatively how much you have learned, called the Kullback–Leibler divergence. Suppose that your prior probability distribution for the parameters $\vec{\theta}$ of a model is $q(\vec{\theta})$ and that your posterior probability distribution is $P(\vec{\theta})$. Then, the Kullback–Leibler divergence is

$$D_{\mathrm{KL}}(P|q) = \int P(\vec{\theta})\log\left(\frac{P(\vec{\theta})}{q(\vec{\theta})}\right)d\vec{\theta}. \tag{A.6}$$

If the base of the log is 2, you get the information gained in bits; if it is a natural log, you get the information gained in nats. For instance, in our coin-flipping example, $q(a) = 1$ and $P(a) = 2a$, where both are normalized. Then, in nats

$$D_{\mathrm{KL}}(P|q) = \int_0^1 2a\,\ln(2a)da = [\ln 2 - 1/2]\ \text{nats} \approx 0.2\ \text{nats}. \tag{A.7}$$

or in bits

$$D_{\mathrm{KL}}(P|q) = \int_0^1 2a\,\log_2(2a)da = \left[1 - \frac{1}{2\ln 2}\right]\ \text{bits} \approx 0.3\ \text{bits}. \tag{A.8}$$

Note, in passing, that a bit is an extremely small amount of information (i.e., in computer science, a bit can only contain one of two values, a one or a zero), which is why we are used to working in bytes (with one byte equaling eight bits, and containing enough information for one ASCII symbol). At the risk of revealing the age of this book, today's computers can store terabytes of information (i.e., 10^{12} bytes)!

A.4 Marginalization

Suppose that your model has multiple parameters, but you're really only interested in the posterior probability distribution of one of the parameters. For example, maybe you are doing a Gaussian fit to a line, and of the three parameters involved in the Gaussian (the centroid wavelength, the width, and the amplitude), you only care about the centroid wavelength. What should you do?

The way this task is performed in a number of analysis packages is that (1) you find the single best fit to the data, then (2) you freeze all of the values of the other parameters at their best-fit values, and finally (3) you vary the parameter of interest, with the other parameter values frozen, and use some criterion to figure out the uncertainty in the parameter of interest (e.g., by effectively doing a Bayesian fit with one parameter, or by doing a $\Delta\chi^2$).

But this procedure is incorrect; it yields the right value only in special circumstances. To understand why, let's think about what we mean by the posterior probability distribution for a single parameter when we have additional parameters.

First, if we go back to a single-parameter model (call that parameter a), suppose that we take the posterior probability distribution $P(a)$ and sample from it. That means that the probability that we choose a given a is proportional to $P(a)$. With lots of samples, we will simply recreate $P(a)$. That's simple enough to be tautological.

So what should we do when there are multiple parameters? The logical and correct generalization is that once we have the full multidimensional posterior $P(a, b, c, \ldots)$ such that the total probability contained in the volume spanned by a and $a + da$, b and $b + db$, etc., is $P(a, b, c, \ldots)da\, db\, dc\ldots$, we imagine picking parameter combinations $(a, b, c, \ldots)$ with probability proportional to $P(a, b, c, \ldots)$ and then just storing the value of a. That makes sense: pick according to the (now multidimensional) posterior probability distribution, and determine the distribution of the values of a that you get as a result.

But it might not be evident why this definition is different from the oft-used definition in the second paragraph above.

To make the difference clear, let's think about an extreme case with two parameters. We'll set this up so that the overall peak in the two-dimensional posterior has only a narrow range in the y parameter with a large amplitude, but at other values of x (the parameter we care about), there is a broad range of values of y that give a significant posterior probability density $P(x, y)$. Let's say the posterior, which we assume to apply in the range $x = 0.05$ to ∞ (i.e., there is a hard lower bound of 0.05 to x) and $y = -\infty$ to $+\infty$, is

$$P(x, y) = 0.23e^{-(x-1.5)^2/2}e^{-x^2y^2}, \tag{A.9}$$

where the factor of 0.23 is so that $\int_{-\infty}^{\infty}\int_{0.05}^{\infty} P(x, y)dxdy = 1$. The overall maximum of $P(x, y)$ is at $x = 1.5$, $y = 0$; that's the only place where the exponent is nonnegative. Thus, if we used the common approach described above, we would expect that the one-dimensional posterior for x, $P(x)$, should peak at $x = 1.5$. Indeed, if we go to the overall maximum in the two-dimensional posterior ($x = 1.5$, $y = 0$) and fix y at its best value ($y = 0$), then what we have is just a standard Gaussian with variance 1, centered on $x = 1.5$.

But as you see in the upper panel of Figure A.1, the contour lines are asymmetric; here we show the locus of points in the (x, y) plane that give $P(x, y) = 0.23e^{-1}$. For

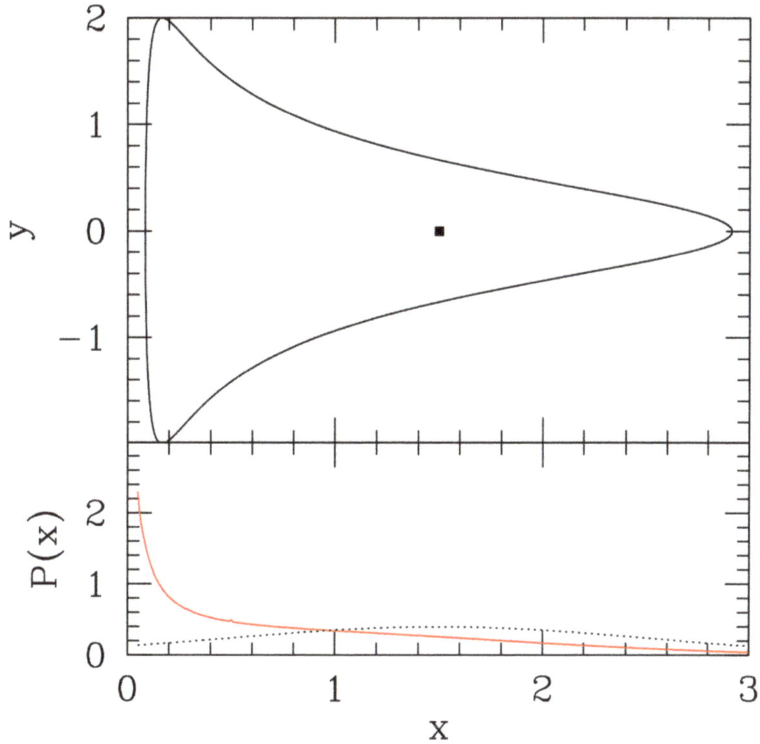

Figure A.1. Demonstration that in general, you cannot find the one-dimensional posterior probability distribution from a multiparameter fit by fixing the other parameters to their best values. The posterior here is $P(x, y) = 0.23e^{-(x-1.5)^2/2}e^{-x^2y^2}$, where the prefactor is chosen so that $\int_{0.05}^{\infty} P(x, y)dxdy = 1$. The solid square in the upper panel shows the location of the maximum of $P(x, y)$, where $x = 1.5$ and $y = 0$. That panel also shows the contour lines where $P(x, y)$ is e^{-1} times its maximum value; the asymmetry is obvious. The lower panel shows the one-dimensional posterior for x computed in two different ways. The black dotted line shows the result from the standard procedure of fixing the other parameters to their best values (here, $y = 0$) and then determining the probability distribution for x. The solid red line shows the correct answer we get by marginalization, i.e., by integrating $P(x, y)$ over all y at a given x. In this case, and in general, the procedure of fixing the other parameters to their best values gives an incorrect answer.

$x < 1.5$, there is a much wider range in y that gives $P(x, y) > 0.23e^{-1}$ than there is for $x > 1.5$.

You may say that this is an extreme example. And it is: we selected the function to make a point. But if you think about it, it is virtually never the case that a posterior has the property that fixing all the parameters but one and looking at that single-parameter cut through the posterior give you exactly the same answer as the correct marginalization procedure. Multidimensional Gaussians, oriented in any direction, do have that property, but posteriors will never have exactly that form. So if your black box code fixes parameters to get one-dimensional posteriors, beware!

To be a bit more formal, suppose that we have parameters $a_1, a_2, \ldots, a_n$ in our model. Our posterior probability distribution is $P(a_1, a_2, \ldots, a_n)$, normalized so that

$$\int P(a_1, a_2, \ldots, a_n)da_1da_2\ldots da_n = 1. \tag{A.10}$$

If we only want to know the posterior probability distribution for parameter a_1, independent of the values of the other parameters, we simply integrate over those other parameters (this integration is called marginalization):

$$P(a_1) = \int P(a_1, a_2, \ldots, a_n)da_2\ldots da_n. \tag{A.11}$$

We then have $\int P(a_1)da_1 = 1$. Similarly, one could find the distribution for the two parameters a_1 and a_2 by integrating P over a_3 through a_n. The parameters you integrate over are called nuisance parameters. In practice, because you will only rarely encounter a posterior that you can integrate analytically, what normally happens is that you sample the posterior with some algorithm and then store the values of the parameters (singly, in pairs, in triples, or whatever you need for your marginalization).

A.5 Parameter Estimation

A.5.1 Discrete Data

Say that you flip a coin 10 times and you get 4 heads and 6 tails. Your model is that the probability of heads coming up in a given throw is a, and thus that the probability of tails coming up in a given throw is $1 - a$. Here we will fix the number of throws at 10 (i.e., the actual number!), which means that in our model we would expect $10a$ heads and $10(1 - a)$ tails; note that our two bins are the number of heads and the number of tails. We will assume that the probability distribution is Poisson; in a Poisson distribution, if we expect m counts (where m is a positive real number), then the likelihood of actually getting d counts (where d is a nonnegative integer) is

$$\mathcal{L}(d|m) = \frac{m^d}{d!}e^{-m} \tag{A.12}$$

(this can be derived from the binomial formula; give it a try!). Given that, the likelihood of the data given the model (with parameter a) is then

$$\mathcal{L}(a) = \left(\frac{(10a)^4}{4!}e^{-10a}\right) \times \left(\frac{(10(1-a))^6}{6!}e^{-10(1-a)}\right), \tag{A.13}$$

which we can rewrite as

$$\mathcal{L}(a) = \frac{10^4}{4!}\frac{10^6}{6!}e^{-10}a^4(1-a)^6. \tag{A.14}$$

Because we will eventually normalize the posterior probability distribution, we can ignore and leave out all of the numerical factors in front and just write

$$\mathcal{L}(a) \propto a^4(1-a)^6. \tag{A.15}$$

The likelihood is only one of the factors that we need to get the posterior probability density $P(a)$. We also have to multiply by the prior probability for a (see the earlier section on priors), i.e., the posterior probability density $P(a) \propto \mathcal{L}(a)q(a)$, where $q(a)$ is the prior probability density for a. Now, by its nature, a has to be somewhere between 0 and 1. For our current purposes, let's use a prior on a of $q(a) = 1$ from $a = 0$ to $a = 1$. Then in this special case, the posterior probability density $P(a)$ is simply proportional to the likelihood $\mathcal{L}(a)$), and because $P(a)$ is a probability density, when it is properly normalized $\int_0^1 P(a)da = 1$ (in the same way that the prior probability density was normalized, $\int_0^1 q(a)da = 1$).

What can we do with that posterior probability density? As a first step, let's determine where we have our maximum probability. We get that by taking the derivative of $\mathcal{L}$ with respect to a and setting it to zero. This gives

$$4a^3(1-a)^6 - 6a^4(1-a)^5 = 0 \rightarrow a = 0.4. \tag{A.16}$$

That's intuitive; with no other information, our best guess is that the true probability exactly reflects the data.

But we almost always want more than just the best-fit value; we also want to be able to say that, with some probability, a is in a particular range. In Bayesian parlance, we would like to know the "credible region" to some level of probability. To get an idea of what this means, we calculate and plot the normalized posterior probability density as a function of a in Figure A.2. Note that the probability density can exceed 1; it is the integral of the probability density that must equal 1.

When we look at the figure we see that the probability density is not symmetric around the peak. For example, at $a = 0.2$ the probability density is about 0.95, whereas at $a = 0.6$ the probability density is about 1.2. This introduces an ambiguity in the definition of the credible region. Should we, for example, start at the peak and move symmetrically to smaller and larger values of a until we get to some total probability? Should we begin from $a = 0$ and find the value of a that gives us an integral equal to a specified probability? Should we find the smallest region that contains the specified probability? The smallest contiguous region that contains the specified probability?

We'll choose the last of these, for illustrative purposes. Suppose that we want a 68.3% credible region (which we choose because this corresponds to the probability between -1σ and $+1\sigma$ for a Gaussian distribution). Then the minimum-width contiguous range that includes this probability goes from $a = 0.264$ to $a = 0.547$, for a total width of $\Delta a = 0.283$.

Captain Obvious: There's a nice way to visualize how we can process a posterior to get a desired fraction of the total probability, in a general case where the posterior is $P(\vec{\theta}|\text{data, prior})$ for parameters $\vec{\theta}$. Take your posterior, which is normalized so that its integral over all parameter values is 1. Now, start with the part of the parameter space with the highest posterior and integrate the probability under the curve as you go to lower and lower values of the posterior (including, possibly, disconnected regions), until you get

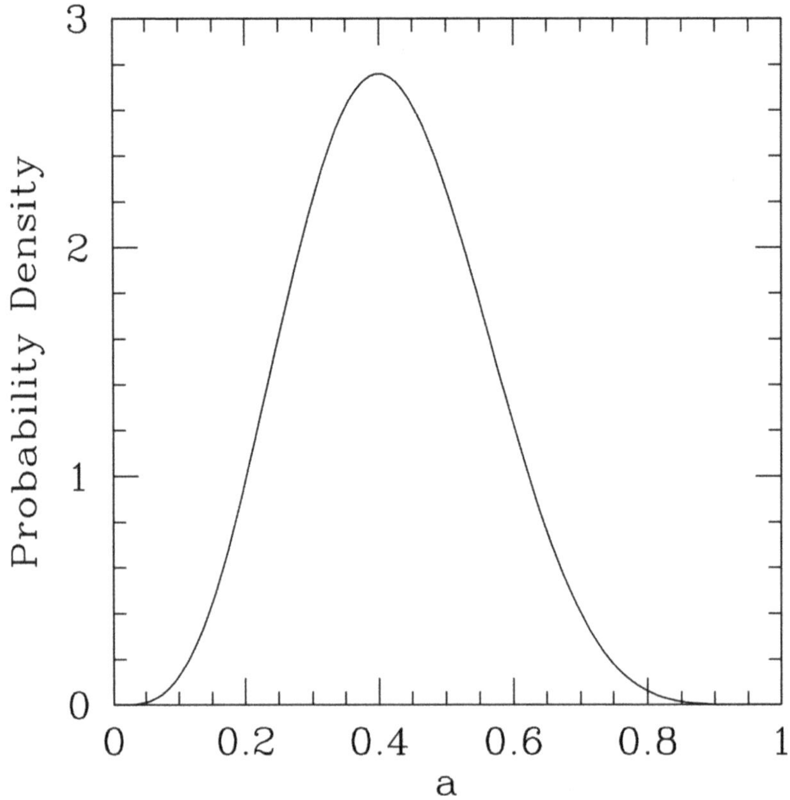

Figure A.2. Posterior probability density for the probability a of heads after 10 flips that produced 4 heads and 6 tails. Here our prior was that any value of a from 0 to 1 was equally likely. Note that, as a result, the posterior probability density peaks at $a = 0.4$ and that the probability density is asymmetric around that peak.

to the desired fraction of 100% of the probability. It's a bit like flipping the posterior over and determining where you'll integrate next by pouring water into the posterior; at any given time, you are integrating at a fixed value of the posterior density, until you get to the total probability that you want.

Even when we only have one or two parameters, there are pathological posteriors that can defeat us. A classic image is a flagpole in the middle of the ocean; it might be the highest point, and if it's high enough it might even dominate the volume, but good luck finding it! Unsurprisingly, things are even tougher when there are many parameters, as there often are in analyses of gravitational-wave data. For such applications, even after we streamline the problem as much as we can (e.g., separating extrinsic from intrinsic parameters in binary coalescences), we still can't do brute-force analyses. Statistical techniques such as MCMC and nested sampling have been developed to provide fast and reliable parameter estimation for multiple parameters, but even they can fail sometimes. To reiterate: it is a terrible idea to use codes blindly. As some wizards with a "mad eye" would say, constant vigilance!

A.5.2 Continuous Data

Suppose that ground-based gravitational-wave detectors have discovered a double neutron star merger and that electromagnetic follow-up observations show that the merger occurred in the nucleus of a dwarf galaxy. We'd like to determine some of the dynamical properties of the nucleus so that we can explore in more detail the possible origins of the system. There are only 10 stars in the nucleus bright enough that we can get good spectra, and we find that, relative to the redshift of the Galaxy, the speeds of the stars along the direction to us in increasing order are −18.4623, −17.6493, −9.26109, −9.09967, 1.10899, 2.96846, 18.0029, 25.5558, 27.8944, and 35.7825, in units of km s^{-1}. We want to fit a zero-centered Gaussian to the data to determine the velocity dispersion of the nucleus, which will in turn allow us to estimate rates of interactions. That is, we want to estimate the standard deviation of the velocities; how do we do this?

It might seem that you would have to bin the measurements because otherwise it doesn't look like a distribution at all (if the bins are narrow, there are either zero points or one point in a bin, so there are no peaks). This is, however, not the case, so let's see how it works.

First, we realize that our Gaussian has the form

$$N(v)dv = A\exp(-v^2/2\sigma^2)dv, \tag{A.17}$$

where σ is the standard deviation and A is a normalization factor. Therefore, it might appear that there are two parameters. However, if we normalize the distribution so that the total number of measurements expected in the model equals the number of data points, then in our case $\int_{-\infty}^{\infty} N(v)\,dv = 10$. This implies that $A - 10/(\sqrt{2\pi\sigma^2})$.

If we imagine that we have divided the data space into an enormous number of narrow bins in velocity where, for bin i, d_i, and m_i are, respectively, the number of observed counts and the number of expected counts in the model, then the Poisson likelihood of the data given the model is

$$\mathcal{L} = \prod_i \frac{m_i^{d_i}}{d_i!} e^{-m_i}. \tag{A.18}$$

The factor $\prod_i (1/d_i!)$ is common to all models, which means that it is a normalization factor that we can ignore. If we normalize our model so that $\sum_i m_i$ equals the number n of data points, then $\prod_i e^{-m_i} = e^{-\Sigma_i m_i} = e^{-n}$ is also a constant that we can ignore. Then, the likelihood becomes

$$\mathcal{L} \propto \prod_i m_i^{d_i} \tag{A.19}$$

and the log likelihood is

$$\ln\mathcal{L} = C + \sum_i d_i \ln m_i, \tag{A.20}$$

where we can ignore the constant C (again, because when we eventually normalize the posterior, the constant will drop out).

From this we see that the bins i without observed counts don't contribute, because $d_i = 0$. Therefore, the sum really only needs to go over the bins that contain counts. Next, what is m_i? It is the expected number of counts in a bin. Suppose that a bin has width dv at velocity v. Then the expected number is $N(v)dv$. This appears to depend crucially on the bin width, but remember that we're just comparing differences of log likelihoods. Therefore, if we use the same bin widths for every value of σ (which we obviously will), the $\ln dv$ values will be common to all models, and hence will cancel. If we make the further assumption that we've done the smart thing and chosen small enough bins that the ones with data all have $d_i = 1$, then we get finally

$$\ln \mathcal{L} = \sum_i \ln[N(v_i)] + \text{const}, \tag{A.21}$$

where the v_i are the measured velocities. The likelihood depends only on the values of the distribution function at the measured velocities.

This is actually a general result for continuous distributions, in any number of dimensions. After you've normalized, the log likelihood is just the sum of the log of the distribution function at the measured locations if you have enough precision that there is at most one count per bin.

For a general model, one would now calculate the log likelihood numerically for a set of parameter values, and then maximize it to get the best fit. In our particular case, we can do it analytically. Dropping the constant,

$$\ln \mathcal{L} = \sum_i [\ln(10/\sqrt{2\pi}) - \ln \sigma - v_i^2/2\sigma^2]. \tag{A.22}$$

This sum is over the 10 measured velocities. We note that the first term is in common between all models, so we drop it. We then have

$$\ln \mathcal{L} = -10 \ln \sigma - (1/2\sigma^2)\sum_i v_i^2. \tag{A.23}$$

The sum of the squares of our velocities is 3870 km ^{2} s $^{-2}$. Taking the derivative with respect to σ and setting it to zero (to maximize) gives

$$-10/\sigma_{\text{best}} + 3870/\sigma_{\text{best}}^3 = 0 \rightarrow \sigma_{\text{best}} = (387)^{1/2} = 19.67. \tag{A.24}$$

Ok, so we have the best-fit value (the mode of the distribution), let's now work on the credible region. When we compute the 68.3% credible region, we need to (1) select a prior on the standard deviation, and (2) decide on how we want to define the credible region. Let's tackle each of these in turn.

When we think about a prior, there are apparently many choices. For example, unlike with our previous discrete case, where a was limited to being between 0 and 1, σ could in principle range from 0 to ∞. What should we choose? We can get an answer to that by looking at the likelihood:

$$\mathcal{L} \propto \sigma^{-10} e^{-3870/(2\sigma^2)}. \tag{A.25}$$

We see that at very small σ the likelihood drops off sharply (because of the $e^{-3870/(2\sigma^2)}$ factor) and similarly at very large σ (because of the σ^{-10} factor). Thus, we actually can take σ to be equally probable in a large range, say 0 to 200, and then have it end abruptly above 200. There is virtually no likelihood above 200, so having it be constant from 0 to 500, or 0 to 1000, will lead to almost identical conclusions. This is an example in which the data are informative enough that reasonable priors will lead to the same conclusion.

As we mentioned before, there are also times when we might not know the scale of σ, in which case perhaps a reasonable prior might be that there is equal probability in equal ranges of $\ln \sigma$ (so that, for example, the prior probability would be the same from 1 km s $^{-1}$ to 10 km s $^{-1}$ as it is from 10 km s $^{-1}$ to 100 km s $^{-1}$). Then, the prior would be proportional to $1/\sigma$ from some minimum to maximum (can you see why?) and the posterior would thus be proportional to σ^{-11} instead of σ^{-10}—not a big difference.

So if we choose a flat prior $q(\sigma) = 1/\sigma_{\max}$ from $\sigma = 0$ to $\sigma = \sigma_{\max}$, with $\sigma_{\max} = 200$, or $\sigma_{\max} = 500$, or $\sigma_{\max} = 1000$, or really any large value, and define the credible region as before (the minimum contiguous region in σ that includes 68.3% of the probability), we find that the credible region runs from $\sigma = 15.6$ to $\sigma = 25.65$. In fact, $\sigma = 25$ was used to generate the data.

There are, of course, times when you simply can't have infinitely fine data. For example, instruments have a finite resolution in energy, angle, and so on. In addition, if you are interested in the parameters of the source, you'll need to use a model of the detector and propagate your intrinsic source model through the detector response to get a prediction in the data space.

As an important aside, always compare your model with the data in data space, which is called forward folding! Don't "backward fold," where you might try to use the gravitational-wave data, divide in some way by the detector noise curve, and figure the properties of the source that way. That leads to ambiguity and pain. In forward folding, you would start with a model of the event (e.g., two black holes with specified masses, spins, orientations, etc.), figure out the gravitational waves they would produce, fold that through the detector response, and then compare with the data.

It is also important to remember that because the value of the likelihood never enters, one can happily calculate maximum likelihoods and credible regions for models that are awful! It's an automatic procedure. That's why Bayesians draw a distinction between parameter estimation and model comparison, which we now discuss.

A.6 Model Comparison

Suppose we have a data set, and two models to compare, such as a general relativity model and a modified gravity model. How do we determine which model is favored by the data? At first glance this may seem easy: just figure out which model matches the data better. But think about models with different numbers of parameters; intuitively, we should give the benefit of the doubt to the model with fewer parameters, based on Ockham's principle. In addition, one could imagine a situation in which the parameters of two models are qualitatively different. Some of the parameters could be continuous (e.g., the mass of a black hole), and some could be discrete (e.g., if one is considering a modified gravity model in which the spin of the graviton is a free parameter but must be an integer or half-integer). For example, suppose we have two models. The first one has one parameter, which can take on any real number between 0 and 1. The second has three parameters, but each of them can only take on the value 0 *or* the value 1, but nothing in between. Which model is simpler, and how would we take that into account?

This is where Bayesian statistics shines. It provides a simple procedure that automatically takes into account different numbers and types of parameters in an intuitively satisfying way. As before we'll give the general principles and then try some examples.

Say we have two models, 1 and 2. Model 1 has parameters $a_1, a_2, \ldots, a_n$, and a normalized prior probability distribution $q_1(a_1, a_2, \ldots, a_n)$. Model 2 has parameters $b_1, b_2, \ldots, b_m$ and a normalized prior probability distribution $q_2(b_1, b_2, \ldots, b_m)$. For a given set of values $a_1, a_2, \ldots, a_n$, let the likelihood of the data given model 1 be $\mathcal{L}_1(a_1, a_2, \ldots, a_n)$, and similarly for model 2. Then the "Bayes factor" for model 1 in favor of model 2 is

$$\mathcal{B}_{12} = \frac{\int \mathcal{L}_1(a_1, a_2, \ldots, a_n) q_1(a_1, a_2, \ldots, a_n) da_1\, da_2 \ldots da_n}{\int \mathcal{L}_2(b_1, b_2, \ldots, b_m) q_2(b_1, b_2, \ldots, b_m) db_1\, db_2 \ldots db_m}, \tag{A.26}$$

where the integration in each case is over the entire model parameter space. Therefore, it is just a ratio of the integrals of the likelihoods times the priors for each model. When you multiply the Bayes factor by the prior probability ratio you had for model 1 in favor of model 2 (you could set that ratio to unity if you had no reason to prefer one model over another), you get the odds ratio $\mathcal{O}_{12}$ of model 1 in favor of model 2. As we mentioned before, each integral (e.g., the numerator of this equation, or the denominator of this equation) is sometimes called the "evidence" for the model in question, so the Bayes factor is the ratio of the evidences.

Captain Obvious: Earlier the authors advocated the use of log likelihoods rather than likelihoods, because logs are easier for computers to handle. But the evidence involves the integral of the likelihood times the prior, so what can we do? Commercial programs handle this just fine, but if you want to write code yourself, there's a nice trick you can use.

The idea is that when you do an integral on a computer you are basically computing a sum. Suppose that your sum, so far, is A. We'll designate by B the thing you want to add (i.e., the value of $\mathcal{L}(a)q(a)da$ that you're adding for some parameter set a). So now you want to compute $A + B$, but both A and B could be huge or tiny. We can nonetheless work entirely with logs, in the following way. We suppose that the log of our running sum is lnA, and we want $\ln(A + B)$. Then,

$$\begin{aligned}\ln(A + B) &= \ln[A(1 + B/A)] \\ &= \ln A + \ln(1 + B/A) \\ &= \ln A + \ln(1 + \exp(\ln B - \ln A)).\end{aligned} \tag{A.27}$$

Now you're working entirely with logs. You can set some threshold; for example, if $\ln B < \ln A - 50$, $\ln(A + B) \approx \ln A$; if $\ln B > \ln A + 50$, $\ln(A + B) \approx \ln B$. Otherwise, you can simply do the calculation. Again, sophisticated codes do evidence calculations without you having to worry about the details, but it is still useful to think about and use tricks like these because it gives you a better sense of the problem.

What does this mean? Don't tell a real Bayesian that we explained it this way, but consider the following. Suppose you and a friend place a series of bets. In each bet, one has two possible models. You compute the odds ratio as above and get $\mathcal{O}_{12}$ in each case. Ultimately, it will be determined (by future data, say) which of the two models is correct (we're assuming these are the only two possible models). If your friend puts down \$1 on model 2 in each case, how much money should you place on model 1 in each bet so that you expect to break even after many bets? You put down \$ $\mathcal{O}_{12}$. That is, it really does act like an odds ratio. The reason a hard-core Bayesian might get agitated about this analogy is that Bayesian statistics emphasizes considering only the data you have before you, rather than imagining an infinite space of data (as happens in more familiar frequentist statistics). Still, this provides some insight.

In practice, different codes use different approaches to compute the evidence for a model. For a nested sampler, it's direct: clever methods have been found for how to calculate the integrals directly. If you're using an MCMC with two different models, some codes allow the sampler to visit one model or the other with a probability that is proportional to the evidence, which means that the Bayes factor is just the ratio of the number of times one model was visited to the number of times the other was visited. As always, you should learn how the codes work so that you can be alert to potential problems.

Dr. I. M. Wrong: What you should be doing is telling readers what a given Bayes factor means. Many brilliant people, including myself, use a scale in which, for example, a Bayes factor of 10 to 30 is strong evidence in favor of one model, and a Bayes factor greater than 100 is decisive.

Major Payne: Not so fast! While it is true that plenty of papers use such a scale, it isn't necessary. The Bayes factor, by itself, tells you all you need to know in a quantitative way. Adding adjectives adds nothing to the discussion, and it risks serious confusion. For example, if a Bayes factor is 150 and we consider that to be "decisive," then we might not bother looking at additional data. In contrast, if we just use the odds ratio, we can place that in the context of later observations.

Why does this automatically take simplicity into account? Think of it like this. If your data are informative, then for a given set of data it is likely that only a small portion of the parameter space will give a reasonably large likelihood. For example, if you are modeling the gravitational waves from a double black hole coalescence, you might have the chirp mass and symmetric mass ratio as parameters; with good-enough data, only chirp masses and symmetric mass ratios close to the right ones will produce a likelihood close to the maximum. Now, think about the priors. For a complicated model with many parameters, the prior probability density is "spread out" over the many dimensions of parameter space. Thus, the prior is comparatively small in the region where the likelihood is significant. If instead you have few parameters, the prior probability density is less spread out, so it's larger where the likelihood is significant and therefore the integral is larger.

If the parameters have discrete instead of continuous values, you do a sum instead of an integral but otherwise it's the same. Note that (assuming that Poisson statistics apply) we have to use more of the full Poisson likelihood here. When we did parameter estimation we could cancel out lots of things, but here we have an integral or sum of likelihoods so we can't do the cancellation as easily. The product $\prod (1/d_i!)$ will be the same for every likelihood, and if your model is normalized so that the total number of expected counts is set to the number of observed counts (which is common, but not universal), then $\prod \exp(-m_i)$ is the same for every likelihood. Thus, those factors can be canceled, but one still has a sum of likelihoods and so taking the log requires some interesting finesses.

Let's try an example. Consider a six-sided die. We want to know the probabilities of each of the six faces. Model 1 is that the probability is the same (1/6) for each face. Model 2 is that the probability is proportional to the number on the face. Normalized, this means a probability of 1/21 for 1; 2/21 for 2; and so on (the 21 is for the normalization). We roll the die ten times and get 5, 2, 6, 2, 2, 3, 4, 3, 1, 4. What is the Bayes factor between the two models?

We're starting with an easy one, in which there are no parameters, so we don't even have to do an integral, just a likelihood ratio. For either model, the normalized model expectation is just the product of the probability times the number of times we roll the dice. So for model 1, the normalized model expectations per bin are $m_1 = 10/6$, $m_2 = 10/6$, and so on. For model 2 we have $n_1 = 10/21$, $n_2 = 20/21$, $n_3 = 30/21$, and so on. Therefore,

$$\mathcal{L}_1 = \left(\frac{10}{6}\right)^1 \cdot \left(\frac{10}{6}\right)^3 \cdot \left(\frac{10}{6}\right)^2 \cdot \left(\frac{10}{6}\right)^2 \cdot \left(\frac{10}{6}\right)^1 \cdot \left(\frac{10}{6}\right)^1 = 165.4 \tag{A.28}$$

and

$$\mathcal{L}_2 = \left(\frac{10}{21}\right)^1 \cdot \left(\frac{20}{21}\right)^3 \cdot \left(\frac{30}{21}\right)^2 \cdot \left(\frac{40}{21}\right)^2 \cdot \left(\frac{50}{21}\right)^1 \cdot \left(\frac{60}{21}\right)^1 = 20.7, \tag{A.29}$$

where we multiply the expectations because the likelihood is a product as in the previous section. Thus, from this data,

$$\mathcal{B}_{12} = \mathcal{L}_1/\mathcal{L}_2 = 7.98. \tag{A.30}$$

Model 1 is favored, assuming that we didn't have strong prior favoritism toward model 2.

Now try another example, with the same data. Model 1 is the same as before, but now model 2 has a parameter. In model 2, the probability of a 1 is $1 - p$, and the probability of a 2, 3, 4, 5, or 6 is $p/5$. Therefore, model 2 encompasses model 1 (or in fancy language, we say the models are nested), so by maximum likelihood alone it will do at least as well. But will it do better enough to be favored? Let's assume as a prior that p is equally probable from 0 through 1. The numerator is the same as before, but for the denominator we need to do an integral. For probability p and our given data, the Poisson likelihood of the data given the model is

$$\mathcal{L}_2(p) = [10(1 - p)] \cdot (10p/5)^3 \cdot (10p/5)^2 \ldots = 10(1 - p)(2p)^9. \tag{A.31}$$

Therefore, the denominator is

$$\int_0^1 5120(1 - p)p^9 \, dp = 46.5 \tag{A.32}$$

and the Bayes factor is

$$\mathcal{B}_{12} = 165.4/46.5 = 3.55, \tag{A.33}$$

so the first model is still preferred. Note that the maximum likelihood for model 2 occurs for $p = 0.9$ (it should—we have one 1 in 10 rolls!) and gives 198.4, so as expected the more complicated model has a higher maximum likelihood; it's just not enough to make up for the extra complication.

Captain Obvious: Although the authors are describing the correct Bayesian way to compare models, the required calculations can be extremely challenging in some cases. That's because to compute the evidence it is necessary to integrate over, in principle, the whole parameter space, although in practice clever methods are often employed to focus only on the parts of the parameter space that contribute significantly to the integral.

It turns out, typically, to be much easier to find the maximum of a function than to integrate the function (although there is never a guarantee of success). This is one reason why analyses sometimes take a shortcut by employing a maximum likelihood ratio test. The idea is that if we are comparing two nested models, which, as the authors have said means that one model contains the other as a special case, then in a way similar to how χ^2 is often used, the maximum likelihood for the more complicated model must be larger than the maximum likelihood for the simpler model by enough of a factor that the extra complexity is justified.

Major Payne: Grumble. Captain Obvious is glossing over some potentially major problems. The maximum likelihood ratio test, like a $\Delta\chi^2$ test, runs into some nontrivial conceptual and practical difficulties. For example, if you do a "by the book" $\Delta\chi^2$ test between nested models, you are supposed to (1) determine the $\Delta\chi^2$ between the best fits of the two models, (2) use as your number of degrees of freedom the number of extra parameters in the more complicated model, and (3) use a χ^2 table to figure out the degree to which you can then favor the more complicated model. For example, suppose that you are comparing two models of the distribution of black hole masses: one that is a pure power law with sharp cutoffs at a minimum and a maximum mass, and the other that adds a Gaussian peak. It has to matter whether the amplitude, centroid, and width of the Gaussian are unconstrained or whether they are required to be in narrow ranges! But the standard $\Delta\chi^2$ or log likelihood ratio test doesn't easily take this into account. I admit that these tests can be useful as a quick check. But if you want my advice (and you should), you should be very cautious about drawing major conclusions from simplified tests such as these.

Model comparison in Bayesian statistics is always between precisely defined models. There is no analog to the idea of a null hypothesis. Hard-core Bayesians consider this to be a strength of the approach. For example, suppose that you try to define a null hypothesis and do a standard frequentist analysis, finding that the null hypothesis can be rejected at the 99% confidence level. Should you, in fact, reject the null hypothesis? Not necessarily, according to Bayesians. Unless you know the full space of possible hypotheses, it could be that there are 10,000 competing hypotheses and of those your null hypothesis did the best.

This is not a fully satisfactory line of argument. It is also important to look at your fit to determine whether or not your model is reasonable. For example, if it is applicable, a standard chi-squared per degree of freedom can give you an idea of whether you need to work a lot harder to get a good model, or if you're nearly there. Indeed, when gravitational-wave data are analyzed, there are a variety of checks to determine whether the model is at least a decent fit to the data. For example, you can subtract the best-fit model from the data to determine whether the residuals are at least roughly consistent with noise.

Where does all of this leave us? As always, when you undertake any problem, be it analytical, observational, or computational, you should think about what you want out of your analysis before you decide on your methods. It is a generally good idea, when evaluating a model or derivation or whatever, that you start with quick, easy methods (e.g., order-of-magnitude estimation) before settling in for more detailed treatments if those are warranted. The same goes for statistics. You could, for example, perform a quick analysis (chi-squared, Kolmogorov–Smirnov test, a Fisher analysis, or whatever) first, to see if your data are informative. If they are, then you may be justified in spending time with a more rigorous method (like an MCMC exploration of the likelihood) to get the most out of your data. In all cases, however, you have to know the limitations of your method! This is one advantage to thinking in the Bayesian way. In many circumstances you can figure out what should be done, and then you have a better sense of how good an approximation a simpler method is.

A.7 Exercises

1. You go to the doctor for what you think is a routine checkup. However, the doctor regrets to inform you that, based on your test results, you have a greatly elevated chance of having a rare disease that will, inevitably, within

no more than two years, turn you into a complete jerk (i.e., not a type of chicken preparation, but rather a contemptibly obnoxious person). Even with your indicators, the probability of you having the disease is just 1/1000. The doctor asks you to take a more specific test that has 90% accuracy. The test comes back positive. What is the probability that you have the disease?

2. You are given a completely ordinary-looking coin and asked to assess the probability P that, in a given flip, you will get heads. Based on your previous experience with coins, your prior probability distribution is

$$q(P) = \frac{1}{\sqrt{2\pi\sigma^2}} e^{-(P-0.5)^2/2\sigma^2}. \tag{A.34}$$

You start flipping the coin and get heads in every flip. If $\sigma = 0.1$, how many flips are needed to convince you that, at 90% probability, the probability of heads is greater than 90%?

3. You are analyzing data on double black hole binaries, and your model is that the chirp mass is equally probable from $\mathcal{M}_{\rm min}$ to 50 $M_\odot$ and cannot be outside that range. Your prior on $\mathcal{M}_{\rm min}$ is a flat probability from 5 $M_\odot$ to 20 $M_\odot$, and zero outside that range. You measure one black hole binary chirp mass with perfect precision and it turns out to be 30 $M_\odot$. Calculate the Kullback–Leibler divergence for your chirp mass distribution, in nats.

4. Suppose that you have three different models for the chirp masses of double black hole binaries. In all three, the probability is zero outside the range 5 $M_\odot$ to 50 $M_\odot$. In Model M_1, the probability is flat from 5 $M_\odot$ to 50 $M_\odot$. In Model M_2, the probability is proportional to $\mathcal{M} - 5\ M_\odot$ from 5 $M_\odot$ to 50 $M_\odot$. In Model M_3, the probability is proportional to $50\ M_\odot - \mathcal{M}$ from 5 $M_\odot$ to 50 $M_\odot$. You measure a single event with a chirp mass of $\hat{\mathcal{M}} = 10\ M_\odot$. Assuming that your prior is that all three models are equally probable, compute the relative posterior odds of M_1, M_2, and M_3.

5. In this problem we want you to ponder an apparent paradox, which we will present in an abstract form although you could imagine applications to gravitational-wave astronomy. You have a selection of one-dimensional data points that you know to be drawn from a unit-variance Gaussian, and you want to compare two models. Model 1 is that the unit-variance Gaussian is centered on 0. Model 2 is that it has an arbitrary center, with an equal probability anywhere between $-\infty$ and $+\infty$. In a frequentist treatment, you might do something like a maximum likelihood ratio test and determine whether Model 2 has a maximum likelihood sufficiently larger than the likelihood of Model 1 (which has no parameters) to justify the extra parameter. What about a full Bayesian model comparison? Is there any finite data set that would give a Bayes factor in favor of Model 2? What is the resolution of the pseudo-paradox?

6. In this problem we ask you to write a code (you can choose your coding poison) to explore the central limit theorem, which is an important principle that very often allows you to treat the average of several measurements as drawn from a Gaussian, even if the underlying distribution is not at all a Gaussian. The theorem is that if you have a distribution with a mean μ and a finite standard deviation σ, and if you take the average of N samples from the distribution, then as $N \to \infty$ the average is distributed like a Gaussian with a mean μ and a standard deviation $\sigma/\sqrt{N}$. Test this out using your code by doing draws from a distribution that you might have for the chirp masses of black hole binaries: $P(\mathcal{M}) \propto \mathcal{M}^{-2}$ from $\mathcal{M} = 5\ M_\odot$ to $50\ M_\odot$, and zero otherwise. Test this out for other distributions as well.

7. Suppose that we have detected gravitational waves from numerous black hole binaries and in each case have measured the chirp mass $\mathcal{M}$ and the effective spin ($\chi_{\rm eff}$, which is the mass-weighted projected spin of the black holes along the orbital axis). If the joint posterior probability distribution between $\mathcal{M}$ and $\chi_{\rm eff}$ is $P(\mathcal{M}, \chi_{\rm eff}) \propto e^{-\chi_{\rm eff}^2/[2(0.1)^2]}\ \mathcal{M}^{-2}$ (in the range $-1 \leqslant \chi_{\rm eff} \leqslant 1$ and $5\ M_\odot \leqslant \mathcal{M} \leqslant 50\ M_\odot$), what are the marginalized posterior probability distributions for $\chi_{\rm eff}$ and $\mathcal{M}$ individually?

8. Suppose that a single example of a special type of gravitational-wave event (you take your pick $\cdots$) has been observed, and during the observation time the total spacetime volume that could have revealed the event is VT. For example, if the sensitivity was constant for an observing time T, during which the volume out to which the event could have been seen was V, then the spacetime volume would be VT; one could also envision an evolving sensitivity with an accessible volume $V(t)$ that changes with time t and integrates to a total spacetime volume VT. The best estimate of the rate density (rate per volume) is then $1/(VT)$. Assuming Poisson statistics, compute the 5th to 95th percentile of the rate density of the special events. In your calculation, assume that the prior probability is the same for any rate density up to extremely large values, i.e., the prior is flat.

9. Suppose that we see two gravitational-wave events. Your task is to calculate the Bayes factor between two hypotheses: (1) they came from a single source, or (2) they came from two distinct sources. Assume that a priori the sources could come from anywhere in the sky with equal probability, and if there are two sources their locations are not correlated a priori. To give you a start, note that the evidence for hypothesis 2 (two sources) can be written as

$$E_2 = \left[\int q_{A1}(\vec{\theta}_1)\mathcal{L}_1(\text{data}_1|\vec{\theta}_1)d\Omega_1\right]\left[\int q_{A2}(\vec{\theta}_2)\mathcal{L}_2(\text{data}_2|\vec{\theta}_2)d\Omega_2\right], \tag{A.35}$$

where $q_{A1}(\vec{\theta}_1)d\Omega_1$ is the prior probability that we use in our analysis that the actual direction to source 1 is within a solid angle $d\Omega_1$ of the direction $\vec{\theta}_1$, and similarly for source 2. In our case, we are assuming that $q_{A1}(\vec{\theta}_1) = q_{A2}(\vec{\theta}_2) = 1/(4\pi)$.

Useful Books

Donovan, T. M., & Mickey, R. M. 2019, Bayesian Statistics for Beginners: A Step-by-step Approach (Oxford: Oxford Univ. Press)

Hoff, P. D. 2009, A First Course in Bayesian Statistical Methods (Berlin: Springer)

Gelman, A., Carlin, J. B., Stern, H. S., Dunson, D. B., Vehtari, A., & Rubin, D. B. 2013, Bayesian Data Analysis (Boca Raton, FL: Chapman and Hall/CRC)

Gregory, P. 2005, Bayesian Logical Data Analysis for the Physical Sciences (Cambridge: Cambridge Univ. Press)

Jaynes, E. T. 2003, Probability Theory: The Logic of Science (Cambridge: Cambridge Univ. Press)

Mertler, C. A., & Reinhart, R. V. 2016, Advanced and Multivariate Statistical Methods: Practical Application and Interpretation (England, UK: Routledge)

Tamhane, A., & Dunlop, D. 1999, Statistics and Data Analysis: From Elementary to Intermediate (England, UK: Pearson)

Appendix B

A Primer on Dynamics

Because many of the paths to binary coalescence involve the dynamics of stellar clusters, here we gather some of the key points. As an overall note, we are thinking about a collisionless cluster, where the stars and compact objects interact with each other gravitationally but do not physically collide. To see that this is reasonable, consider a star cluster in the nucleus of a galaxy, where at the center there might be thousands of stars at a number density of $n \approx 10^6$ pc $^{-3} \approx 3 \times 10^{-50}$ cm^{-3}. Suppose that the velocity dispersion (i.e., the statistical dispersion of velocities about its mean) is $\sigma = 100$ km s^{-1} and that a typical star has an effective cross section of $\Sigma \approx 10^{23}$ cm^2 (taking into account gravitational focusing; see below). Then, the rate of interactions for a single star is $\tau^{-1} = n\Sigma\sigma \approx 10^{-12}$ yr^{-1}. Thus, even in this relatively high-density environment, only $\sim$1% of stars will have interactions over the current age of the universe ($H_0^{-1} \sim 10^{10}$ years). This is what allows us to make the collisionless approximation. With that assumption, we now describe briefly some of the key aspects of collisionless gravitational dynamics.

B.1 Gravitational Focusing

When we think about interactions and rates, we have to make calculations such as the one above: the rate is the number density times the relative speed times the effective cross section. The effective cross section, however, is more than the area of the target. To understand why, let us suppose that we are interested in a situation in which two objects come within some minimum distance r_p of each other. As you will demonstrate in the exercises, if the total mass of the two is m and the relative speed at a very large distance (which we often colloquially call "infinity") is v, then the cross section we need to put into the rate formula is

$$\Sigma = \pi r_p^2 \left(\frac{2Gm}{r_p v^2} + 1 \right). \tag{B.1}$$

By itself, the second term in the parentheses gives the expression we might have expected: πr_p^2. But if the speed at infinity is small enough, then gravity deflects the path and makes the interaction more likely than it would have been. You can see that the first term is the square of the ratio of the mutual escape speed at a distance r_p (because the escape speed is $\sqrt{2Gm/r_p}$) to v. This makes sense: if the speed is much larger than the escape speed, the trajectory is barely deflected, whereas if the speed is much slower, then gravity is dominant. Of course, any inverse power of v would satisfy that constraint; in the exercises, you can derive the inverse square.

One initially counterintuitive point is that if gravitational focusing is dominant (i.e., the first term in the parentheses is much greater than unity), then the rate of interactions decreases with increasing speed, because then $\Sigma \sim v^{-2}$ and $\tau^{-1} = n\Sigma v \sim v^{-1}$. This is because in that limit the effective cross section increases rapidly with decreasing speed as lower-speed trajectories can be bent more by gravity. In the opposite limit, i.e., when $2Gm/(r_p v^2) \ll 1$, then $\tau^{-1} \sim v$, and the rate increases with increasing speed at infinity.

B.2 Heavy Objects Sink

"Heavy objects sink" is obvious for a dense thing in a fluid, but on reflection, it is not obvious for stellar dynamics. We are thinking about a situation in which every point mass (which is how we're treating stars and compact objects in this appendix) has a mass that is tiny compared with the mass of the cluster as a whole. As a result, each object is orbiting in the potential of the cluster. If that potential were time independent and spherically symmetric, then (by Noether's theorem) the orbital energy and all three components of the orbital angular momentum would be conserved and therefore neither the semimajor axis nor the eccentricity of any orbit would be changed, although the orbits could still precess.

What alters this situation is gravitational interactions between objects. If two objects pass by each other, the trajectory of each is deflected. This changes the energy and angular momentum of each object relative to the center of mass of the cluster; of course, in an isolated interaction, the combined energy and angular momentum of the pair remains constant. The typical result is that the heavier of an unequal-mass pair will lose energy and therefore sink in the potential, i.e., toward the center of the cluster, which means that the lighter of the pair will gain energy and therefore rise in the potential, i.e., away from the center of the cluster. Over time this leads to what is called mass segregation, in which the heavier objects tend to settle toward the center of the cluster.

Major Payne: Let's perform a simple calculation to motivate the idea that in two-body encounters, heavy objects tend to transfer energy to lighter objects. Suppose that a star of mass m_1 and a star of mass m_2, both with initial speed v_0 as measured from the center of a cluster, approach each other nearly head on but with enough of an impact parameter that their gravitational interaction reverses the direction of motion of both, i.e., their motion is turned around 180° after they pass by each other. Call the final speeds of the two objects v_1 and v_2, respectively. Then the equations of conservation of momentum and energy are

$$\begin{aligned} m_1 v_1 - m_2 v_2 &= m_1 v_0 - m_2 v_0 \\ \frac{1}{2}m_1 v_1^2 + \frac{1}{2}m_2 v_2^2 &= \frac{1}{2}m_1 v_0^2 + \frac{1}{2}m_2 v_0^2, \end{aligned} \tag{B.2}$$

where we have ignored the potential energy because we assumed they both start and end very far from each other (at infinity), and of course, we have ignored all relativistic effects. After the 180° turn, the solution of these equations is that the final speeds are

$$v_1 = \frac{m_1 - 3m_2}{m_1 + m_2} v_0, \qquad v_2 = \frac{m_2 - 3m_1}{m_1 + m_2} v_0. \tag{B.3}$$

If $m_1 = m_2$, then $v_1 = -v_0$ and $v_2 = -v_0$, which means that the stars simply turn around but go at the same speed as before. But if, say, $m_1 > m_2$, then $|v_1| < v_0$ and $|v_2| > v_0$. Thus, as advertised, energy is transferred from the heavier to the lighter star in this encounter.

Captain Obvious: More generally, there is a productive analogy with thermodynamics: the heavy and light object, being in roughly the same place in the overall potential, typically have about the same speed. Thus, the heavier object has greater kinetic energy. If we think of this as a temperature, then the transfer of energy of the heavier object to the lighter object is similar to our understanding that heat flows from hotter regions to cooler regions. A critical difference is that because, in a gravitational potential, objects with less energy (i.e., more negative energy, so closer to the center) move faster, the effective heat capacity of gravity is negative. That is, loss of energy leads to faster motion. That has major consequences for many areas of astronomy.

Another important thermodynamic analogy with self-gravitating clusters involves the increase of the entropy of the system. Suppose, for example, that we start for simplicity with an isolated spherical cluster of stars, with a sharp edge, of stars that all have the same mass. Suppose also that initially the number density of stars is uniform throughout the cluster. If you take two nearby stars and reduce the energy of one (so that it moves closer to the cluster center) and correspondingly increase the energy of the other (so that it moves farther away), you will find that the total phase space (which is a product of the spatial volume and momentum volume) occupied by both stars combined is larger than it was before. We can motivate this with an extreme example: if we have two stars with the same orbital gravitational binding energy, and we double the orbital binding energy of one of the stars, then to conserve the total energy of the two stars, the other star has to become unbound, i.e., the spatial volume it explores is then infinite! More quantitatively, because the characteristic speed of an orbit of radius r is $v \sim r^{-1/2}$, increasing r increases the spatial volume ($\sim r^3$) faster than it decreases the momentum volume ($\sim r^{-3/2}$). This movement of stars therefore effectively increases the entropy of the cluster. Hence, the drive toward higher entropy takes a uniform-density system and drives it toward a concentrated core and a tenuous, larger, outer region. Although you might think at first that the close to uniform-density initial state of the universe already maximized its entropy, the special properties of gravity mean that systems that interact only via gravity are never in equilibrium because structure formation increases the entropy further.

B.3 Two-body Relaxation

More generally, in addition to heavy objects tending to sink and light objects tending to float in a gravitationally bound cluster, two-body interactions change the three-dimensional motions of stars and therefore also change the eccentricities of their orbits in the cluster. For an object of mass m moving in an environment with an overall mass density of ρ and a velocity dispersion of σ, the characteristic timescale needed to halve or double the object's semimajor axis is called the energy relaxation time because the semimajor axis is related to the orbital energy. We can produce a Fermi estimate of this time by noting that it is roughly the time needed for the object to interact significantly with its own mass. Here, "interact significantly" turns out to mean that the matter needs to pass close enough to the object that the escape speed from the object at that distance equals the speed of the matter. Thus, for an object of mass m, the radius for significant interaction is $r \approx Gm/\sigma^2$, so if the density is ρ, then the rate at which matter passes through this radius is $\dot{m} \sim \rho r^2\sigma \sim \rho G^2m^2/\sigma^3$, and the time needed to interact with mass m is $T \sim m/\dot{m} \sim \sigma^3/(G^3\rho m)$.

In more detail,

$$
\begin{aligned}
t_{\rm rlx,ener} &\approx \frac{1}{3\ln\Lambda}\frac{\sigma^3}{G^2\rho m} \\
&\approx 2\times 10^7\ {\rm yr}\left(\frac{10}{\ln\Lambda}\right)\left(\frac{\sigma}{10\ {\rm km\ s^{-1}}}\right)^3\left(\frac{m}{M_\odot}\right)^{-1}\left(\frac{\rho}{10^5\ M_\odot\ {\rm pc^{-3}}}\right)^{-1}.
\end{aligned}
\tag{B.4}
$$

Here we have scaled to typical parameters in the centers of dense globular clusters, and $\ln\Lambda$ is the Coulomb logarithm, which comes out of a detailed study of the dynamics of a particular system (in ways that are echoed by, for example, a study of bremsstrahlung). The logarithm means that details matter little; note, for example, that $\ln 100 \approx 5$ and $\ln 10^9 \approx 20$, from which we get the astrophysicists' rule of thumb that "all logarithms are 10," at least to within a factor of 2 or so.

Major Payne: The Coulomb logarithm is a highly important concept with many applications and does not deserve to be brushed aside in this way! Luckily, I am here to fill that gap in your knowledge. We'll start with the electromagnetic analogy and then comment on an important difference between that and the gravitational case.

Suppose we have an electron of charge $-e$ and mass m moving past an ion of charge Ze, with a speed at infinity of v and an impact parameter of b; that means that if the path were a perfectly straight line, the closest the electron would come to the ion would be a distance b. We consider a path that is deflected only slightly, which means that to lowest order, we can consider just the change in velocity normal to the path and can do our integration over the original straight-line path. Because the magnitude of the electrical force on the electron at a distance r from the ion is Ze^2/r^2, the component of the acceleration normal to the path at a time t from the closest passage is $Ze^2b/[m(b^2+v^2t^2)^{3/2}]$. Therefore,

$$
\Delta v = \frac{Ze^2}{m}\int_{-\infty}^{\infty}\frac{b\,dt}{(b^2+v^2t^2)^{3/2}} = \frac{2Ze^2}{mbv}. \tag{B.5}
$$

If we suppose that we have an infinite uniform sea of ions that deflect the electron, then because of symmetry there is no net direction of deflection (hence $\langle\Delta v\rangle = 0$), but there is a net change of the square of Δv, as with any random walk. Thus, the expected change in $(\Delta v)^2$ for a single interaction with impact parameter b and speed at infinity v is $4Z^2e^4/(m^2b^2v^2)$. To figure out the expected rate of change over all interactions we need to integrate over (1) the distribution of speeds v, and (2) the distribution of impact parameters b. The integration over speeds is not a problem; for example, if we assume a Maxwell–Boltzmann distribution $P(v) \propto v^2 e^{-mv^2/(2kT)}$, then at low speeds $P(v) \propto v^2$ and at high speeds $P(v)$ dies exponentially, so when multiplied by v_2 the integral is convergent. But b provides a problem. If the perturbers (ions in this case) are distributed uniformly at large distance, then the number $N(b)db$ between an impact parameter b and an impact parameter $b + db$, with $db \ll b$, is $N(b)db \propto bdb$. When multiplied by $(\Delta v)^2$, our integral is then $\propto \int_0^\infty (db/b)$, which diverges at both limits!

We note, though, that the divergence is "only" logarithmic. For example, if we set the lower limit of b to be $b_{\min}$ (for example, at too small a b the electron will actually hit the ion) and the upper limit to be $b_{\max}$ (for example, because the system isn't actually infinite) then that integral becomes $\ln(b_{\max}/b_{\min})$. Because logarithms change only slowly with their argument, the details of $b_{\min}$ and $b_{\max}$ don't matter much and the factor is called $\ln \Lambda$, the Coulomb logarithm. An interesting difference between the electromagnetic application (e.g., bremsstrahlung) and the gravitational application (to two-body relaxation or dynamical friction) is that because there are positive and negative electric charges, a given charge is surrounded by a slight excess of the opposite charge (and is roughly cancelled at the Debye radius). Thus, the deflection angle dies away faster with increasing distance than Coulomb's Law would suggest for a bare charge. In contrast, there are no negative masses, which means that the gravitational effect would extend indefinitely without a cutoff in the system. However, as we indicated above, at large-enough distances you run out of stars; the number density isn't constant throughout the universe! We can be casual about this because logarithms vary so slowly.

The energy relaxation time should then be compared with the age of the system to determine whether there has been enough time to undergo significant dynamical evolution. For example, globular clusters are roughly the age of the universe ($\sim 10^{10}$ years), which means that they have had plenty of time to evolve. In contrast, a massive galaxy with $\sigma = 200$ km s^{-1} and an average $\rho = 10^4\ M_\odot$ pc^{-3} has an energy relaxation time of $\approx 10^{12}$ years for 1 $M_\odot$ objects. We learned earlier that two-body interactions tend to transfer energy from heavier to lighter objects, which means that heavier objects settle toward the center of the cluster over time until their self-interactions become important. Equation (B.4) shows that the energy relaxation time goes inversely with the mass of the object, so when the time is much longer than the age of the universe for an average-mass star, only objects such as black holes will have had time to sink to the center.

Let us now focus on the orbital evolution of stars of roughly the average mass of the system. In such cases, it makes sense (and is true!) that the orbit has the character of a random walk. In a random walk, the average distance from the starting point increases like the square root of time. Now, it turns out that the time needed for energy relaxation is also roughly the time needed to change the angular momentum by roughly the angular momentum of a circular orbit of the initial semimajor axis. Because an orbit with eccentricity e has an angular momentum that is $(1 - e^2)^{1/2}$ times the angular momentum of a circular orbit of the same semimajor axis, this means that the angular momentum relaxation time is

$$t_{\rm rlx,ang} \approx (1 - e^2) t_{\rm rlx,ener}. \tag{B.6}$$

Because the pericenter distance of an orbit of semimajor axis a and eccentricity e is $r_p = a(1 - e)$, this means that highly eccentric orbits can be perturbed rapidly to decrease their pericenter distance. That's because when $e \sim 1$, Equation (B.6) shows that the angular momentum can halve or double its value much more rapidly than the energy relaxation time, which is the time to change the semimajor axis a by of order itself. As we discuss in Section 5.1.3, this is thought to be the main mechanism that drives tidal disruption events and extreme mass ratio inspirals.

In summary, two-body relaxation causes stars of average mass to increase or decrease their semimajor axes and eccentricities randomly. For stars or compact remnants heavier than average, the net effect is to cause them to sink toward the center of the cluster; conversely, lighter-than-average stars tend to move away from the center. Relaxation operates more rapidly for more massive objects, which means that if the energy relaxation time for a given type of object is shorter than the timescale of the system then heavier things are expected to be overrepresented in the center (globular clusters are examples of such systems). The high number densities that are expected close to the centers of such systems therefore promote interactions, particularly interactions involving binaries (which have effective areas the size of the orbit). The stochastic wandering of eccentricities leads to losses of objects that get close enough to a central massive black hole (if the system has a massive black hole) via tidal disruption or inspirals.

B.4 Resonant Relaxation

A fundamental assumption in the theory of two-body relaxation, which leads to the analogy with random walk motion, is that each two-body gravitational encounter is uncorrelated with every other such encounter. But in 1996 Kevin Rauch and Scott Tremaine followed an idea originally credited to Jerry Ostriker to show that there is another potentially important regime. They noted that, for stellar orbits close enough to a massive black hole such that most of the mass contained in the orbit belongs to the hole, the stars are in a potential that is dominated by the hole, and therefore, they move in slowly changing Keplerian orbits. If the orbits were exactly Keplerian, then they would simply trace over themselves indefinitely; indeed, the "resonant" in "resonant relaxation" refers to the equality of the orbital period (how long it takes to go 2π in phase) and the radial epicyclic period (how long it takes to go from apocenter to pericenter and back), which produces the fixed orientation. Over time, therefore, it would be as if their mass were smeared over the orbit, like a wire with greater linear mass density where the orbit spends more time. That is, the linear mass density would be greatest near the apocenter and least near the pericenter. If we consider the net gravitational effect of two wires on each other, the time-averaging means that (by Noether's theorem again) their energies are constant. Because the energy of an orbit depends only on the semimajor axis, the constancy of energy means that their semimajor axes are constant. There is, however, usually a net torque. Thus, the angular momentum and therefore the eccentricity can change. Moreover, as long as the wires maintain their orientation in space, the sign of the torque remains constant. Thus, in many circumstances, this effect can change the angular momentum, eccentricity, and pericenter distance much more effectively than can two-body relaxation.

The constancy of orientation is broken by precession, which can be classical (the long-term effect of other stars) or general relativistic. Resonant relaxation is being actively explored to determine its net effect on dynamics and rates, but the current feeling is that its net effect is small. One reason for this is that the rate of encounters of, say, stellar-mass black holes with supermassive black holes is driven by the rate at which they flow from larger distances, rather than details of the processes that occur relatively close to the massive black hole.

B.5 Dynamical Friction

Now suppose that the object of interest is much more massive than most of the other objects with which it interacts. An example would be a stellar-mass black hole ($\sim 10\ M_\odot$) among ordinary stars (which might have an average mass of $\sim 0.5\ M_\odot$, particularly for an old stellar system). An even more extreme example is a supermassive black hole (mass $\sim 10^5 - 10^{10}\ M_\odot$) in a field of stars. In these cases, we can approximate the interactions of the object not as individual, stochastic gravitational scatterings, but as a smoother process called dynamical friction.

Dynamical friction is often visualized in the following way. Suppose that our heavy object is moving relative to a field of stars. The trajectories of the stars will be deflected by the gravity of the object (see Figure B.1), and therefore, the density will be enhanced in a wake behind the object. This enhanced density exerts a greater gravitational pull than the pull of the stars in front of the object, with the result that the object is slowed down. Suppose that the stars have an overall mass density ρ (this is over a volume that contains many stars) and a Maxwell–Boltzmann velocity dispersion σ, and consider an object of mass M that moves at speed v_M relative to the center of momentum of the stars. If $v_M \gg \sigma$, then as we have argued before the rate of interaction of the heavy object with the stars is proportional to v_M^{-3}. The momentum per star relative to the heavy object

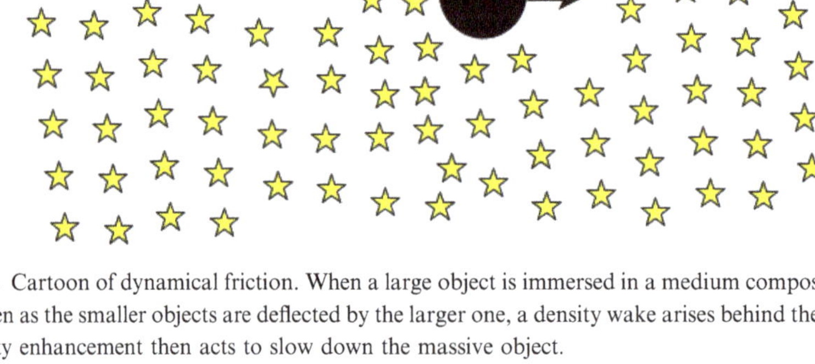

Figure B.1. Cartoon of dynamical friction. When a large object is immersed in a medium composed of smaller objects, then as the smaller objects are deflected by the larger one, a density wake arises behind the large object. This density enhancement then acts to slow down the massive object.

is $\propto v_M$, so the drag force $F_{\rm drag} = dp/dt$ is proportional to $v_M^{-3} v_M = v_M^{-2}$. In the opposite limit $v_M \ll \sigma$, things are more subtle, because stars from any direction can overtake the black hole and thus the drag from different stars acts in different directions. The result is that if $v_M \ll \sigma$, the drag force is proportional to v_M, which is the same drag law as you would have for a ball moving slowly through a high-viscosity medium such as honey.

Major Payne: Let's at least give the details! If we define $X \equiv v_M/\sigma$, and we assume that the stellar velocities are isotropic and that their speeds follow a Maxwell–Boltzmann distribution, then a careful treatment of the dynamics gives a drag force of

$$F_{\rm drag} = M\frac{dv_M}{dt} = -\frac{4\pi(\ln\Lambda)G^2\rho M}{v_M^3}\left[\mathrm{erf}(X) - \frac{2X}{\sqrt{\pi}}e^{-X^2}\right]v_M. \tag{B.7}$$

If $X \gg 1$ (i.e., the relative speed is much larger than the velocity dispersion), then the term in brackets approaches 1 and we see that $F_{\rm drag} \propto v_M^{-2}$. Thus, in this limit, again, the force is lower when the relative speed is higher, though of course v_M cannot be chosen to be too small, as then the condition $X \gg 1$ would be broken. If $X \ll 1$ then $\mathrm{erf}(X) \to 2X/\sqrt{\pi}$ and the term in brackets is proportional to $X^3 \propto v_M^3$, so $F_{\rm drag} \propto v_M$.

Captain Obvious: Dynamical friction has important analogs. For example, it turns out that if the motion of the massive object relative to gas or a field of stars is much faster than the sound speed (for a background of gas) or velocity dispersion (for a background of collisionless objects such as stars), then only the density of the background matters rather than whether it is gas or stars. When the motion is much slower than the sound speed or velocity dispersion, gas can flow around the massive object, and thus drag by gas is much less efficient than drag by stars. When the motion is comparable to the sound speed or velocity dispersion, then there can be a resonance that significantly enhances gas drag compared with stellar drag. Moreover, there are studies that show that, surprisingly, there are cases in which there could be negative dynamical friction. The idea is that if a massive object moves supersonically through gas, the gas can pile up in a shock in front of the object (this is called a bow shock). That density enhancement could in principle be enough that its gravity can pull the object forward. Another related effect is the Bondi–Hoyle–Lyttleton accretion of gas, where the deflected gas shocks with itself, loses energy, and falls toward the massive object. You can show that when gravitational deflection is important (i.e., when the escape speed from the massive object is much larger than the relative speed at infinity), this mode of accretion of gas is much more effective than accretion due to direct impact on the massive object.

B.6 Binary–Single Interactions

As we discussed at the start of this section, even in dense stellar systems, stars are very unlikely to collide. However, if stars are in binaries then their target size is vastly increased, and binaries can "collide" with single stars or other binaries. This has important consequences for the production of coalescing double compact object systems in dense stellar systems.

To see this, we note that an orbit might have a radius of 1 au $\approx 10^{13}$ cm, so that the area of the orbit is a few $\times\ 10^{26}$ cm^2. This is several thousand times the effective area of a star, which means that although even in a dense stellar cluster only a small fraction of stars will have a direct collision in the age of the universe, a typical 1 au binary will have tens of close interactions in such a cluster.

Close three-body interactions between objects of comparable mass are chaotic, so one might at first think that any kind of general analysis is hopeless. But studies starting in the 1960s showed that there are some trends in binary–single interactions that can guide our insight. Among these, the following are particularly important:

1. *Soft binaries soften and eventually ionize.* If the total energy of the system is positive (i.e., the relative kinetic energy of the single exceeds the gravitational binding energy of the binary—the binary is then called "soft"), then the tendency is that, after the interaction is resolved, the final binary is less bound than the original binary. Eventually, for a soft-enough binary, an encounter with a single star disrupts the binary and the result is three single stars.
2. *Hard binaries harden.* The flip side of this is that if the total energy of the system is negative (which means that the binary is "hard"), the tendency is that after the interaction is resolved, the final binary is more bound than the original binary.
3. *Massive stars get together.* After a binary–single interaction, it is likely that the two most massive of the three objects in the interaction are the members of the final binary. This likelihood increases with greater mass contrast.

These trends have multiple important consequences. First, for a system that is old enough for there to have been many binary–single interactions, we do not expect many soft binaries. This is because a softened binary is larger and therefore has a shorter interaction time, so there is "runaway softening" and ultimately ionization. Second, hardening does not run away. A hardened binary is smaller and therefore has a longer interaction time than before. The net result is that for a very hard binary, its binding energy grows at a constant rate because encounters are rarer, but each individual interaction takes away, on average, a constant fraction of the original energy. Third, relatively heavy objects such as black holes can start their lives single and then exchange into binaries. As a result, such heavy objects become progressively more likely to reside in binaries. Thus, dense stellar systems are highly efficient per stellar mass at producing coalescing compact binaries.

Captain Obvious: Earlier, Major Payne explained the Coulomb logarithm. You will see from those equations that, when it comes to the evolution of the orbits of single objects, the net effect is equal for equal logarithmic intervals of the impact parameter. That is, if we think of some fiducial impact parameter b_0, the net effect of perturbers between b_0 and $2b_0$ is the same as that from $2b_0$ to $4b_0$, or $4b_0$ to $8b_0$, and so on. But this is not the case when we think about the effect of a perturber on the intrinsic properties of a binary (its semimajor axis or eccentricity, for example)! We can understand this by realizing that perturbers with the larger closest distances r_p take longer for their encounters (on the order of r_p/v for a speed v at infinity). If that time is many binary orbital periods, then the net effect of the encounter on the binary averages out. Thus, to affect a binary with a perturber of order the mass of the binary, the closest distance to the center of mass cannot be more than a few times the semimajor axis of the binary.

B.7 von Zeipel–Lidov–Kozai mechanism

The final dynamical effect we will highlight has received dramatically increased attention over the past several years because of potential applications to exoplanets. In the early 1960s, Mikhail Lidov (who was interested in the orbits of artificial satellites) and Yoshihide Kozai (who investigated comets interior to Jupiter) studied three-body hierarchical systems with an inner binary and an outer tertiary. They demonstrated that with a sufficiently large relative inclination between the binary plane and the tertiary plane, and over many orbits of both the binary and tertiary, the relative inclination drops, then increases, then drops, and so on. Moreover, they also showed that while the inclination is doing this dance, the eccentricity of the inner binary also increases, then drops, then increases, and so on, exactly out of phase with the inclination oscillations! Recent historical work has shown that in the early 1900s, Hugo von Zeipel achieved many of these same insights. In the simplest treatment, such that the outer orbit's semimajor axis and eccentricity are constant, the semimajor axis of the binary remains constant during these oscillations, which means that in some circumstances, the binary pericenter distance can be driven to small-enough values that gravitational radiation plays an interesting role.

Given that binary–binary scattering can result in stable hierarchical triples in tens of percent of comparable-mass interactions and that the massive stars that evolve to neutron stars and black holes are commonly in triple or higher-order systems, this mechanism has promise to drive mergers at a rate that might be competitive with other channels. Estimates and simulations are being made of the rates and other properties, but there are challenges that make this nontrivial. For example, suppose that an isolated triple system (meaning one that is not close enough to any other stars to have significant interactions) evolves to a state in which there are three compact objects with a large-enough mutual inclination between the inner binary and outer tertiary that the von Zeipel–Lidov–Kozai mechanism could push the inner binary to a high eccentricity and produce a merger. The problem is that if the original three main-sequence stars had that configuration, then we would in many cases have expected the stars to collide on the main sequence or during the giant phase of one or more of the stars. Such an early collision would mean that the system would never evolve to be three compact objects. A similar difficulty attends the idea that two white dwarfs in a triple system might collide because of von Zeipel–Lidov–Kozai oscillations. There are ways out, e.g., it could be that the time for a von Zeipel–Lidov–Kozai cycle is longer than the main-sequence lifetime of the stars. However, it is unclear how common that is, and realistic calculations depend strongly on observational inputs that are difficult to obtain. Another scenario in which von Zeipel–Lidov–Kozai oscillations could be important is in galactic centers, where the triple could consist of a double compact object binary as the inner binary and the central supermassive black hole as the tertiary. Overall, this is an interesting channel and it might have better opportunities than other channels to result in, for example, palpably eccentric coalescences in the frequencies observed with ground-based detectors.

B.8 Exercises

1. In this problem, you will derive Equation (B.1). You will start by assuming that a test particle of some very small mass starts at effectively infinite distance from a compact body of mass M. The initial speed of the test particle is v with respect to M, and if it traveled in a straight line, then the closest it would get to M is a distance b; b is called the impact parameter of the trajectory. As a result, the specific angular momentum of the trajectory relative to M is bv and the specific energy of the trajectory is $\frac{1}{2}v^2$. By conserving energy and angular momentum, determine the value of b such that the closest approach to M is r_p. Hint: at the closest approach, the velocity vector is perpendicular to the direction to M. The effective cross section for an interaction of closest approach r_p or closer is then $\Sigma = \pi b^2$, so use this to derive Equation (B.1).

2. In a box, an extreme example was given of how, when one star becomes more bound to the center of a gravitational potential, the movement outward of the other star increases the momentum-volume space entropy. Here you will examine a less extreme case. Suppose you have two stars of the same mass m orbiting around a much larger mass M. Initially, both are in a circular orbit with the same radius r. One of the stars then moves to a circular orbit of radius $r(1 - \epsilon)$, with $\epsilon \ll 1$. Assuming that the total angular momentum (but not the total energy!) of the two stars combined is fixed, your task is then to
   a) Calculate the new (assumed circular) orbital radius of the other star.
   b) Calculate the total momentum-space volume of the two stars after the movement and compare that with the total momentum-space volume of the two stars before the movement.

For the second part, call the spatial volume $\frac{4}{3}\pi r^3$ for an orbital radius r and the momentum volume $\frac{4}{3}\pi(mv)^3$ for mass m and orbital speed v. The momentum-space volume is the product of the momentum volume and the spatial volume.

3. In a box, a head-on interaction was considered in which two stars reverse their direction and heavier objects transfer energy to lighter objects.
   a) Write a code (choose your poison!) to follow a more general gravitational interaction between two objects, with arbitrary initial masses, velocities, and impact parameters.
   b) Run this code many times for different choices of speeds (say, drawn from a Maxwell–Boltzmann distribution), with isotropic velocities and random impact parameters. If the average speeds of the two masses are the same, is the net average result that energy is transferred from the heavier to the lighter object?

c) For a given mass ratio, what is the ratio of the average speed that is needed so that there is no net energy transfer when averaged over isotropic encounters and the Maxwell–Boltzmann distribution of speeds?

4. In this problem, we will perform calculations related to the production of a binary by direct capture during a two-body encounter. The scenario is that in a dense stellar system, two black holes, which are initially unbound with respect to each other, pass close enough to each other that the gravitational radiation released during the encounter binds the black holes into a binary. Because the relative speed at great distances (typically ~tens of km s^{-1}) is tiny compared with the speed at pericenter in such encounters (typically tens of thousands of km s^{-1}), we can approximate the orbit as parabolic. The energy released in gravitational waves for a parabolic encounter between two masses m_1 and m_2 with closest approach r_p is

$$\Delta E = \frac{85\pi G^{7/2} m_1^2 m_2^2 (m_1 + m_2)^{1/2}}{12\sqrt{2}\, c^5 r_p^{7/2}}. \tag{B.8}$$

Given this:

a) For an initial relative speed at a large distance of v_∞, calculate the closest approach r_p such that ΔE is equal to $\frac{1}{2}\mu v_\infty^2$, where $\mu = m_1 m_2/(m_1 + m_2)$ is the reduced mass. Thus, encounters with closest approach distances of r_p or smaller will result in a bound binary.

b) Calculate the effective cross section of such encounters, assuming that at r_p the relative speed is much larger than v_∞. Hint: gravitational focusing is dominant in this limit.

c) Suppose that the core of a globular cluster has $v_\infty = 10$ km s^{-1}, and 100 black holes each with mass 10 $M_\odot$, at a number density of 10^5 pc^{-3}. Given the cross section that you found in part (b), compute the expected number of double black hole mergers you would expect in the globular in 10^{10} years. Here, the assumption (which you should check) is that once black holes are captured into a binary by this mechanism, coalescence is rapid.

5. Dr. Wrong doesn't understand all this focus on binary compact object mergers. Instead, direct collisions of single neutron stars in clusters with each other will make wonderful burst sources. Dr. Wrong has requested that you work out the numbers. Suppose that you consider a dense globular cluster, such that in the center the number density of neutron stars is 10^6 pc^{-3} and there are 1000 total neutron stars per cluster. Suppose that each neutron star has a radius of 10 km and mass of $1.5\ M_\odot = 3 \times 10^{33}$ g, and that the typical random speed in the cluster is 10 km s^{-1}. To within an order of magnitude, calculate how often two neutron stars in a given cluster will hit each other. If there are 10^{10} such clusters in the universe, how often will this happen in the universe? Hint: be careful when you calculate the cross section for collisions because gravitational focusing is important.

6. When we talked about relaxation time, we assumed that there was no dominant object in the middle. More generally, the local relaxation time is

$$t_{\rm rlx}(r) = \frac{1}{\ln \Lambda}\frac{\sigma^3(r)}{G^2M^2n(r)}, \tag{B.9}$$

where $\ln \Lambda \sim 10$ comes from the Coulomb integral, $\sigma(r)$ is the local velocity dispersion, M is the typical mass of an object, and $n(r)$ is the local number density of objects. Consider a region $r < r_{\rm infl}$, where $\sigma(r)$ is given by the Keplerian orbital speed.
   a) If $n(r) \propto r^{-3/2}$ (a typical profile), how does the relaxation time depend on r?
   b) In contrast, for $r \gg r_{\rm infl}$, assume that $n(r) \propto r^{-2}$ and $\sigma(r)$ is constant. Then how does the relaxation time depend on r?

6. Here we will investigate further the concept of the loss cone. Suppose that the loss cone as seen from radius r involves orbits of angular momentum between $J = 0$ and $J = J_{\rm LC}$. Then, as orbits with $J < J_{\rm LC}$ are eliminated, orbits with $J > J_{\rm LC}$ move in to fill them. If the typical change of J in one orbital time $t_{\rm orb}$ is $\Delta J \gg J_{\rm LC}$, then the loss cone is refilled in a dynamical time, and this is the full loss cone regime. If $\Delta J \ll J_{\rm LC}$, then the loss cone has to be filled diffusively, which takes much longer than one orbit. This is the empty loss cone regime. Given this, answer the following questions:
   a) Because motion in angular momentum space is a random walk, how long does it take to diffuse by $J_{\rm LC} \ll J_{\rm circ}$, where $J_{\rm circ}$ is the angular momentum of a circular orbit with the same energy? Call this time t_J. Remember that $t_{\rm rlx}$ is basically the time needed to change the angular momentum by $J_{\rm circ}$. More specifically, given that for a Keplerian orbit, the angular momentum scales as $(1 - e^2)^{1/2}$, how does t_J scale with e for $(1 - e) \ll 1$?
   b) The capture of a single 10 $M_\odot$ black hole by emission of gravitational radiation near a 10^6 $M_\odot$ massive black hole requires a pericenter distance of about 0.1 au. The standard relaxation time (time required to change by $\sim J_{\rm circ}$) is about 10^9 years at 1 pc (roughly equal to $r_{\rm infl}$). Given this, how does t_J compare to $t_{\rm orb}$ at 1 pc? If $n(r > r_{\rm infl}) \propto r^{-2}$, how does t_J compare to $t_{\rm orb}$ at $r > r_{\rm infl}$? If $n(r < r_{\rm infl}) \propto r^{-3/2}$, how does t_J compare to $t_{\rm orb}$ at $r < r_{\rm infl}$?

7. As we have discussed, one of the ways that black holes can acquire mass, and possibly grow into supermassive black holes, is through Bondi–Hoyle–Lyttleton accretion. In this process, gas that moves at a speed v relative to the black hole (we assume here that v is much larger than the sound speed of the gas) is gravitationally deflected by the hole, heats itself, shocks as a result, releases energy, and if it is close enough to the hole it is then bound and eventually accretes into the hole. For a black hole of mass M, the cross section for this type of accretion is $\Sigma_{\rm BHL} = \pi(GM/v^2)^2$ times a numerical factor close to unity that depends on the details of the flow.
   But what if the matter does not interact with itself? An example would be dark matter. Then, for the matter to accrete it needs to hit the hole directly. For

a nonrotating black hole (rotating black holes have slightly different numbers), capture requires that the angular momentum per unit mass is less than $4GM/c$. Given this, compute the ratio of the cross section for direct-impact accretion to the cross section for Bondi–Hoyle–Lyttleton accretion and comment on the implications for the rapid growth of black holes by accretion of dark matter. Hint: assume that at a great distance from the black hole, the dark matter is moving at the same speed v relative to the hole as is the gas, and note that in galaxies v is typically a few hundred kilometers per second.

Useful Books

Binney, J., & Tremaine, S. 2008, Galactic Dynamics (Princeton, NJ: Princeton Univ. Press)

Merritt, D. 2013, Dynamics and Evolution of Galactic Nuclei (Princeton, NJ: Princeton Univ. Press)

Thorne, K. S., & Blandford, R. D. 2017, Modern Classical Physics: Optics, Fluids, Plasmas, Elasticity, Relativity, and Statistical Physics (Princeton, NJ: Princeton Univ. Press)

Appendix C

General Relativistic Calculations of I, Q, and λ

Here we fill in a number of the calculational details that we omitted for brevity in Chapter 7. In particular, we show how to compute the moment of inertia, the rotational quadrupole, and the tidal deformability, for small rotation rates or small tidal perturbations, consistently in general relativity.

C.1 The Moment of Inertia I

The way to find the moment of inertia in general relativity is as follows. First, you pick a value for the stellar rotation frequency Ω_*, and for this value, you solve the first-order-in-rotation Einstein equations in the interior and the exterior of the star, which requires the choice of two constants of integration. Then, you find the values of these constants such that the solution is continuous and differentiable at the surface. It turns out that one of these constants is the angular momentum S, so with the right value (i.e., the one that leads to a continuous and differentiable solution at the surface) and the value we chose for Ω_*, we can then compute the moment of inertia via $I = S/\Omega_*$. Thus, for us to compute the moment of inertia we need to solve the Einstein equations to first order in rotation.

When all is said and done, it turns out that the moment of inertia in general relativity can be written as

$$I = \frac{8\pi}{3\Omega_*} \int_0^{R_*} \frac{e^{-(\lambda+\nu)/2} r^5 (\epsilon + p)\omega}{r - 2M(r)} dr, \tag{C.1}$$

where ω is not an angular velocity; instead, it is an unknown metric function of radius (see Equation (7.12)) that controls the amount of frame dragging. This expression reduces to the Newtonian one in Equation (7.16) when $\epsilon \gg p$, $M \ll r$, $\omega \sim \Omega_*$ and $\lambda + \nu \sim 0$.

Major Payne: Wait a minute! Where did this come from? I hate it when professors just pull equations out of thin air. There is actually a beautiful way to derive this expression, and it's not that hard, so let's do it. The calculation starts by using the metric and the stress–energy tensor I presented earlier (see Equations (7.12) and (7.13)) in the Einstein equations, and expanding to first order in rotation. Doing so, one finds a differential equation for the metric function ω, which, as it turns out, can be written as

$$\frac{1}{r^4}\frac{d}{dr}\left(r^4 j \frac{d\omega}{dr}\right) + \frac{4}{r}\frac{dj}{dr}\omega = 0, \tag{C.2}$$

where we have introduced $j = \exp[-(\lambda + \nu)/2]$.

One must now solve the above equation in the interior and in the exterior of the star, ensuring both continuity and differentiability at the stellar surface (you now need two such conditions because it's a second-order differential equation). In the exterior, this equation can be solved exactly to find $\omega = \Omega_* - 2S/r^3$, where $S = I\Omega_*$ is an integration constant, and we have fixed the other integration constant so that $\omega \to \Omega_*$ as $r \to \infty$. In the interior, we have to solve the differential equation numerically, and this requires a boundary condition at the center of the star. As before, this can be found through a local asymptotic analysis, namely by using the ansatz

$$\omega = \omega_0 + \omega_1 r + \omega_2 r^2 + \mathcal{O}(r^3). \tag{C.3}$$

Inserting this ansatz into the above differential equation, one finds that $\omega_1 = 0$, $\omega_0 = \omega_c$ with ω_c the central value of the rotation function, and $\omega_2 = (8\pi/5)(\epsilon_c + p_c)\omega_c$.

Given a choice of Ω_*, one can then find the values of ω_c and I that lead to an interior solution that is continuous and differentiable at the stellar surface. When one poses the problem in this way, it becomes a "shooting problem" in numerical analysis. Alternatively, one can choose a value of Ω_* and ω_c and then integrate the differential equation in the interior up to the surface. This solution can then be made continuous and differentiable at the surface by solving for the correct I. But there is yet another way in which we can find the correct moment of inertia. This can be obtained by multiplying Equation (C.2) by r^4 and integrating to find

$$\frac{d\omega}{dr} = -\frac{4}{r^4 j}\int_0^{R_*} r^3 \frac{dj}{dr}\omega dr. \tag{C.4}$$

To proceed further, we now realize that the above expression must be valid both in the interior and in the exterior of the star. In the exterior, the solution is such that $d\omega/dr = 6I\Omega_*/r^4$, while $j = 1$. Using this and solving for the moment of inertia, one then finds the expression the authors presented in Equation (C.1).

No matter which way one calculates this, at the end of the day, given a choice of ϵ_c and for a given equation of state, one can compute not just the mass of the star and its radius from the zeroth-order equation, but also the moment of inertia. Note that the moment of inertia does not depend on the rotation frequency Ω_* (as long as the star is rotating slowly), as this scales out in the calculation of I. This then means that one can construct curves of the moment of inertia as a function of the stellar mass, or the moment of inertia as a function of the stellar compactness $C = GM_*/(c^2R_*)$. This curve is perhaps not as interesting as the mass–radius curve, as it is simply a decaying function of mass or compactness.

We can nonetheless Fermi estimate the magnitude of the moment of inertia. The only relevant scales here are the mass of the star and its radius, so $I \propto M_* R_*^2$. But the radius is just $R_* = M_*/C$, so we can rewrite this as $I \sim C^{-2} M_*^3$. For a solid sphere in Newtonian physics, for example, we then have that $I = (2/5) M_* R_*^2 = (2/5) C^{-2} M_*^3$. For an $M_* = 1.44 M_\odot$ star with a radius of 13 km, the compactness is about $C \sim 0.16$, and thus, our Fermi estimate is $I \sim 1.9 \times 10^{45}$ g cm^2. Recently, the NICER telescope was able to measure the equatorial radius of a given pulsar, and from this, researchers were able to infer its moment of inertia to be $I_{\rm PSR} \sim 1.7^{+0.6}_{-0.5} \times 10^{45}$ g cm^2 by first inferring the compactness (from a pulse profile model), and then using the I–C relations to infer the moment of inertia. We see that this is less than 14% off from our Fermi estimate!

C.2 The Rotational Quadrupole Moment Q

As in the case of the moment of inertia, it turns out there is an integral expression for the quadrupole moment, namely

$$Q = \frac{1}{8\pi} \int S(r, \theta) P_2(\cos\theta) r^4 dr d\Omega, \tag{C.5}$$

which was found by Fintan Ryan in the 1990s. Here, $S(r, \theta)$ is a complicated function that depends on the zeroth, first, and second order in rotation metric functions. Clearly then, in order to find the relativistic version of the quadrupole moment, we must first solve the Einstein equations to second order in rotation.

Is there another way to get the quadrupole moment? But of course! From Newtonian gravity, recall that the gravitational potential Φ far away from a source of mass M_* behaves as

$$\Phi = -\frac{GM_*}{c^2 r} - \frac{GQ}{c^2 r^3} P_2(\cos\theta) + \mathcal{O}(1/r^4), \tag{C.6}$$

where r is the distance from the center of mass to a field point located at some angle θ. But the gravitational potential is nothing but the time–time component of the metric tensor! Indeed, one can expand this component of the metric far away from a gravitating source to find

$$g_{tt} = -1 - 2\Phi + \mathcal{O}(G^2) = -1 + \frac{2GM_*}{c^2 r} + \frac{2GQ}{c^2 r^3} P_2(\cos\theta) + \mathcal{O}(1/r^4, G^2/c^4). \tag{C.7}$$

Clearly then, if we are able to find the second order in rotation time–time component of the metric tensor, then we should be able to read out the quadrupole moment from its far-field behavior.

Major Payne: Ok, but what are those equations at second order and how do we solve them? Let's try to bring back some level of detail, although this time, I will not present all of the equations because these are, even for me, much more complicated than those at first order in rotation. The procedure, however, is quite similar to what I've already described at first order. One inserts the metric ansatz in Equation (7.12) and the stress–energy tensor in Equation (7.13) into the Einstein equations and expands them to second order in rotation. In doing so, one will find equations for the metric functions (h_2, m_2, K_2), as well as for the perturbed energy density (or the perturbed pressure), but there is a complication.

As the authors already indicated, at second order in rotation, the energy density and the pressure should not just be functions of radius but should also be functions of polar angle. Indeed, the star will be oblate, so its polar radius will be slightly smaller than its equatorial radius. As a consequence, a sphere with radius equal to the equatorial radius will actually be outside of the star at the poles! This creates a serious technical complication in solving the equations, because for perturbation theory to be valid, the perturbation to the energy density should be always much smaller than the background energy density, and this cannot be true at the poles. The solution is to introduce a new radial coordinate R, related to our old radius via $r = R + \xi_2(R)\alpha Y_{2m}(\theta, \phi)$, where, as before, α is a normalization constant chosen for convenience and Y_{2m} are spherical harmonics. The function $\xi_2(R)$ is defined such that $\epsilon[r(R, \theta, \phi)] = \epsilon^{(0)}(R)$, i.e., $\xi_2(R)$ is such that the energy density is just the unperturbed energy density evaluated in the new coordinate R. In doing so, the perturbed Einstein equations no longer give us an equation for the perturbed energy density, as this field is zero in the new coordinates, but instead we find an equation for the function ξ_2.

That complication aside, the procedure to solve the perturbed Einstein equations proceeds just as in the case of the zeroth- and first-order problem. The equations reduce to two coupled differential equations for the metric functions K_2 and h_2, and two coupled algebraic equations for ξ_2 and m_2. We thus first solve these equations in the exterior, and as it turns out, an exact closed-form solution can be found, which will depend on an unknown exterior constant. Next, we carry out a local asymptotic analysis to find the boundary conditions at the center of the star, which will depend on another unknown interior constant. We then solve for the interior solution numerically with our boundary conditions at the center up to the surface and find the interior and exterior unknown constants that guarantee that K_2 and h_2 will be continuous at the stellar surface. Finally, by expanding the analytic exterior solutions far from the star, we can read out the coefficients of the term that falls off as $1/r^3$, which gives us the quadrupole moment. As it turns out, this quantity will be a sum of the spin angular momentum squared divided by the mass and the exterior unknown constant times the cube of the mass.

Given a central density and a rotation rate, for a fixed equation of state, one can then calculate the quadrupole moment for a star of a given mass, or a given radius, or a given compactness. In general, the quadrupole moment is a rather boring decaying function of any of these quantities, just as the moment of inertia is, but, as in the moment of inertia case, we can Fermi estimate its magnitude. First, we notice that the quadrupole moment has the same units as the moment of inertia, but this time we have three quantities we can play with: M_*, R_*, and Ω_*. In fact, we know that the quadrupole moment must scale with the spin squared, so we must have that $Q \propto \Omega_*^2$. We can restore the units by multiplying the previous expression by some product of M_* and R_* that has dimensions of length to the fifth power, but what combination should we pick? From the Newtonian expression for the quadrupole moment in Equation (7.17), we know that we must integrate an integrand that goes as r^4 over the entire star, so the right combination is simply R_*^5; indeed, the Newtonian expression for the quadrupole moment in Equation (7.20) already told us that $Q \sim (1/2)\Omega_*^2 R_*^5$ for a solid sphere. Rewriting this slightly, we have $Q \sim 3\chi_*^2 M_*^3 / C_*$, where we recall that $\chi_* \equiv (c/G)(S/M_*^2)$. For a rapidly rotating neutron star at $\Omega_* = 2\pi$ 200 Hz with a mass of $1.44 M_\odot$ and a moment of inertia of 1.9×10^{45} g cm^2 (as we found in the previous section for a solid sphere), we then have that $\chi_* \sim 0.13$ and thus $Q \sim 3\chi_*^2 M_*^3/C_* \sim 6 \times 10^{43}$ g cm^2. Recently, researchers have used the measurements of the radius of a neutron star from NICER data (together with the I–C and the I–Q relations) to infer the quadrupole moment of that pulsar to be $Q_{\rm PSR} = 1.5^{+0.6}_{-0.5} \times 10^{43}$ g cm^2, which is only about a factor of 4 off from our Fermi estimate.

C.3 The Tidal Deformability Λ

How do we go about finding this tidal deformability? Unlike in the moment of inertia and the quadrupole-moment case for slowly rotating stars, we cannot simply express λ as an integral over the source. Instead, we must solve for the perturbation to the star's metric as induced by the external field, and then "read out" λ from the far-field expansion of the metric (similar to what we did for the quadrupole moment). When we include this external field, we can expand the time–time component of the metric as

$$g_{tt} = -1 - 2\Phi + \mathcal{O}(G^2) = -1 + \frac{2GM_*}{c^2 r} + \frac{2GQ}{c^2 r^3} P_2(\cos\theta) - \frac{2Gr^2\mathcal{E}}{3c^2} P_2(\cos\theta) + \mathcal{O}(1/r^4, G^2, r^3), \tag{C.8}$$

where $\mathcal{E}$ is related to the trace of the electric-type, quadrupole tidal tensor, and where here we have simultaneously expanded both in $r \gg R_*$ and $r \ll r_{12}$, where r_{12} is the distance to the companion (the orbital separation). The region where both of these expansions are valid is called the "buffer zone." The procedure is then simple: solve for the metric perturbation in the interior of the star and in the exterior (ensuring it is continuous and differentiable at the surface), and then read out the coefficient of the term that goes as $1/r^3$ and the one that goes as r^2 in the buffer zone, because their ratio essentially gives you λ.

Major Payne: Ok, let's try to provide a bit more detail here. As the authors have mentioned, the tidal deformabilities depend on the metric perturbations, which in turn are determined by the Einstein equations. To obtain this, one can simply take Equation (7.12) for the metric and Equation (7.13) for the stress–energy tensor, but with $\omega_1 = 0 = \Omega_*$, because our star is not rotating. This only leaves three metric perturbation functions (h_2, m_2, K_2) and one function ξ_2 for the coordinate transformation (just as in the quadrupole-moment case, a transformation here is also necessary). But this time the

Einstein equations can be fully decoupled to get a single ordinary differential equation for the metric perturbation h_2, which is the one that determines the behavior of g_{tt}. This equation is

$$\frac{d^2h_2}{dR^2} = -\left\{\frac{2}{R} + \left[\frac{2M}{R} + 4\pi R(p - \epsilon)\right]e^{\lambda}\right\}\frac{dh_2}{dR} + \left\{\frac{6e^{\lambda}}{R^2} - 4\pi\left[5\epsilon + 9p + (\epsilon + p)\frac{d\epsilon}{dp}\right]e^{\lambda} + \left(\frac{d\nu}{dR}\right)^2\right\}h_2. \tag{C.9}$$

With this equation at hand, the calculation now follows the same recipe as for the quadrupole moment. First, it turns out the above equation can be solved exactly in the exterior of the star in terms of two constants of integration. Unlike in previous cases, we cannot set one of these constants to zero by requiring asymptotic flatness, as we wish to find a solution that is valid in the buffer zone (and this does not extend all the way to spatial infinity). Moreover, as the above equation is homogeneous, we can rescale h_2 by one of these constants, so that the exterior solution only depends on their ratio. Then, we carry out a local asymptotic analysis about the center of the star to find the appropriate boundary conditions, which in turn will depend on another unknown constant. We then solve numerically for h_2 in the interior starting from our boundary condition, and with choices of the constants such that the metric perturbation is continuous and differentiable at the stellar surface. The exterior constants then yield the tidal deformability for a given choice of central density.

Just as before, we can now compute the tidal deformabilities for a bunch of different central densities, given a particular equation of state, and thus, calculate λ as a function of the stellar mass, the stellar radius, or the compactness. As in the case of the quadrupole moment, though, the λ–C relation is rather boring, decaying with compactness, because more massive stars are more difficult to deform. We can estimate though the magnitude of the tidal deformability using our Fermi tools. Recall that the definition of tidal deformability is through the ratio of the induced quadrupole moment to the quadrupole tidal tensor, $Q_{ij} = -\lambda \mathcal{E}_{ij}$, and then $\lambda \propto M_* R_*^2/(M_*/r_{12}^3) \sim r_{12}^3 R_*^2$, where we recall that r_{12} is the distance to the body that is producing the external field. Notice here that the quadrupole moment that enters is the induced one and not the rotational one, which is why we do not force Q to be proportional to Ω_*^2. If that body is at least a few R_* away from the neutron star, then $\lambda \sim R_*^5 \sim M_*^5/C^5$. We see then that the tidal deformability is enhanced by compactness to a large power, because for neutron stars C is small compared to unity. The maximum tidal deformability then is something like $\lambda/M_*^5 \sim 10^3$ for a star with a compactness of $C \sim 0.2$. The gravitational waves emitted in the merger event LIGO show that $(\lambda/M_*^5)_{1.4} \sim 190^{+390}_{-120}$ at 90% confidence for a neutron star with $M = 1.4 M_{\odot}$, assuming these waves were emitted in a binary neutron star coalescence. We see that this is not that far off from our Fermi estimates!

Index

www.ingramcontent.com/pod-product-compliance
Ingram Content Group UK Ltd.
Pitfield, Milton Keynes, MK11 3LW, UK
UKHW060959120926
8556IPUK00033B/915

9 780750 330497